SUPERCONDUCTIVITY SOURCEBOOK

V. DANIEL HUNT
President, Technology Research Corporation

WILEY

A Wiley-Interscience Publication

JOHN WILEY & SONS

New York • Chichester • Brisbane • Toronto • Singapore

Library of Congress Cataloging in Publication Data:

Hunt, V. Daniel.

  Superconductivity sourcebook.
  Bibliography: p.
  1. Superconductivity. I. Title.

QC611.92.H86 1989   537.6'23   88-27676
ISBN 0-471-61706-7

Printed in the United States of America

10 9 8 7 6 5 4 3 2

*To Ellen O. Feinberg, John R. Clem, and the staff at
the Ames Laboratory, Iowa State University for their
continuing excellent efforts in providing the primary
superconductivity technology transfer information
source in the form of "High-Tc Update."
We appreciate the assistance and information made
available via High-Tc Update which is a fundamental
resource used in this sourcebook.*

CONTENTS

PREFACE

The recent discovery of superconductivity at temperatures above 95°K is one of the more important scientific events of the past decade. The sheer surprise of this discovery, as well as its potential scientific and commercial importance, largely underlie the degree of excitement in the field. Because our previous understanding of superconductivity has been so fundamentally challenged, a door has been opened to the possibility of superconductivity at temperatures at or above room temperature. Such a development would represent a truly significant breakthrough, with implications for widespread application in modern society.

While the base of experimental knowledge on the new superconductors is growing rapidly, there is as yet no generally accepted theoretical explanation of their behavior. Applications presently being considered are largely extrapolations of technology already under investigation for lower-temperature superconductors. To create a larger scope of applications, inventions that use the new materials will be required. The fabrication and processing challenges presented by the new materials suggest that the period of pre-commercial exploration for other applications will probably extend for a decade or more.

Near-term prospects for applications of high-temperature superconducting materials include magnetic shielding, the voltage standard, superconducting quantum interference devices (SQUID), medical imaging systems, infrared sensors, microwave devices, and analog signal processing. Longer-term prospects include large-scale applications such as microwave cavities; power transmission lines; and superconducting magnets in generators, energy storage devices, particle accelerators, rotating machinery, levitated vehicles, and magnetic separators. In electronics, long-term prospects include computer applications with semiconductor superconducting hybrids, Josephson devices, or novel transistor-like superconducting devices.

The United States has a good competitive position in the science of this field, and U.S.

researchers have contributed significantly to the worldwide expansion of scientific knowledge of the new materials. International competition is intense. Several leading industrialized countries have mounted substantial scientific and technological efforts, especially Japan, the USSR, and a number of Western European nations.

The short-term problems and long-term potential of high-temperature superconductivity may both be easily underestimated. Given this potential and the current limited understanding of the new superconducting materials and their properties, it is essential that government, academic institutions, and industry take a long-term, multidisciplinary view. Since science and technology in this field are strongly intertwined, progress must occur simultaneously in basic science, manufacturing/processing science, and engineering applications. It is also important to maintain an open and cooperative international posture.

V. Daniel Hunt

Springfield, Virginia
December, 1988

ACKNOWLEDGMENTS

The information in the *Superconductivity Sourcebook* has been compiled from a wide variety of authorities who are specialists in their respective fields.

The following publications were used as the basic technical resources for this book. Portions of these publications may have been used in the book. Those definitions or artwork used here have been reproduced with the permission to reprint of the respective publisher.

High-T$_c$ Update, published by the Office of Basic Energy Sciences, U.S. Department of Energy, under contract with the *Ames Laboratory, Iowa State University*, Ellen O. Feinberg, Editor; John R. Clem, Technical Advisor.

Supercurrents, Volume 1, 2, and 3, Donn Forbes, Editor, *Supercurrents Magazine*, P.O. Box 889, Belmont, California 94002.

Research Briefing on High-Temperature Superconductivity, Research Briefing, *National Academy Press*, Washington, D.C., 1987.

Commercializing High-Temperature Superconductivity, U.S. Congress, *Office of Technology Assessment*, OTA-ITE-388, GPO #052-003-01112-3, Washington, D.C., June 1988.

Selection Guide for Superconducting Magnets and Magnetic Systems, *American Magnetics Inc.*, 1988.

McGraw-Hill Dictionary of Physics, Sybil P. Parker, Editor-in-Chief, *McGraw-Hill Book Company*, 1984.

University Physics, Francis W. Sears, Mark W. Zemansky, and Hugh D. Young, *Addison-Wesley Publishing Company*, Seventh Edition, 1987.

Glossary of Chemical Terms, Clifford A. Hampel and Gessner G. Hawley, *Van Nostrand Reinhold*, 1976.

High Technology Business, *Infotechnology Publishing Corporation*, Volume 8, No. 1, January 1988 issue.

Dictionary of Electronics, Carol Young, *Penguin Books Ltd.*, 1979.

Dictionary of Physics, Valerie H. Pitt, *Penguin Books Ltd.*, 1975.

We thank IBM, AT&T, Applied SuperConetics Inc., GA Technologies, Argonne National Laboratory, Department of Energy, Department of Defense, DARPA, Electric Power Research Institute, CSAC, American Magnetics Inc., Newport News Shipbuilding, Intermagnetics General Corporation, General Electric Medical Systems, Supercon, and Hypres Inc. for the permission to utilize their photographs and line drawings in this book.

The preparation of a book of this type is dependent upon an excellent staff, and I have been fortunate in this regard. Special thanks to William Heyman and Don Keehan for research assistance, and to Margaret W. Alexander for the word processing of the manuscript.

CHAPTER 1

OVERVIEW OF SUPERCONDUCTIVITY

INTRODUCTION

Perhaps the most remarkable feature of the discovery of high-temperature superconductivity is the fact that it was so unexpected. The sheer surprise of this discovery, as well as its potential scientific and commercial importance, largely underlie the degree of excitement and fervor of the field. Superconductivity, in the past, has always been a challenging fundamental and technological problem, for which understanding and application have come slowly. Because our previous understanding of superconductivity has been so fundamentally challenged, there is hope that the progress that has been achieved so dramatically in the past two years can continue.

High-temperature superconductivity offers an important opportunity for our nation's scientific and technological community. The opportunity merits a substantial thrust in fundamental research; at the same time, enough is already known to encourage commercial development efforts with the newly discovered materials.

Superconductivity has been found in at least 26 metallic elements and in thousands of alloys and compounds. Each of these materials superconducts as long as its temperature stays low enough, it does not carry too much current, and the magnetic field surrounding it is not too strong. If any of these conditions are not met, the material goes normal or quenches, regaining normal resistance to electricity.

HISTORICAL BACKGROUND

Kamerlingh Onnes, a Dutch physicist, discovered superconductivity in 1911 at the age of 58. Onnes had devoted his career to exploring the limits of coldness. In 1908 he made helium a liquid by cooling it to 452 degrees below zero Fahrenheit (4° Kelvin). Liquid helium enabled him to chill other materials nearly to absolute zero (0°K), the coldest possible temperature.

In 1911, Onnes began to investigate the electrical properties of extremely cold metals. He passed a current through a mercury wire, then measured the electrical resistance of the wire as he chilled it. At 4.2°K the resistance suddenly vanished. (See Figure 1-1.) "The experiment left no doubt about the disappearance of the resistance of mercury," he wrote. "Mercury has passed into a new state, which because of its extraordinary electrical properties may be called the superconductive state."

He soon found superconductivity in several other metals. "It is of great importance that tin (Sn) and lead (Pb) were found to become superconductive also. Tin has its step-down point at 3.8°K, a somewhat lower temperature than the vanishing point of mercury. The vanishing point of lead may be put at 6°K. Tin and lead being easily workable metals, we can now contemplate all kinds of electrical experiments with apparatus without resistance."

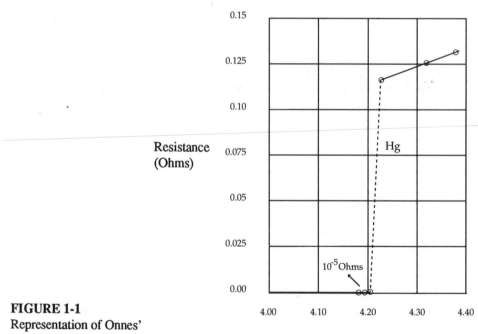

FIGURE 1-1
Representation of Onnes'
Original Temperature/
Resistance Graph

Source: Supercurrents,
December 1987

Onnes knew his discovery had vast commercial potential as well as scientific importance. An electrical conductor with no resistance could carry current any distance with no losses, or carry current in a loop for the age of the universe. In one key experiment, Onnes started a current flowing in a loop of lead wire cooled to 4°K. A year later the current was still flowing undiminished.

Since an electrical current creates a magnetic field, Onnes believed coils of superconducting wire could be formed into large industrial magnets. With no resistive losses, these superconducting magnets could operate without the expense of a continuous energy source. But his own experiments were unsuccessful: the coil would lose its superconductivity if either the current or the magnetic field became too high.

Nevertheless, he remained optimistic: "In the future much stronger and more extensive fields may be realized than are now reached in the strongest electromagnets. As we may trust in an accelerated development of experimental science, this future ought not to be far away."

The fulfillment of this dream took longer than Onnes expected—he died in 1926 without seeing any practical application of superconductivity. However, his discoveries stimulated other scientists to establish centers for superconductivity research, seeking scientific understanding and new materials with better performance characteristics.

People soon began to invent applications for superconductors—for example, to reduce losses in electric power systems. However, in the 1920s it was found that superconductivity disappeared in these metals when rather low electric currents were passed through them. As a result, power applications were abandoned. Electric utility equipment typically carries thousands of amperes and has associated magnetic fields of up to 20,000 gauss or 2 tesla.

The next major discovery occurred in 1933, in Walther Meissner's superconductivity laboratory in Berlin. Meissner and R. Ochsenfeld learned that a superconductor is more than a perfect conductor of electricity—it also has surprising magnetic properties, as shown in Figure 1-2. A superconductor will not allow a magnetic field to penetrate its interior. If a superconductor is approached by a magnetic field, screening currents are set up on its surface.

FIGURE 1-2
Illustration of phenomenon known as the Meissner effect. A permanent magnet floats above a superconductor bathed in liquid nitrogen because its magnetic field is completely repulsed at the surface of the superconductor.

Source: IBM Research

These screening currents create an equal but opposite magnetic effect, thereby cancelling the magnetic field and leaving a net of zero inside the superconductor. Reversing the sequence gives the same result: by first placing the material in a magnetic field and then cooling it to the superconducting state, it sets up screening currents that expel the magnetic field. This phenomenon is known as the Meissner Effect.

As IBM researcher Paul Grant explains, "A magnet placed near a superconductor will literally see its mirror image and, since like poles repel, both superconductor and magnet will try to move away from each other." If you place a magnet above a superconductor, the superconductor will levitate the magnet.

The Meissner Effect occurs only if the magnetic field is relatively small. If the magnetic field becomes too great, it penetrates the interior of the metal and the metal loses its superconductivity.

Superconductivity was a mystery for many years. Einstein, Bohr, and Heisenberg tried unsuccessfully to explain it. In 1935, the brothers Fritz and Heinz London devised 2 equations describing zero resistance and the Meissner effect. Fifteen years later Vitaly Ginzburg and Lev Landau, Russian scientists, described superconductivity in terms of quantum mechanics. These were phenomenological theories, in other words, they described the observable phenomena without explaining what was occurring on a microscopic level.

Although the London Equations and the Ginzburg-Landau Theory helped to guide experimental researchers, a comprehensive microscopic theory of superconductivity did not appear until 1957, 46 years after Onnes' discovery. John Bardeen, Leon Cooper, and Robert Schrieffer, then at the University of Illinois, won the Nobel Prize for this theory, now known as the Bardeen, Cooper, Schrieffer (BCS) Theory. According to Schrieffer, "all the electrons condense into a single state, and they flow as a totally frictionless fluid." The electrons, which normally repel each other, develop a mutual attraction and form twosomes called Cooper pairs. Schrieffer described the theoretical issue as "how to choreograph a dance for more than a million million million couples." D.N. Langenberg extended the dance analogy:

> Consider an enormous ballroom packed with dancers, shoulder to shoulder. Suppose each dancer is vigorously doing his or her own individual dance. The dancers will collide with each other and with any other objects which may be scattered about the dance floor. If there is some pressure on the whole group to move toward one side of the ballroom, dancing all the while, the collective motion will be random and chaotic, and a lot of energy will be lost in collisions. This represents the electrons in a normal metal, colliding with each other and with irregularities or impurities in the crystal lattice.

> Now suppose the dancers are paired in couples, each pair dancing together. The partners comprising each couple are not dancing cheek to cheek, but are separated by a hundred other dancers. Consequently, if every couple is going to dance together, it is clear that everybody must dance together. The result is a single coherent motion, with order extending all the way across the ballroom. The superconducting state is something like that.

The theory of Bardeen, Cooper, and Schrieffer defined, the interaction between electrons and phonons (vibrational modes in the lattice of atoms making up the material) leads to a pairing of electrons. At low temperatures, these so-called Cooper pairs condense into an electrical superfluid, with energy levels a discrete amount below those of normal electron states (known as the superconducting energy gap). In the same period, new materials were discovered that displayed superconductivity at temperatures as high as 20 K, almost 5 times higher than the temperature of superconductivity in mercury.

These scientific discoveries had two important consequences. First, in a direct experiment to verify the energy gap, Giaever at the General Electric Research Laboratories observed electron tunneling between superconductors—that is, electrons passing from one superconductor to another through a thin insulating barrier. The observation of normal electron tunneling led Josephson in England to speculate that Cooper pairs could also tunnel through a barrier, a prediction that was soon verified by Rowell and Anderson at Bell Laboratories. These discoveries laid the foundation for a whole new superconducting electronics technology.

Second, Kunzler and coworkers at Bell Laboratories established that a group of superconducting compounds and alloys (the Type-II superconductors), could carry extremely high electric currents (up to a million amperes per cm^2 of conductor cross-section) and remain superconducting in intense magnetic fields (up to 30 tesla [T] or 300,000 gauss). These materials, offering the prospect of very high magnetic fields and current-carrying capacity at a much lower cost than before, revived interest in superconducting magnets and electric power components.

That same year, 1957, a Russian physicist named Alexei Abrikosov predicted the existence of new superconductors with better performance characteristics. These materials would superconduct even in very high magnetic fields. Tiny filaments of magnetic flux would penetrate the materials, but the areas between the filaments would carry currents with no resistance. Abrikosov's analysis of these materials, which he called Type II superconductors, provided a theoretical basis for the development of commercial superconducting magnets.

The theoretical basis for electronic applications appeared five years later, in 1962. Brian Josephson, then a graduate student at Cambridge University, analyzed what would happen at the intersection of two superconductors separated by a thin insulating barrier (now called a Josephson Junction). His predictions defied common sense. He predicted that a supercurrent would tunnel right through the barrier and that there would be no voltage across the barrier as long as the supercurrent did not exceed a critical value. He also predicted that a voltage across the barrier would create a high-frequency alternating supercurrent. These predictions evoked considerable skepticism, but John Rowell, Phil Anderson, and S. Shapiro verified them experimentally in 1963.

EXPERIMENTAL DEVELOPMENTS[1]

During the 1950s and 1960s, while the theoreticians were catching up with the experimentalists, researchers at the University of Chicago and Bell Labs were exploring new materials in a search for higher transition temperatures. Ted Geballe, a superconductivity researcher at Bell Labs in the early 1960s, described this period in an interview with *Supercurrents:*

The field really wasn't going anywhere. Right after Heike Kammerlingh Onnes and his colleagues discovered superconductivity, they thought "We can make magnets, we can make motors, we can do all these wonderful things, it's zero resistance!" But as soon as they did, the

[1] *Supercurrents,* The Superconducting Magazine, Volume 1, January 1988.

field or the current drove the superconductor normal again, and they couldn't really do anything with it. So the initial hope of technology, which was bright back in 1915 to 1920, turned into disillusionment—and that disillusionment carried all the way through until 1960.

In the early 50s John Hulm and Bernd Matthias at the University of Chicago began looking for high-temperature superconductivity in new materials, mainly alloys and compounds. In 1952 Hulm and George Hardy discovered that V_3Si (vanadium-3-silicon) had a transition temperature of 17°K. Then at Bell Laboratories Matthias made Nb_3Sn (niobium-3-tin), and we found it superconducting at 18°K, which was then the highest temperature. When we reported this development at the international conference the people in Physics Today called it Schmutzphysics, from the German word Schmutz, meaning dirt. At that time the physics community was interested in pure aluminum and pure tin. Nobody had expected high-temperature superconductivity out of such dirty material!

Seven years later, in 1960, Gene Kunzler and Jack Wernick found the very high critical fields associated with Nb_3Sn. We had just gotten a copper magnet at Bell Labs that would create a field of 88 kilogauss (8.8 tesla)—that was the state of the art. Kunzler put this Nb_3Sn in the magnetic field and found that it stayed superconducting. Nobody dreamed that a superconductor would be superconducting at 88 kilogauss! That was a milestone, almost as exciting as today. From the standpoint of technology, maybe it was more epochal—that remains to be seen.

At any rate, the discovery confirmed this theory that had been written about Type II superconductors by Abrikosov in Russia in 1957, as just a model, with very little data. It explained that some alloys would act like pure metal superconductors until they reached a certain magnetic field. At that point, lines of magnetic flux would penetrate them, but they'd go on superconducting, and they'd carry huge amounts of current. The discovery of Type II superconductivity opened up the technology once again.

John Hulm, now Director of Corporate Research at Westinghouse, offers this perspective on the high currents carried by Type II superconductors:

The critical current density of niobium-tin and niobium-titanium alloys usually lies somewhere between a hundred thousand and a million amperes per square centimeter. Putting a million amperes of current through an area the size of your fingernail is truly impressive. You can't do that with a normal conductor such as copper unless you go to the most elaborate high-speed cooling devices—and even then, you would need a stream of cooling water the size of the Charles River to carry away the heat.

The next 20 years saw no major scientific advances in superconductivity. In 1973, John R. Gavaler of Westinghouse, found a transition temperature of 23° in Nb_3Ge (niobium-three-germanium), but Nb_3Ge has seldom, if ever, been used in a commercial device.

During the 1960s and 1970s, companies such as Supercon, Inc., Intermagnetics General Corporation, and Oxford Instruments achieved major engineering advances necessary for development of commercial superconducting wires and magnets. It took nearly 10 years, after the discovery of Type II superconductors, to develop reliable products. The driving force behind this development was the high-energy physics community's interest in magnetic fields for particle accelerators.

The necessary engineering achievements included the following:

• Developing liquid helium refrigeration systems that would operate reliably in a commercial environment.

- Configuring Type II superconductors into wires, a particularly difficult problem with Nb_3Sn because it has a high critical field but is very brittle.
- Devising protection systems to keep the wires from melting if they suddenly lost their superconductivity.
- Creating support structures to withstand the mechanical forces on the magnets.
- Minimizing heat leaks between the supercooled components and the room temperature components.

In the years between 1960 and 1986, several hundred materials were found to be superconducting at sufficiently low temperatures. However, the highest critical temperature (i.e., the temperature below which a material becomes superconducting, T_c) achieved in this period was 23 K, which still required either liquid helium or liquid hydrogen cooling.

The workhorse of the high-field magnet technology has been niobium-titanium (NbTi), a ductile alloy that can be made into wires. A second material of great promise, niobium-tin (Nb_3Sn), can support even larger electric currents and remain superconducting in higher magnetic fields, but it has found much less use because of its brittle nature. Other materials have found more limited uses—for example, pure niobium in radiofrequency cavities and niobium nitride (NbN) in electronics.

RECENT DISCOVERIES

Alex Mueller and Georg Bednorz (see Figure 1-3), scientists at the IBM Zurich Laboratory, sparked the recent excitement in 1986. They had spent much of 1984 and 1985 looking for high T_c superconductivity in a class of metal oxide ceramic materials called perovskites. In January of 1986 they found indications of superconductivity at approximately 30°K in barium lanthanum copper oxide (see Figure 1-4). In April they reported the findings to a German journal, Zeitschrift fur Physik, which published their paper in September. The title of the paper was very controlled and restrained: *"Possible High T_c Superconductivity in the Ba-La-Cu-O System."*

Malcolm Beasley, a professor at Stanford, reflects on the developments which followed:

Bednorz and Mueller saw the resistance drop sharply at about 30°K. It didn't even go to zero, but they had courage to say maybe this was superconductivity in this wild new set of materials. A lot of us read their article, but I can't say it made us stand up straight—you see hints of high T_c superconductivity in the literature all the time. The excitement in the United States began during a meeting of the Materials Research Society in Boston in the fall of 1986, when Koitchi Kitazawa and Shoji Tanaka from the University of Tokyo not only confirmed that the material found by the Zurich researchers had a T_c near 30°K, they went further: they showed that it had the true hallmarks of superconductivity, both zero resistance and the Meissner effect, and they also told us the structure of the material, how its atoms were arranged. Their presentation was so convincing that nobody doubted it. It was at this point that everybody in the U.S. went scurrying back to their labs and said, "My God, we have a new high temperature superconductor!"

Soon researchers around the world confirmed Bednorz and Mueller's findings. Paul Chu's group (see Figure 1-5) at the University of Houston raised the T_c of the material to

FIGURE 1-3
Photograph of J. Georg
Bednorz (left) and K. Alex
Mueller at IBM's Zurich
Research Laboratory

Source: IBM Research

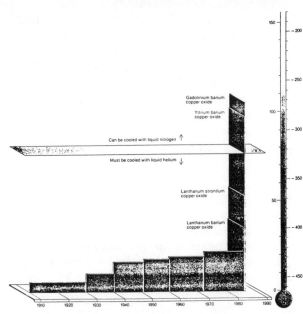

FIGURE 1-4
Illustration of High
Temperature Superconduc-
tivity Development

Source: EPRI

FIGURE 1-5
Photograph of Dr. Paul
Chu

Source: University of
Houston

57°K in December. Then Chu called a February 16 press conference to announce that his team, including M.K. Wu at the University of Alabama at Huntsville, had found a new compound with a T_c of 93°K! At first he wouldn't identify the compound, but his article in the March issue of *Physical Review Letters* revealed that he had substituted yttrium for lanthanum. Since then researchers worldwide have substituted 10 other rare earth elements to produce compounds that superconduct at temperatures over 90°K. All this work derives directly from the conceptual breakthrough of Bednorz and Mueller, who won a Nobel Prize in physics for their discovery.

A larger number of so-called high-temperature superconductors are now known to exist. In general these are variations of two basic types (the so-called 40 K and the 95 K [or 1-2-3] materials). Those with T_c greater than 77 K are based on only one structure, with copper (Cu) and oxygen (O) a constant feature. The new materials present an enormous scientific opportunity and open new vistas for potential applications. Because our understanding of superconductivity has been challenged in so fundamental a fashion, and with the present theoretical understanding of superconductivity being insufficient to explain the properties of the new materials, there is hope that what has been achieved in such a short time can be extended. The excitement surrounding the field has caught the imagination of policymakers, the media, and the public at large.

FUNDAMENTALS OF SUPERCONDUCTIVITY

Electric currents travel through superconductors with no resistance, hence no losses (provided the current is steady—alternating currents meet resistance even in superconductors). When current flows in an ordinary conductor, say a copper wire, some power is lost. In a light bulb or an electric stove, resistance creates light and heat, but in other cases the energy is simply wasted. With no resistance, magnets wound with superconductors can create very high fields without heating up and dissipating energy. Motors and generators with superconduct ing windings could be smaller, lighter, and more efficient than those built with copper. Very high magnetic fields might be used to fire projectiles, to float molten metal in a steel mill, or to levitate trains.

Above its transition (or critical) temperature, a superconductor exhibits electrical resistance like any other material. Below the transition temperature, the material has zero resistance to direct current (DC): a steady electric current will circulate in a superconducting coil forever, so far as anyone knows. Variation in the flow of current does lead to electrical losses; thus a superconductor dissipates energy when turned on or off, or when carrying an ordinary alternating current (AC losses). However, these losses are much less than those in a good normal conductor (e.g., copper) at the same temperature.

Until 1986, the highest known transition temperature was 23°K (in niobium-germanium). Then the IBM scientists in Zurich discovered a new class of materials that showed superconductivity at 35–40°K. Shortly thereafter, the compositions now termed the 1-2-3 superconductors were found, with transition temperatures in the vicinity of 95°K. The first of the 1-2-3 materials announced contained yttrium, barium, and copper oxide. More recently, copper-oxide ceramics containing bismuth or thallium have been found. These have superconducting critical temperatures in the range, respectively, of 110°K and 125°K. In April 1988, the first high-temperature superconductor (HTS) compositions containing no copper were announced, with transition temperatures up to 30°K.

A major advantage of the new materials, of course, is the potential for simpler, less expensive cooling. For technical reasons, superconductors must be operated well below their transition temperatures—as a rule of thumb, at half to three-quarters the transition value. In fact, then, liquid nitrogen temperatures will be marginal for the 1-2-3 compositions (although the practical advantages of operating in the range of 40°K rather than 4°K can be great). If the more recently discovered ceramics, with critical temperatures of 110°K and up prove to have useful properties, liquid nitrogen cooling will almost certainly prove adequate.

All the high-temperature superconductor compositions so far discovered are ceramics, rather than metals. They are new materials and are poorly understood. All ceramics are brittle and they require processing and fabrication methods very different from metals and alloys.

Superconducting Materials

The high-temperature superconductors are mixed metal oxides that display the mechanical and physical properties of ceramics. A key to the behavior of the new materials appears to be the presence of planes containing copper and oxygen atoms chemically bonded to each other. The special nature of the copper-oxygen chemical bonding gives rise to materials that conduct electricity well in some directions, in contrast to the majority of ceramics, which are electrically insulating.

The first of high-T_c oxides discovered was based on the chemical alteration of the insulating ternary compound La_2CuO_4 by replacement of a small fraction of the element lanthanum (La) with one of the alkaline earths barium (Ba), strontium (Sr), or calcium (Ca). The substitution led to compounds with critical temperatures of up to 40 K. In these materials, an intimate relation between superconductivity and magnetic order is presently under intensive study and has inspired one of the many classes of theories that attempt to explain high-temperature superconductivity.

In a second class of compounds, based on $YBa_2Cu_3O_x$ (where Y is yttrium, a rare earth), the metallic atoms occur in fixed proportions. These are the so-called 1-2-3 compounds, which are highly sensitive to oxygen content, changing from semiconducting, at $YBa_2Cu_3O_{6.5}$, to near $YBa_2Cu_3O_7$, without losing their crystalline integrity. The high sensitivity of their properties to oxygen content is due to the apparent ease with which oxygen can move in and out of the molecular lattice.

The 40 K and 1-2-3 (or 95 K) materials have similar structures but differ significantly in other respects. In both compounds, the rate earth and alkaline earth atoms provide a structural framework within which the chains of copper and oxygen atoms may be hung.

Surprisingly, the substitution of other rare earths, even magnetic ones, for yttrium in the 95 K compounds results in very little change in superconducting properties. Various substitutions are under study, both to understand the present materials and to achieve higher critical temperatures in new ones.

Status of Theoretical Understanding

In the microscopic theory of Bardeen-Cooper-Schrieffer, the presence of a net attractive interaction between conduction electrons, which would normally repel each other because of their like electrical charges, is essential to the occurrence of superconductivity. In conventional superconductors this attraction originates in the dynamic motion of the crystal lattice, which leads to an attractive "electron-phonon-electron" interaction. But the recent appearance of superconductivity in a class of materials quite different from the conventional superconductors, and with extremely high transition temperatures as well, has led physicists to explore a very wide spectrum of possible new pairing mechanisms involving, for example, spin fluctuations, acoustic plasmons, and excitonic processes. The physical origin of the pairing "glue" remains an open and to some extent crucial question. There is a wide range of theoretical possibilities, and the ultimate explanation may involve a combination of mechanisms. Indeed, some theorists have discarded conventional Bardeen-Cooper-Schrieffer theory and have suggested that there may not even be the traditional close relationship between energy gaps and basic superconducting properties.

Given the wealth of puzzling experimental features in a variety of different materials, it may take a considerable effort, with a diverse theoretical program, to unravel fully the secrets of these compounds. In the meantime, the fact that they obviously do exist can form the basis of immediate commercial exploitation. But the prospect of even more promising materials has led to a substantial theoretical effort aimed at elucidating the principles underlying the phenomenon. Typical of the questions currently under active consideration are the role played by oxygen, the nature and scope of dynamic mechanisms and the resulting electron pairing, whether this coupling is weak or strong, and whether the anisotropic nature of the materials is a truly important feature. The appearance of superconducting coherence lengths 1 or 2 orders of magnitude smaller than those previously encountered, the very low carrier concentrations, and the role of copper and oxygen will probably require a considerable extension of our current understanding of superconductivity. The fact that the superconducting interaction mechanism in the new materials is likely to be very different from that in low-temperature superconductors certainly enhances the prospect that other high-temperature superconducting materials may be discovered. Looking for materials with higher transition temperatures remains largely a matter of trial-and-error, guided by intuition—laboratory research that is time-consuming and expensive.

Physical Properties Important for Technology

Maintaining the superconducting state requires that both the magnetic field and the electrical current density, as well as the temperature, remain below critical values that depend on the material. Figure 1-6 shows this schematically, while Table 1-1 gives the critical values of temperature and field for a number of superconducting materials. Practical applications, in general, require that both the transition temperature and the critical current density be high. In some cases relatively high magnetic fields are necessary as well.

The most important physical properties for applications are the superconducting critical temperature (T_c), the upper critical magnetic field (H_{c2}), and the maximum current-carrying capacity in the superconducting state (J_c). Also important are the mechanical, chemical, and electromagnetic properties: physical and thermal stability, resistance to radiation, and alternating-current loss characteristics. Each is discussed more fully below.

Critical Temperature, T_c. A rule of thumb for general applications is that materials must be operated at a temperature of 3/4 T_c or below. At about 3/4 T_c, critical fields have reached roughly half their low-temperature limit, and critical current densities roughly a quarter of their limit. Thus, to operate at liquid nitrogen temperature (77 K), one would like T_c near 100 K, making the 95 K material just sufficient. To operate at room temperature (293 K) one requires a material with T_c greater than 400 K, well above the highest demonstrated value. Higher T_c materials would be superior across the board for applications, other properties being acceptable, and materials with T_cs above 400 K would have a truly revolutionary impact on technology. In this temperature domain one could consider mass market applications.

Upper Critical Magnetic Field, H_{c2}. $YBa_2Cu_3O_7$ samples generally exhibit extremely high upper critical fields. Preliminary measurements indicate that for single crystals H_{c2} is anisotropic, that is, dependent upon field direction relative to the *a*-, *b*-, or *c*-axes of the orthorhombic lattice. Values ranging from 30 T (*c*-axis) to 150 T (*a*- or *b*-axes) are reported

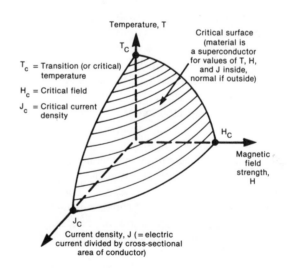

FIGURE 1-6

Dependence of the Superconducting State on Temperature, Magnetic Field, and Current Density.

Source: Office of Technology Assessment, 1988

T_c = Transition (or critical) temperature

H_c = Critical field

J_c = Critical current density

Temperature, T

T_c

Critical surface (material is a superconductor for values of T, H, and J inside, normal if outside)

H_c

Magnetic field strength, H

J_c

Current density, J (= electric current divided by cross-sectional area of conductor)

TABLE 1-1 Critical Values for Superconducting Materials

	Temperature (degrees)	Magnetic field (gauss Kelvin)	Current density[a] (amps per square centimeter)
Aluminum	1.2	105	
Mercury	4.2	410	
Lead	7.2	800	
Niobium	9.2	0.4×10^4 [b]	
Niobium (75%)—			
titanium (25%)	10	14×10^4 [b]	$\sim 10^5$ [c]
Niobium—tin	18	23×10^4 [b]	$\sim 10^7$ [c]
1-2-3 ceramic			
($YBa_2Cu_3O_{6.9}$)	93	$100 + x\ 10^4$ [b]	10^3 to $> 10^6$ [d]

[a] At zero magnetic field.

[b] Upper critical field (Type II superconductor).

[c] At 4.2°K.

[d] At 77°K. The highest values are reached with oriented single-crystal films.

Source: Office of Technology Assessment, 1988

at 4.2 K. The mechanical stresses associated with the confinement of such high magnetic fields in typical compact geometries are frequently beyond the yield or crushing strengths of known materials. Hence, improving these intrinsic H_{c2} values is less important than increasing T_c or J_c values. In fact, materials with higher T_cs should exhibit higher H_{c2} values if the performance of known materials is any guide. However, developing materials that can practically be fabricated into magnets and that retain useful J_cs at fields approaching H_{c2} even at 77 K is an important challenge.

Critical Current Density, J_c. For practical applications, J_c values in excess of 10^3 amperes per square millimeter (A/mm^2), are desirable both in bulk conductors for power applications and in thin film superconductors for microelectronics.

Bulk ceramic conductors of $YBa_2Cu_3O_7$ have achieved about $10^2\ A/mm^2$ at 4.2 K and 6 T. However, J_c falls off very steeply to levels around 1-10 A/mm^2 at 77 K and 6 T. These J_c values are determined from magnetization measurements; J_c values derived from transport measurements are usually lower. There is no clear understanding of these reduced J_c levels at the present time, but achieving acceptable values for J_c in bulk high-temperature superconductors is of critical importance and should be a principal focus of research on fabrication processes.

Based on current experience, a reasonable target specification for a commercial magnet conductor would be J_c of $10^3\ A/mm^2$ at 77 K and 5 T, measured at an effective conductor resistivity of less than 10^{-14} ohm-m,[2] with strain tolerance of 0.5 percent, and availability at prices comparable to or less than those of conventional low-temperature superconductors.

Preliminary measurements on epitaxially grown single-crystal thin films indicate J_c values in excess of 10^4 A/mm^2 at 77 K and zero magnetic field. These values seem adequate for microelectronic applications.

Mechanical Properties

Present ceramic high-temperature superconducting materials can be strong, but they are always brittle. Hence, it may be that high-temperature superconductors wire will be wound into magnets prior to the final high- temperature oxidation step in its fabrication, after which it becomes very brittle. Other conductor fabrication techniques might be feasible, however. For example, those used for producing flexible tapes of Nb_3Sn. An elastic strain tolerance of 0.5 percent may be achieved in a multifilamentary conductor by a fine filament size and by induced compressive stresses.

Currently available ceramic technology allows the fabrication of the kinds of complicated pieces that may be needed for such applications as radiofrequency cavities. There are some indications that the new materials may be deformable above 800 C and can then be shaped. The development of a mechanical forming process, however, is constrained by the parallel need for the process to optimize J_cs, both by aligning anisotopic crystal grains and by increasing the strength of the intergranular electrical coupling.

Life testing will also be necessary to understand the performance of materials under realistic conditions such as temperature cycling and induced stresses due to transient fields. The adhesion of high-temperature superconductors to other materials is important in microelectronics, in which temperature cycling results in thermal expansion and contraction that cause stresses at the interface.

Chemical Stability

The 1-2-3 compounds readily react with the ambient atmosphere at typical ambient temperatures. These problems seem to be less severe, however, as the purity and density of the materials are improved. Both water and carbon dioxide participate in the degradation through the formation of hydroxides and carbonates. Further study of the nature of this degradation is needed to develop handling procedures or protective coatings that will ensure against impairment of superconducting properties by atmospheric attack.

Chemical stability is also limited because oxygen leaves the structure under vacuum, even at room temperature. Surface protection techniques need to be developed to allow satisfactory performance and lifetime of the materials under various conditions of storage and operation. These concerns are heightened in thin films, in which, for some applications, the chemical composition of the outer atomic layers near the surface must be maintained through many processing steps, and in which diffusion into the substrate interface could degrade super-conducting properties.

---

[2] The sensitivity of many of the measurements reported on the new ceramics is poor, and "zero resistance" often means 10^{-10} to 10^{-7} ohm-m, 4 to 7 orders of magnitude greater than values required for practical application.

Radiation Effects

High-temperature superconductors appear to be somewhat more sensitive to radiation than conventional superconductors. High sensitivity to radiation damage could pose a difficult, although not insurmountable, problem for application to magnetic fusion machines. For electronic applications the substitution of either conventional or high-temperature superconducting devices for those employing semiconductors would result in an improvement of several orders of magnitude in resistance to radiation damage.

Alternating Current Losses

Conventional superconductors exhibit losses in alternating current applications, such as in 60 Hertz power transmission or in microwave devices. Although little is known about the alternating current characteristics of the new high-temperature superconductors, there is no reason to expect that the new materials will exhibit lower alternating current losses than other superconducting materials. Recent measurements on thin films in parallel applied fields show the presence of a large surface barrier for the entry of flux, which indicates that hysteresis losses would be small.

Synthesis and Fabrication

Tailoring high-temperature superconductor materials for applications either in electronics or where high currents and/or fields are needed (e.g., electric power) will entail design and processing at size scales from the atomic level on up. Engineers and scientists engaged in applications developments, as well as materials processing, will have to concern themselves with electronic structure (energy gaps), crystal structure (the arrangement of atoms in the material), microstructure (grain boundaries), and the fabrication of films, filaments, tapes, and coils (see Figure 1-7).

FIGURE 1-7
Coil of superconducting yttrium-barium-copper oxide.

Source: Argonne National Laboratory

The high-temperature superconductor ceramics are not only brittle, and chemically reactive, but highly anisotropic—meaning that properties vary with direction within a grain of the material. The 1-2-3 ceramics, for instance, show differences of as much as 30:1 in critical current density depending on grain orientation. Some of the current density limitations can be traced to anisotropy, but grain boundaries seem to be the primary culprit.

Some of the processing and fabrication techniques familiar from work with electronic and structural ceramics hold promise for the new superconductors. Bulk samples of high-temperature superconductor material can be made by hot pressing, extrusion, and tape casting, among other methods. The anisotropy in the 1-2-3 materials has led many research groups to seek processes for aligning the grains—e.g., extruding a slurry of single crystals in a high magnetic field to create a tape. Semiconductor fabrication techniques, likewise, can in some cases be adapted for making thin films.

Past work on niobium-tin, a brittle intermetallic compound, may also hold lessons for high-temperature superconductor materials, which, like other ceramics, cannot deform plastically. Because they break easily and without warning—like glass, being very sensitive to small imperfections (hence the scribed lines used to "cut" glass)—practical applications may require specialized in-situ processing, as well as careful design to minimize strain. Magnets wound with niobium-tin are made starting with strands of niobium in a copper-tin alloy matrix—flexible and ductile. Heat treatment after the wires have been drawn and

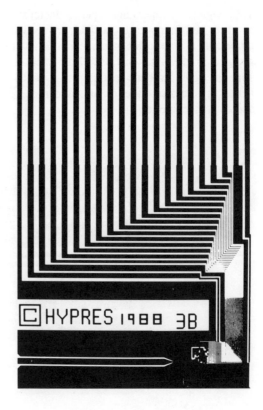

FIGURE 1-8
HYPRES Superconducting
Integrated Circuit.

Source: HYPRES, Inc.

wound into coils causes the tin to combine with the niobium, forming the superconducting compound, with its vastly different properties. Some R&D groups have pursued similar processes for ceramic superconductors.

Progress has been faster with thin films (see Figure 1-8), which can be created via a wide range of well-known techniques—e.g., sputtering, and evaporation by molecular or electron beams. Finding good substrates on which to deposit the high-temperature superconductor layer has been the primary problem. The high-temperature superconductor compounds react chemically with many otherwise desirable substrate materials, including those used for integrated circuits (silicon, sapphire). Strontium titanate gives high current densities compared to other choices, but is expensive and has otherwise undesirable properties. Silicon would be ideal as a step toward combining semiconductor and superconducting electronics. While the temperatures so far required for creating the proper high-temperature superconductor composition has posed difficulties, many research groups have been working on the problems, with encouraging results.

Two steps are required to synthesize the 95 K superconducting materials. First, the basic structure must be formed at temperatures above 600–700 C. The tetragonal structure so formed is deficient in oxygen and does not possess superconducting properties. Accordingly, the second part of the synthesis involves annealing under oxygen at a temperature below 500 C. The arrangement of this additional oxygen in the lattice causes a conversion from tetragonal to orthorhombic symmetry that supports high-temperature superconductivity.

For the future development of high-temperature superconducting materials, we require a much better understanding of how synthesis conditions relate to the structure of the 1-2-3 compounds on the atomic and nanometer scales. We need to know, further, how this structure relates to superconducting properties and to other important properties such as chemical stability and mechanical strength.

The fabrication of many high-temperature superconducting ceramics involves grinding of prereacted starting materials, which can result in contamination from grinding media. At the present time there is no evidence that impurities introduced by grinding degrade superconducting properties; however, further work is required to optimize this process. Also, the rate of oxygen uptake during the oxygen anneal depends on the available surface area of the sample; for large particles or very dense ceramics, this critical oxygen uptake reaction can be slow. A recently announced fabrication technique, in which materials in bulk (see Figure 1-9) are made by melting the ingredients, may make the manufacture of wires and specially shaped pieces much easier and may eliminate the need to work with sintered materials. Bulk superconductors made by the melt-textured growth process greatly increase the current-carrying capacity of the bulk material.

In addition, processes are required for the commercial production of high-quality thin films on useful substrates. What is needed is to compare the various ways that have been used to produce thin films—electron beam, planar magnetron sputtering, pulsed laser evaporation, molecular chemical vapor deposition—and establish the strengths and weaknesses of each method. Epitaxial growth methods also need to be studied.

A further requirement is the achievement of reliable, low-resistance ohmic contacts to the new materials. A better understanding of phase equilibria, solid solutions, and intermetallic compounds is needed to find stable ohmic contacts that do not degrade superconducting behavior.

New Superconducting Materials

Finally, we must not neglect the search for new compounds with intrinsically superior superconducting properties. Cryogenic systems that operate below 77 K should be investigated, and the compatibility of high-temperature superconductors with refrigerants other than liquid nitrogen (e.g., liquid neon) should be tested. Further, and perhaps most importantly, the events of the past two years have shown that surprises do occur, and it may be that superconductivity at or above room temperature may be detected at some future date in compounds not yet studied.

ASSESSMENT OF TECHNOLOGY

Basic research in high-temperature superconductivity is being actively pursued in all of the developed nations and in several developing nations. In most cases scientists have switched spontaneously from other scientific activities into high-temperature superconductivity research. To this point, however, little new money has gone into basic research efforts. Plans are being prepared for 1989, but at present no major new U.S. government resources have been committed. The prevailing attitude appears to be that of waiting to see how the science progresses.

In Japan the scientific and technical community has responded vigorously, they have reprogrammed and identified new funds for immediate additional action by government agencies. The Japanese have also been quite active in formulating plans for the next fiscal year (which begins in April). Private industrial corporations are said to be heavily investing their own funds in research on high-temperature superconductors, with the Japanese government intervening to establish industry consortia to pursue prototyping and other early development activities. Japan offers perhaps the strongest long-range competitive threat to the U.S. position.

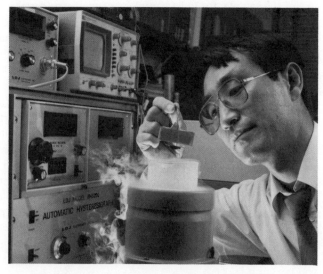

FIGURE 1-9
Sungho Jin tests resistivity
of bulk superconductor.

Source: AT&T Bell
Laboratories

In Europe, historical strengths in basic research and industrial development are being applied to the new superconductors. National and cross-national efforts are in the early organizational stages at best (again, with the exception of the reprogramming of research funds), and major project goals to drive technical problem solving are not yet in place. On the other hand, a variety of industrial corporations are involved in research, and precompetitive collaborations appear to be at advanced planning stages.

In the USSR, traditional scientific strengths in superconductivity theory and basic experimental approaches are being applied to the new materials. In addition, work is being carried out on the susceptibility of high-temperature superconducting materials to radiation damage.

The National Academy of Sciences, in their report entitled "Research Briefing on High-Temperature Superconductivity," made several technology recommendations as summarized below:

- The discovery of materials that exhibit superconductivity at temperatures up to 95 K is a major scientific event, certainly one of the more important of the last decade. Meeting the complex challenge of understanding the phenomenon will improve fundamental knowledge of the electronic properties of solids.

- Although a large number of promising theories are being explored, there is as yet no generally accepted theoretical explanation of the high critical temperature behavior. Current theoretical understanding does not preclude T_cs above 95 K.

- The base of experimental knowledge on the new superconductors is growing rapidly. The intrinsic properties that can guide theory are still being determined. A number of investigators have reported superconducting-like transitions at temperatures above 95 K, in some cases even above room temperature; at present those effects have not been firmly established.

- The prospect exists for applying the new superconductors to both electrical and electronic technology. The nature of the new materials (quaternary ceramic oxides) suggests that a substantial materials engineering effort will be required to develop bulk conductors for power applications or thin films for electronic applications.

- The applications currently being considered are largely extrapolations of technology already under investigation for lower temperature superconductors. To create a larger scope of applications, inventions that use the new materials will be required. Given the materials engineering problems already mentioned, the period of precommercial exploration of the new superconductors for other applications will probably last for a decade or more. Although it is too early to make a sound engineering judgment about most of the possible high- temperature superconductivity applications, the potential impact could be enormous, especially if operation at room temperature can be achieved.

- Near-term prospects for high-temperature superconductivity applications include magnetic shielding, the voltage standard, SQUIDs, infrared sensors, microwave devices, and analog signal processing. Longer-term prospects include large-scale applications such as microwave cavities; power transmission lines; and superconducting magnets in generators, energy storage devices, particle accelerators, rotating machinery, medical imaging systems, levitated vehicles, and magnetic separators. In electronics, long-term prospects include computer applications with semiconducting superconducting hybrids, Josephson devices, or novel transistor-like superconducting devices. Several of these technologies will have military applications.

• The complexity of the materials technology and of many of these applications makes a long-term view of research and development essential for success in commercialization. The infectious enthusiasm in the press and elsewhere may have contributed to premature public expectations of revolutionary technology on a very short time scale. Over-reaction in either direction could be detrimental to achieving the true long-term potential of high-temperature superconductivity.

Further Progress: The Next Steps

The short-term problems and long-term potential of high-temperature superconductivity may both be easily underestimated. Given this potential and today's limited understanding of the new superconducting materials and their properties, it is essential that government, academic institutions, and industry take a long-term, multidisciplinary view. Because science and technology in this field are strongly intertwined, progress must occur simultaneously in basic science, manufacturing processing science, and engineering applications. It is also important to maintain an open and cooperative international posture.

The National Academy of Sciences' report entitled *"Research Briefing on High-Temperature Superconductivity,"* has identified eight major scientific and technological objectives for a national program to exploit high-temperature superconductivity. They are:

1. To improve understanding of the essential properties of current high-temperature superconducting materials (especially T_c, H_{c2}, J_c and alternating current losses) through the acquisition of additional experimental data.
2. To develop an understanding of the basic mechanisms responsible for superconductivity in the new materials.
3. To search for additional materials exhibiting superconductivity at higher temperatures by the synthesis of new compositions, structures, and phases.
4. To prepare thin films of controllable and reproducible quality from present high-temperature superconducting materials and to establish preferred techniques for growing films suitable for electronic device fabrication.
5. To develop bulk conductors from current high-temperature superconducting materials, with special emphasis on enhanced electric current-carrying capacity.
6. To advance the understanding of the chemistry, chemical engineering, and ceramic properties of the new materials, focusing on synthesis, processing, stability, and methods for large-scale production.
7. To fabricate a range of prototype circuits and electronic devices based on superconducting microcircuits or hybrid superconductor/semiconductor circuits, as suitable thin film technologies become available.
8. To fabricate a range of prototype high-field magnets, alternating and direct current power devices, rotating machines, transmission circuits, and energy storage devices, as suitable bulk conductors are developed.

Dr. Erick Bloch, Director of the National Science Foundation, urges us to temper our optimism with patience and persistence. "What has been accomplished so far is largely basic research. We are rapidly learning more about the materials and their characteristics and about the nature of superconductivity itself. We still don't know the limits of the new superconductivity, nor do we know how to make materials that will both superconduct and have desirable engineering properties. Much scientific and engineering work will be necessary to fill these gaps in our knowledge."

CHAPTER 2

SUPERCONDUCTIVITY APPLICATIONS

INTRODUCTION

Virtually all of the applications currently envisioned for high-temperature superconductors are extrapolations of devices already operated at liquid helium temperatures. The most important applications, however, may well involve devices that have yet to be contemplated, much less invented.

Much of the excitement[1] over high-temperature superconductors has been stirred by speculation concerning applications as low-loss electric power transmission or magnetically levitated trains. In some of these cases, development will depend more on system costs and progress in competing technologies than on the specifics of high-temperature superconductors. Both transmission lines and levitated trains have been demonstrated with low-temperature superconductor materials. Superconducting transmission lines, which must be run underground because of the cooling requirements, may eventually prove cost-effective relative to conventional underground transmission; thus far, however, these applications have not moved out of the test stage. Magnetic levitation trains could be built by the end of the century in Japan and West Germany.

More than likely 5 to 10 more years of research and development lie ahead before significant applications of high-temperature superconductors emerge. Those that come earlier are likely to be highly specialized—perhaps in military systems, perhaps targeted on very demanding civilian needs.

---

[1] Commercializing High-Temperature Superconductivity, U.S. Congress, *Office of Technology Assessment*, OTA-ITE-388, GPO #052-003-01112-3, Washington, D.C., June 1988.

1. Basic research, both theoretical and experimental, aimed at explaining high-temperature superconductors, finding new materials and exploring their properties, and understanding structure-property relationships.
2. Applied research, focused particularly on development of processing methods and optimization of material properties through manipulation of processing variables. A great deal of research and development is still needed before routine production of tapes and multifilamentary conductors can begin.
3. Applications engineering (for high-temperature superconductors)—e.g., development of prototype chips containing many Josephson junctions—including extensive testing under realistic operation conditions (environmental exposure, thermal cycling, mechanical vibrations, electrical surges, loss of temperature control). Joining techniques for conductors will also be needed, as well as repair methods.
4. Process engineering—manufacturing methods for routine (rather than laboratory) production. Problems here will include yields and reliability in superconducting circuits, and methods for producing long continuous lengths of superconducting cable. Inspection, testing, and quality control procedures will need a good deal of attention.
5. Systems engineering—design, development, and demonstration of applications in which superconducting components are integrated into such end products as computers, electrical generators, and coil or rail guns. For instance, without further progress in transition temperatures, high-temperature superconductor's interconnects in computers will require cooling to liquid nitrogen temperatures. Fortunately, these temperatures also offer performance advantages for semiconductor chips.

Many of these activities can go forward in parallel. In some cases it makes sense to proceed sequentially. For instance, applied research aimed at increasing current density can and should proceed in conjunction with process research and development, because processing affects microstructure, and microstructure affects current density. But work on production scale-up could be developed until the effects of processing variables can be reasonably well understood. On the other hand, research intended to discover whether a particular processing technique—e.g., laser annealing—compromises some properties that will be needed early in the production process.

As shown in Table 2-1, present and potential applications fall into several distinct classes. Present applications include high-field magnets, radiofrequency devices, and electronics. Superconductivity brings unique advantages to high-field applications because resistive conductors such as copper dissipate large amounts of energy as heat when carrying large currents. Superconductors are also useful in high-Q cavities because of their low alternating current losses at high frequencies compared to those in normal metals.

Electronic applications for superconductors usually involve low electric currents (although high current densities) and low magnetic fields. The core element has been a unique bistable device, the Josephson junction. Superconductors may also eliminate resistive current losses in electronic lines and device interconnections. In addition, various kinds of superconducting sensors have been produced. All of these applications, including the assembly of superconducting electronic components into larger devices, will be reconsidered with the new compounds.

TABLE 2-1. Principal Applications of Superconductivity

PRESENT APPLICATIONS

Magnets
- Commercial and industrial uses
  - Medical diagnostics and research (magnetic-resonance imaging and spectroscopy)
  - Radiofrequency devices (gyrotrons)
  - Ore refining
  - R&D magnets
  - Magnetic shielding
- Physics machines (super colliders, magnetic fusion machines, radiofrequency cavi ties)

Electronics
- Defense Systems
- Sensitive accurate instrumentation (superconducting quantum interference devices (SQUIDs), infrared sensors, oscilloscopes)
- Electromagnetic shielding

POTENTIAL APPLICATIONS
(Proven superconducting technology, but no current market adoption.)

- Computers
  - Semiconducting superconducting hybrids
  - Active superconducting elements
- Power utility applications
  - Energy production (magnetohydrodynamics, magnetic fusion)
  - Large turbogenerators
  - Energy storage
  - Electrical power transmission
- Transportation
  - High-speed trains (magnetic levitation)
  - Ship drive systems

Application of high-temperature superconductors divide into those relevant to currently available materials with critical temperatures near 95 K, and those relevant to possible future materials with higher critical temperatures. The most exciting possibilities, of course, arise with materials with critical temperatures above room temperature.

MAGNET APPLICATIONS

Early applications of superconductivity have involved the design and construction of powerful magnets wound with low-temperature superconductor materials and cooled with liquid helium. Such magnets have been used in scientific experiments (e.g., the Tevatron), and in magnetic resonance imaging. Learning to design and build magnets helps with more demanding applications, such as rotating machinery.

Almost all the power consumed by a superconducting magnet goes to operate the cooling system. For a big magnet wound with copper, resistive losses far outweigh the refrigeration costs for an equally powerful low-temperature superconductor magnet. Indeed, large copper-wound magnets need their own cooling systems just to carry off the heat generated through resistance. The cost comparison below, for a bubble chamber magnet at Argonne National Laboratory (see Table 2-2)—a typical early scientific application—shows that a conventional magnet would cost five times more to operate.

Superconducting magnets have other advantages compared with conventional magnets. Stability is easier to achieve, for instance. In a conventional magnet, the field strength varies as the windings heat up and expand. The stability characteristics of low-temperature superconductor magnets give them advantages both in scientific apparatus and in magnetic resonance imaging.

With low-temperature superconductor magnet technology well in hand, high-temperature superconductor designs will have to perform at least as well (in terms of characteristics such as stability) before their simpler cooling systems and lower operating costs will make them competitive. Fabricating the conductors will be difficult. Stable operation and protection against overheating in the event of refrigeration failures require multifilamentary cables, just as for low-temperature superconductors, with filament diameters of a few microns. Given the brittleness of the new ceramics, methods for producing filaments and for fabricating cables are just beginning.

Once high-temperature superconductor wire and cable becomes available, applications-specific requirements will come to the fore. Magnetic resonance imaging, for example, while requiring highly stable fields for good image quality, does not otherwise make heavy demands on the magnet system. Still, joining methods that eliminate resistive imperfections will be needed for image quality comparable with that already achieved using low-temperature superconductors. Magnetic resonance imaging systems are expensive, and savings from simpler cooling will not make that much difference for

TABLE 2-2. Annual Operating Costs

	Superconducting Magnet (actual)	Conventional Magnet (estimated)
	(thousands of dollars)	
Electrical power	$ 17.5	$ 550
Cooling	81.3	4
Maintenance	5.2	?
	$ 104	$ 554 +

Source: Argonne National Laboratory

commercial competition. Magnetic separation is another story. Here, for instance, cheap but powerful magnets could be used to sort scrap metal for recycling, in refining ores, purifying chemicals, removing sulfur from pulverized coal, and cleaning up waste water. In all these applications, cost, reliability, and ease of use (including maintenance) by a largely blue-collar labor force become significant design considerations. Design considerations for magnetic levitation trains likewise include cost, reliability, longevity, and safety. But the political and economic questions loom ever larger than for, say, desulfurizing coal. In the United States, investments in fixed-rail transportation would have to clear obstacles ranging from opposition by airlines to high costs for rights-of-way. In Japan, where the needs and constraints differ, research and development on high-temperature superconductor-based magnetic levitation is much more likely to go forward.

Superconducting magnets using liquid helium technology have been successfully applied for a number of years in engineered systems and development projects in hospitals, mines, industrial plants, laboratories, and transportation systems. Most of these applications require multiple technologies, with superconductivity playing a critical role.

In medicine, superconducting magnets have been a significant factor in the development of a new market. In high-energy physics, superconductors have led to machines of unprecedented and previously inconceivable energy. In electric power, potential applications in energy storage and power transmission equipment (see Figure 2-1) promise to extend the capacity and range of current technology. Superconducting magnets (see Figure 2-2) are essential components in experimental systems for magnetic fusion and magnetohydrodynamics (MHD). The extremely high power-to-weight ratio possible for superconducting machines makes them particularly attractive for space applications.

FIGURE 2-1
Illustration of Energy
Storage Plant

Source: EPRI

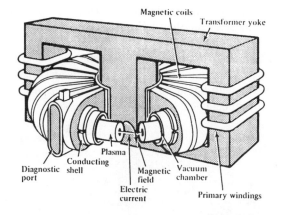

FIGURE 2-2
Superconducting magnets
are essential components
for magnetic fusion and
magnetohydrodynamics.

For magnet and power applications, the higher the critical temperature, the smaller the scale at which commercial viability will be achieved. As an example, the power level at which motors and generators become competitive will be much lower than with the present low-temperature superconductors, when compared to nonsuperconducting machines. A liquid nitrogen-cooled motor, for instance, operating at modest current and magnetic field, might well be smaller, more efficient, and more reliable for the same power output than many present-day motors.

For most applications the switch from liquid helium to liquid nitrogen technology is not revolutionary but will lead to improvements. The continued need for refrigeration is a disadvantage and will reduce market penetration. Of course, the reconsideration of applications held to be impractical at liquid helium temperatures might lead to new products. A hollow conductor cooled with liquid nitrogen is easy to visualize in practical use, for instance. It may not be necessary to demand that the technical specifications of new materials compete with the best commercial superconducting materials of today: a conductor of modest specifications may have value in a wider context than conventional low-temperature superconductors. The new materials, in short, may not so much replace present-day superconductors as extend the applications of superconductivity to a larger circle of users.

MEDICAL APPLICATIONS

Magnetic resonance imaging (MRI) has been possible for some years now. Based on the principle of powerful magnets, the imagers (see Figure 2-3) are capable of providing soft tissue images (see Figure 2-4) that are obtained without the effects of its predecessor, the x-ray. The machines are large, extremely expensive and require liquid helium to cool the magnets to their superconductive state. Their size, initial cost, and maintenance are prohibitive to some medical facilities.

More powerful magnets, cooled to only nitrogen temperatures will make the magnetic resonance imager less expensive to buy and reduce the costs of cooling to a small percentage of the helium system. In time, it could become a replacement in total for the x-ray machine of today, eliminating the dangers of ionization to the patient.

Magnetic resonance imaging (MRI) and spectroscopy (MRS) constitute radically new techniques in medical diagnosis and treatment, and their full impact is yet to be realized. Much more widespread availability of MRI and MRS systems can be anticipated, with reductions in cost and enhanced features. The use of high-temperature superconducting materials would likely bring further reductions in the costs of manufacture and operation. The redesign of MRI and MRS systems with liquid nitrogen cooling would also make them more user-friendly and reliable by reducing cooling system complexity.

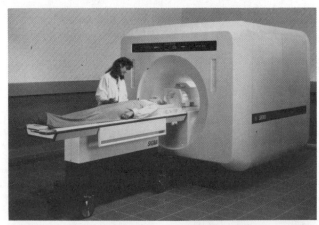

FIGURE 2-3
Magnetic-resonance imaging (MRI) is major medical application of superconducting materials.

Source: General Electric Medical Electric Systems Group

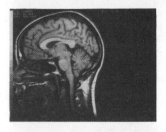
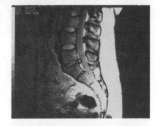
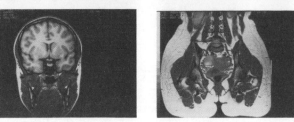

FIGURE 2-4
Soft tissue image analysis provided via magnetic resonance imaging.

Source: General Electric MR Development Center

(upper left) Image of a head demonstrating excellent soft tissue contrast and fine anatomic detail obtainable at high magnetic field strengths. Using 3mm slice thickness, this image was acquired on a 1.5 Tesla Signa™ system.

(upper right) Image of lumbar spine in the sagittal plane showing a herniated disc at the L5-S1 level. The 3mm thin slice was acquired with a surface coil on a 1.5 Tesla Signa system.

(lower left) A 3mm image of the heat in the coronal plane showing pituitary fossa. The ability to achieve thin slices allows visualization of the pituitary stalk and small tumors of the pituitary gland. This image was obtained on a 1.5 Tesla Signa system.

(lower right) Image of the abdomen showing the psoas muscle, uterus, bladder, and pelvic bone marrow. This 3mm thin slice image was acquired on a 1.5 Tesla Signa system. The bright area shows a pelvic cyst.

SUPERCONDUCTING SUPER COLLIDER (SSC)

Superconductivity's most profound impact on society may stem from its role as an enabling technology for scientists asking fundamental questions about the universe. What are the basic building blocks of matter? What forces act on them? How did the universe begin? When will it end? What is the origin of matter?

The most sensitive probe ever designed for answering these questions is a particle accelerator scheduled to operate in 1996: the Superconducting Super Collider (SSC).

The SSC will explore new realms inaccessible to today's instruments. Lying in a 53-mile underground tunnel, it will accelerate two beams of protons in opposite directions to over 99.9% of the speed of light. Superconducting magnets will bend and focus the proton beams. The protons will collide in six interaction halls, creating intense concentrations of energy, which will transform into matter in the form of new particles. Electronic detectors will record data about the particles, and computers will analyze the data. Its high energy (20 TeV, or trillion electron volts, per beam) and its high collision rate (100 million per second), will make the SSC the most powerful accelerator ever built. The schematic provided in Figure 2-5, shows the layout for the 53-mile ring. The injector, drawn to scale, is an accelerator 4 miles in circumference—the size of the Tevatron at Fermilab.

Data from the SSC will enhance the interchange between high-energy physicists and cosmologists. The flow of information goes both ways—cosmologists teach high-energy physicists valuable things about elementary particles. For example, the question of whether the universe is open or closed bears on the mass of neutrinos. Chris Quigg of the SSC Central Design group explains it this way: "The universe is expanding—galaxies are moving away from us in all directions. But is the universe open or closed? In other words, will it expand forever, or will it fall back on itself at some point and demolish everything in the Big Crunch?"

Superconducting magnets reduce the capital costs as well as the operating costs of high-energy accelerators. The highest energies can be achieved in colliders—accelerators in which two particle beams are routed in opposite directions and then made to collide. (It's also possible to accelerate a single beam of particles toward a stationary target, but

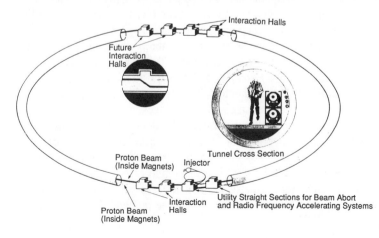

FIGURE 2-5
Illustration of
Superconducting
Super Collider
(SSC) Concept

Source: Supercurrents,
March 1988

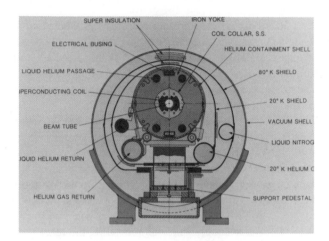

FIGURE 2-6
Cross Section of
Superconducting Super
Collider Magnet

Source: DOE

then part of the energy has to go into motion of the particles after the collision to conserve momentum, and therefore is not available to create new phenomena.) The most economical way to build a collider is to create a ring in which the particles accelerate over many revolutions. Colliders need magnetic fields to bend and focus the particles through a curved path. By generating higher magnetic fields than ordinary iron magnets, superconduct ing magnets bend the proton beams more strongly, and thus minimize the size of the ring.

The SSC will contain a total of 10,000 magnets (see Figure 2-6), over 8000 of them being dipole magnets 17 meters long. The project has required several years of research and development because the SSC will operate at a higher field—6.6 tesla, as compared to the 4.5 tesla at Fermilab.

Its array of superconducting magnets will create a project of the magnitude unsurpassed by any other application. The SSC will presently rely on technology currently available, i.e., it will use supercooled magnets using helium as a refrigerant. The –452 degree F refrigeration system required by the SSC will be the most costly element of the system and when the warmer liquid nitrogen components become available, they will be immediate candidates to replace the older superconducting magnets.

There is a great degree of discussion in the scientific community about the efficacy of the SSC project in that its $3.2 billion price tag (see Table 2-3) will require a significant part of the budget of the National Science Foundation (NSF) for three years, leaving the funding of other projects proposed by the academic community an uncertainty. The NSF presently is tasked to fund all basic research across all the sciences.

Countering the arguments of the scientific community of draining the funding pot, many view the SSC project as a focal point for industry to develop and perfect still newer applications for superconductors. They cite the Tevatron effort, the nations currently operating billion volt electron accelerator, as an example of what can happen to new markets. The fine wire filaments currently used in today's superconductors was created and perfected here and several companies exist on the demand created for this product today.

TABLE 2-3. SSC Construction Cost Breakdown

(Millions of FY 1988 Dollars)

TECHNICAL COMPONENTS		$ 1519
Magnets	(1068)	
Cryogenics	(129)	
Other	(322)	
CONVENTIONAL FACILITY		614
Collider Facilities (Tunnel)	(370)	
Other	(244)	
SYSTEMS ENGINEERING & DESIGN		307
MANAGEMENT AND SUPPORT		205
CONTINGENCY		565
		$ 3210

Source: DOE

The SSC, like the Tevatron, will create a commitment by government and a demand for industry to fulfill for many yet perfected components as nitrogen superconductors and their refrigeration systems come on line, both in the initial production and in maintaining the SSC for years to come. This demand is exactly what business will need to see if it is to devote both effort and resources. The SSC will also attract larger business than the Tevatron because its need to provide larger air liquification systems will far exceed the latter's helium systems.

Superconductors are giving us the opportunity to expand our knowledge. What new technologies might we develop if we understand the basic units of matter and the forces— or even the single force—affecting them? This is a question the superconducting super collider may help us answer.

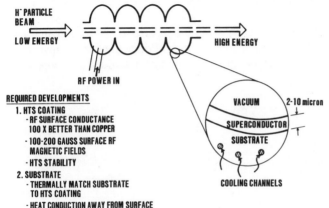

FIGURE 2-7

High-Temperature Superconductors for RF Cavity Applications

SUPERCONDUCTING RADIOFREQUENCY CAVITIES

If microwave alternating current loss characteristics are tolerable, the new superconductors may greatly improve the performance of superconducting radiofrequency cavities (see Figure 2-7) by allowing them to operate at higher fields. Indeed, the potential impacts embrace all of microwave power technology, especially in the promising millimeter-wave region. Accelerator technology might also be significantly advanced by the availability of liquid nitrogen-cooled superconducting cavities. The applicability of superconducting technology to recirculating linear accelerators, on the other hand, is an accepted fact. In addition to providing high-quality beams for nuclear physics research, these machines are natural candidates for continuous beam injectors used in free-electron lasers. As technology matures and industrial-applications develop for high-power, high-efficiency tuneable lasers in biotechnology, fusion plasma heating, and other fields, superconducting radiofrequency devices will proliferate.

DEFENSE SYSTEMS APPLICATIONS

Defense applications of superconductivity range from shielding against nuclear blasts to high-speed computers and motor-generators for ships. Conceptually, there may be little difference between military and commercial applications. But in practice, differences will be pervasive at levels all the way from devices and components (e.g., radiation hardening) to the system configuration itself (cost-performance tradeoffs much different than for commercial markets). Computing requirements for smart weapons—for example, real-time signal processing—tend to be quite different from those important in the civilian economy. Thus, as development proceeds, military uses of superconductivity will diverge in many respects from civilian applications. Some of the military applications could be compelling. Submarine (see Figure 2-8) detection with SQUID-based sensors, for instance, offers at least a factor of 10 improvement over current methods. Conventional electric generators for shipboard or vehicle use, or for producing electric power under battlefield conditions, produce about 2 horsepower per pound; prototype low-temperature superconductor generators have already reached 25 horsepower per pound. Superconducting coil or rail guns promise increases in projectile velocities of 5 to 10 times.

The U.S. Department of Defense (DOD) has funded superconductivity research and development since the early 1950s, contributing to the development of large, high-field magnets, electrical machinery, low-temperature superconductor sensors, and supercon-

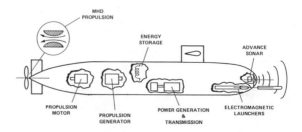

FIGURE 2-8
Potential Uses of
Superconductivity in
Submarines

ducting computers. DOD (and the Department of Energy) also supported much of the materials processing research and development that proved necessary to achieve high current densities in low-temperature superconductor wire and cable. Since 1983, the research and development objectives of DOD programs in low-temperature superconductivity have been redirected, and the programs have grown, as a result of the Strategic Defense Initiative (SDI).

For the Strategic Defense Initiative, high-temperature superconductor shielding, waveguides, and sensors (for use in space) hold obvious attractions, while low-temperature superconductors work also continues. Early in 1988, Bechtel and Ebasco began an SDI-funded design competition on low-temperature superconductor magnetic energy storage for powering ground-based free-electron lasers. SDI has also targeted very high-frequency communication systems, where low-temperature superconductors could offer substantial improvements in performance and extended frequency range. Here, the 1-2-3 ceramics seem to offer theoretically promising electronic characteristics (i.e., larger energy gaps). They would also avoid the many practical problems that liquid helium cooling poses in a military environment.

DOD has also renewed its attention for two of the prospective high-field, high-power applications—ship propulsion, and coil/rail guns. Military funding of research and development on low-temperature superconductor machinery began in the middle 1960s, with a 3000-horsepower prototype completed several years ago. Magnetohydrodynamic (MHD) thrusters offer a wholly different alternative, doing away with propellers, as well as shafts and gearing.

DARPA Superconductivity Program

In 1978, the Defense Advanced Research Projects Agency began funding research and development on electromagnetic launchers, or coil/rail guns. The initial goal, apparently, was a cannon for the Navy. With the advent of SDI, much of the DOD work has been redirected toward higher velocity systems, capable of launching a projectile into space. Like the commercial applications, the requirements for machines or for coil/rail guns, start with good conductors. DARPA's program to push superconductivity faster in the research phase and to move products into the market place, began with an Industry/University Briefing held on 26 June, 1987, in Arlington Virginia. The meeting was held to add emphasis to a Broad Agency Announcement (BAA) indicating the intent to spend $50 million over several years in support of superconductivity initiatives. The announcements broad technical scope included synthesis, processing and fabrication, complete characterization, and component demonstrations of superconductors for use in advanced manufacturing scenarios.

Because of the importance of superconductivity to the services and the country in general, DARPA expects to expand its high temperature superconductor effort.

The objective of the DARPA program is to develop materials processing and fabrication approaches to producing thin film and bulk superconductors with transition temperatures at or above 90 degrees Kelvin.

The DARPA program (see Figure 2-9) hopes to accomplish its objective by:

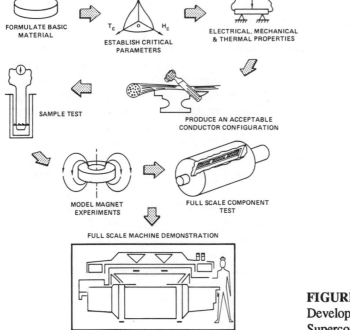

FORMULATE BASIC
MATERIAL

ESTABLISH CRITICAL
PARAMETERS

ELECTRICAL, MECHANICAL
& THERMAL PROPERTIES

SAMPLE TEST

PRODUCE AN ACCEPTABLE
CONDUCTOR CONFIGURATION

MODEL MAGNET
EXPERIMENTS

FULL SCALE COMPONENT
TEST

FULL SCALE MACHINE DEMONSTRATION

FIGURE 2-9
Development of High T_c
Superconducting Systems

- Exploiting research results which include the intellectual basis for understanding the properties of superconducting materials.
- Developing manufacturing science to support fabrication of prototype applications for DOD and NASA.
- Concentrating efforts to develop as quickly as possible an industrial technology base for the processing, fabrication and manufacture of the new superconducting ceramics.
- Synthesizing, processing, and fabrication of materials in engineering shapes and sizes to include complete characterization of the material, component demonstration, and a determination of its viability for use in advanced manufacturing scenarios.
- Demonstrating viability of fabricating high-temperature superconducting materials in component forms, i.e., films, filaments, fibers, tapes, single crystals, dense monoliths, composites, etc.
- Proving the feasibility of superconductive materials in conceptual applications.
- Shortening research and development time to production.
- Contributing to a viable industrial base for fabrication of superconductivity components.
- Enhancing U.S. Defense capabilities.

The superconductors developed would have appropriate properties and forms useful in both large and small applications scenarios. The superconductors would then be demonstrated and hopefully lead to the rapid manufacturing of devices, machines and products useful in the Department of Defense (DOD) arena.

Superconductors for Large Scale Applications

Some of those potential applications of interest to the DOD included:

- Magnetic submarine detection devices capable of increasing detection ranges at least 10 times present capability. Current technologies are adequate; applications are required here.
- Superconductivity based computer systems, to provide tremendous levels of computation and storage in a small space. The reference point of "a Cray in one cubic foot" has been given as a goal. The usefulness of such a computer would be immeasurable in improving the fighting capabilities of vehicles and weapon systems now unable to benefit from today's technology due to their size, weight and power requirements.
- Gyrotrons for precision guidance applications.
- Traveling wave tubes for electronic applications.
- Free electron lasers for research and for military weapon applications.
- Superconductive motors for a variety of applications requiring reduced power drain.
- Electric power generation using high performance superconductive magnets and windings.
- Energy storage devices, of all sorts to store and later provide power to equipment on demand.
- Fusion applications for potential use in fusion power generation.
- Field power supplies, presently relatively inefficient batteries; but for the future, high quality, small size, large capacity cells capable of use in all temperature extremes.
- Switches and other energy storage devices. The Josephson Junction, a high speed superconductive switch and its derivatives would fit into this category of development.

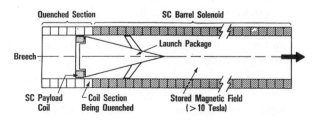

FIGURE 2-10
Superconducting Quench Gun

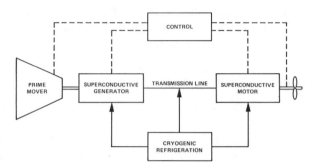

FIGURE 2-11
Block Diagram of
Superconductive Electric
Drive Propulsion System

- Electromagnetic quench guns (see Figure 2-10) and launchers using superconductive magnets and zero resistance windings. Developing rapid fire guns firing high velocity projectiles, useful in military applications on ships, tanks and fixed installations. The goal would be to eliminate explosive propellants and utilize novel new energy sources. Large magnets and high-power switching devices would also be needed.

- New power plant designs using the superconductive motors and generator (see Figure 2-11) technology that might be combined with the nuclear power generators of the present sub. Goals would be to reduce the size of the power plant by a factor of two while maintaining the present speed. The work would require innovations into radiation hardened high current carrying devices.

- Propulsion systems, such as ion generators, for deep space power applications, or by the development of high efficiency, high power, small size motors for drive applications.

- Magnetic propulsion systems to create a new power source for low noise propulsion even at speeds in excess of 20 knots.

- Applications for ship, aircraft, and land vehicle power distribution. The goal would be to expand the range of current systems by 20 percent and would require the development of more efficient power systems as well as the means to apply that power to propulsion systems.

- Current approaches for submarine propulsion involves the use of steam reduction gears to transmit power from the power source to the drive mechanisms. Proposed superconductivity systems would include a nuclear-steam-electric motor design that could not only simplify mechanical designs, but create increases in speed of better than 10 percent with lower signatures being projected by the submarine. The reduction in size and weight of the power plant would allow for the potential increase in weapons carrying capability and/or more computational power in the form of new superconductive computers.

- Magnetic shielding, to provide for an environment useful for making precision magnetic measurements, in an environment unaffected by earth field, and to create possible shields for penetration of unwanted directed magnetic energy sources.

- Bearing systems created from opposing electromagnetic fields, creating an air bearing between surfaces. Possible applications might be high speed transportation systems.

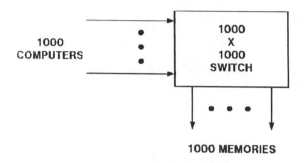

1000 MEMORIES

SIZE: 6" CUBE (APPROXIMATELY)

FIGURE 2-12
Superconducting Crossbar
Switch Concept

Superconductors For Small Scale Applications

Smaller scale applications of interest to DOD include:

- Optical and infrared detectors, novel light modulators, superconducting/semiconducting mirrors, electronics for display, detection, beam steering, Q switching and multiplexing.
- Microwave and millimeter wave components including high-efficiency transmission lines, detectors, mixers, amplifiers, waveguides, phase shifters, antenna arrays, and high frequency chips, (high Ghz region) array transform processors and other analog devices. The long antennas now required are an impediment to operations at the lower frequencies. Suitable application of superconductor high sensitivity receiving components could make the antenna more efficient and possibly smaller or the increased sensitivity of receivers might permit the use of less extensive antenna systems.
- Magnetic components and detectors similar to the SQUID magnetometer.
- Digital devices for low power, high performance computers and signal processing as a basis for a new generation of supercomputer.
- Hybrid computer logic, crossbar switches (see Figure 2-12), connectors, and interconnects to provide for a ten-times increase in speed and a ten-times size reduction. This will require both new devices and the development of new computer architecture to handle and make maximum usage of the new speed. Parallel Processors would benefit from this initiative.
- Analog to digital conversion having the properties of high speed, low power, wide dynamic range, highly linear conversion over the operating range. The Josephson Junction and its derivatives might play a role here.
- Hybrid infrared sensors permitting a resolution improvement of five times present capability. This would allow for not only resolving targets within an area, but permit better pointing accuracy as well. The device would require better understanding of the interface between superconductor device and semiconductor devices. Much of the thin film work on-going might have direct application to this problem.

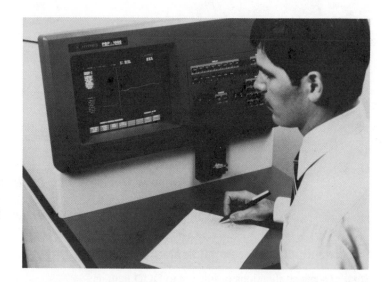

FIGURE 2-13
Hypres PSP-1000
Picosecond Signal
Processor

Source: Hypres Inc.

• Superconducting infrared sensors capable of greater sensitivities to extend the operations of present battlefield sensing devices and weapon systems and to allow for better sensors in space applications. New materials and device fabrication will be required.

• Analysis equipment such as high frequency (Multi-Ghz) sampling oscilloscopes (see Figure 2-13) and broad band network analyzers.

Many of the devices in this category are related in some way to better receiving, processing, storage and transmission of energy, all of which can be improved by superconductive devices. Others devices allow for the laboratory development and field measurement and analysis of these new systems.

ELECTRONIC APPLICATIONS

Some of the most promising applications of high-temperature superconductors are electronic systems involving thin film lines or Josephson elements. Applications in computers would have the largest commercial impact, but may take longer because of their complexity. Sensor and instrument applications are simpler and are likely to be commercialized within a few years. Simplest of all is the use of high-temperature superconductors for low-field electromagnetic shielding.

From the beginning, Josephson junctions have been the basis for many superconducting electronic devices. Superconducting quantum interference devices (SQUIDs), simple circuits incorporating Josephson junctions, have extremely high sensitivity levels, which have led to a considerable range of practical uses for low-temperature superconductor SQUIDs. The Josephson effect can also be exploited for computer logic and memoy. Although a number of practical problems stand in the way, Josephson junctions could in principal replace semiconductor chips in powerful digital processors.

Computers and Logic Devices

Dr. Fernand Bedard, a research physicist at the National Security Agency, says that Josephson electronics offer the theoretical possibility of a computer 6 inches on a side which could outperform a Cray supercomputer. There are two main limits to developing faster computers: switching speeds and wiring delays. A Josephson Junction can switch from a zero voltage to a non-zero voltage stage in around 4 picoseconds, faster than any semiconductor device. (A picosecond is a trillionth of a second.) "The low power dissipation of the Josephson devices is also rather important," says Hisao Hayakawa of the Electrotechnical Laboratory in Ibaraki, Japan. "More than 40% of the cycle time in a conventional computer system is due to wiring delays—the processor must wait for signals to come from the other parts of the computer. Thus, to reduce the cycle time one must reduce wiring delays by packing the devices more densely. However, the need to dissipate the power generated by the devices imposes severe limits on their density. In Josephson devices the power dissipation is typically a few microwatts, which is about three orders of magnitude lower than that of semiconductor devices . . . one can integrate more than 10^4–10^5 devices on a 5-mm square chip."

In 1969, IBM began developing a computer with logic and memory chips made from Josephson Junction switches. They abandoned the project in 1983 after investing $300 million in development. The main technical difficulties IBM encountered were: a) building chips that would operate reliably despite repeated cycling between room temperature and 4°K, b) designing a cache memory, and c) developing a way to cool the chips. IBM apparently decided that other chip technologies involving gallium arsenide would offer an earlier return on their investment. It may also be relevant that IBM's corporate strategy involves serving the largest markets—not pushing the limits of technology.

Work on the Josephson computer has continued in Japan as part of the Ministry of International Trade and Industry's supercomputer project. Presentations at the 1987 International Superconducting Electronics Conference in Tokyo revealed that memory is still the primary focus of research—memory cells are complex and difficult to miniaturize. T. Imamura of Fujitsu displayed an integrated circuit composed of two niobium-aluminum oxide-niobium Josephson Junctions, one mounted above the other. This device may represent a major step toward large superconducting vertically integrated circuits.

Much work has already been carried out on computer subsystems based on liquid helium superconductors. Semiconductor technology is still advancing rapidly, however, and continues to dominate the computer field. The discovery of high-temperature superconductors may change this situation.

One possible role of superconductors in such systems is simply to interconnect the semiconducting devices with superconducting microcircuit transmission lines. This possibility is already interesting at 77 K, because certain semiconducting devices switch faster at this lower temperature. However, 77 K copper lines present significant competition because of their decreased resistivity compared to that at room temperature.

The most exciting opportunities would use room temperature superconductors, offering compatibility with the entire line of semiconductors, including the highest performance bipolar devices. In the most promising scenario, the use of room temperature superconductors could affect the full range of data processing systems, which form the largest high-technology industry in the world today.

Although the implications of high-temperature superconductors for semiconducting computer systems have yet to be assessed, the reduced losses compared to normal conductors offer many possible advantages. System performance (i.e., switching speed) can be increased by reducing the RC time constant associated with the interconnect line. Narrower lines can be used, saving space on the chip. The elimination of power losses and voltage drops permits miniaturization of power busses and potentially, the entire system.

The high-temperature superconductors have also been proposed for computer applications using Josephson junctions. High-temperature superconductors may offer higher device switching speeds, higher bandwidth transmission, and the possibility of using semiconducting memory to supplement ultra-high-speed superconducting logic. The disadvantages of high-temperature superconductors are increased thermal noise and switching power losses at 77 K compared to earlier liquid-helium temperature designs.

A variety of superconducting transistor-like devices have been proposed, among them superconducting field-effect transistors (FETs), several nonequilibrium devices, and optically switched FETs. These devices are at early stages of development, even using the conventional low-temperature superconductors; but although there are still considerable materials and fabrication problems, the potential performance of some of these devices might be enhanced by higher switching speeds and output voltage changes stemming from the larger energy gaps of the high-temperature superconductors.

The Terahertz Initiative, a division of the Strategic Defense Initiative, has begun sponsoring research in superconducting electronics (see Figure 2-14) to develop high-frequency devices for communications, radar, and signal processing in outer space. Dr. Dallas Hayes of Hanscom Air Force Base, Massachusetts, director of the program, says "The objective of the Terahertz Initiative is to replicate today's microwave communication technology at millimeter and submillimeter wave lengths. Superconducting electronics devices can transmit and receive information at much higher frequencies than conventional microwave devices with semiconductor technology—they can operate at gigahertz frequencies (billions of cycles per second), whereas conventional devices generally operate in the megahertz range (millions of cycles per second). Because of its higher frequencies, superconducting communication equipment can transmit data faster than

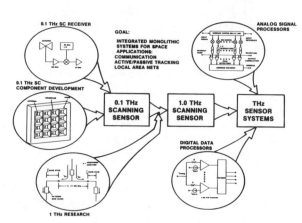

FIGURE 2-14
Superconductive (SC)
Terahertz Technology
Program

conventional equipment, and it will also be much harder to jam or intercept." Superconducting imaging radar devices may yield detailed information about the size and shape of objects because their narrow, submillimeter wave length allows finer resolution. Improved radar developed in this program could also enhance the safety of commercial airliners. But since the earth's atmosphere tends to absorb high-frequency signals, the main application of these devices will be in space-to-space communications.

The Terahertz Initiative has established research contracts with Arnold Silver at TRW in Los Angeles; Richard Blaugher at Westinghouse in Pittsburgh; Ted Van Duzer at the University of California, Berkeley; James Lukens at the State University of New York, Stonybrook; R.A. Buhrman at Cornell University in Ithaca, New York; and Malcolm Beasley at Stanford University.

Superconducting electronic devices have also found an application in astrophysics—most major radio telescopes observing distant galaxies at millimeter or submillimeter frequencies now use superconducting receivers because of their sensitivity and their low noise level.

COMPUTER SYSTEMS

Josephson junction (JJ) based electronic devices promise switching speeds 10 times faster than the very best compound semiconductors. Because the energy losses are several orders of magnitude smaller, JJ-based integrated circuits could be packed much more densely. However, the practical problems of making JJ-based chips far exceed those of SQUIDs.

Even if the practical problems were solved, Josephson computers might not be commercialized. The competing technologies extend well beyond silicon and gallium arsenide chips: a good deal of research and development has been going into alternative computer architectures such as massively parallel processors. Much of this work seeks increases in processing power without major advances in components. Still, faster chips will always promise faster machines. But, in a further contrast with SQUIDs—which are the most sensitive magnetic field detectors known—the theoretical limits of JJ-based logic devices fall well short of what might eventually be possible, for example, using optical switching. Thus the window for opportunity for JJ-based computing may never open. (It may never open for optical computing, either.) On the other hand, advances in device design—and, in particular, a practical three-terminal device that would erase the primary drawback of JJ chips, low gain—could open a broad new frontier. It is simply too early to say.

Research and development in the United States and Japan on low-temperature superconductor-based JJ computing illustrates some of the problems that designers of high-temperature superconductor logic and memory would face. IBM was able to build logic chips with 5,000 junctions reliably, but had trouble with cache memory. IBM's prototype memory chips, with over 20,000 JJs, proved susceptible to errors caused by slight variations in control current—a good example of the kind of problem that a 3-terminal device would help solve. More recently, Japanese companies have built several kinds of low-temperature superconductor chips incorporating niobium JJs. Fujitsu's 4-bit microprocessor, 25 times faster than a gallium-arsenide microprocessor, consumes only 0.5 percent as much power as either. NEC has produced a 1,000 bit dynamic memory, containing 10,000 JJs; access time is a factor of 200 better than for silicon.

The first applications of high-temperature superconductors in computers may be inter-connects—electrical pathways joining otherwise conventional chips. Signal dispersion and other problems associated with transmitting electrical pulses within the processor limit performance; practical means for incorporating high-temperature superconductor interconnects should find ready application in large and powerful machines.

Moreover, at liquid nitrogen temperatures, superconductors and semiconductors could operate compatibly in hybrid designs. Ordinary semiconductors cannot be used at liquid helium temperatures; even if they could be made to operate in otherwise satisfactory fashion, semiconductors would dissipate too much heat, overwhelming the cooling sys-tem. Given that hybrid low-temperature superconductor-semiconductor systems are not feasible, past work on Josephson computing has involved either all-superconducting chips, or unique designs with controlled temperature gradients. The Hypres data sampler, for example, uses an integrated circuit cooled to liquid helium temperature on one end only—that end holding about 100 low-temperature superconductor JJs.

Three-terminal devices could be a big step forward in superconducting electronics, making possible logic designs at the chip level much like those now used with semicon-ductors. It could well be, however, that major advances in high-temperature supercon-ductor electronics would come only with devices that departed in a major way from currently known electronic devices. The first requirement, in any case, is mastery of thin-film fabrication technology.

SENSORS APPLICATIONS

SQUIDs

(SQUID is an acronym for Superconducting Quantum Interference Device.) SQUIDs can detect the very faint signals produced by the human heart (10^{-6} gauss) and brain (10^{-9} gauss). These simple circuits can also measure a wide variety of other electromagentic signals (anything with an associated magnetic signature from DC up to microwave frequencies). SQUIDs are about 1,000 times more sensitive than the next best magnetic field detectors. They can sense the disturbances in the Earth's magnetic field caused by a submarine deep in the ocean, or the field distributions caused by geologic formations holding oil or mineral deposits. Requiring, in simplest form, only one or two Josephson junctions (rather than the large numbers required in computer applications), low-tem-perature superconductor SQUIDs—typically fabricated from niobium—are now made routinely.

To minimize thermal noise, SQUIDs should be operated at the lowest possible tem-perature, and in any case at less than half to two-thirds of the superconducting transition temperature. At liquid nitrogen temperatures, for instance, sensitivity will be 20 times poorer than at liquid helium temperature. Even so, a high-temperature SQUID would still be a more sensitive magnetic field detector than any of the alternatives except a low-temperature superconductor SQUID. If they can be built successfully, high-temperature superconductor SQUIDs will quickly find a considerable range of applications (though none of these are likely to be high-production-volume applications).

Arnold Silver, now at TRW in Los Angeles, observed in 1964 that two Josephson Junctions connected in parallel could serve as exquisitely sensitive detectors of magnetic fields.

Since 1964 researchers have applied the phenomenal sensitivity of SQUIDs to a wide variety of scientific research. SQUIDs can measure infinitesimally small voltages, currents, and resistances. Geologists have used SQUIDs to find oil deposits and geothermal energy sources, and also to study seismic activity. Defense researchers have used SQUIDs to detect submarines.

At Stanford University a SQUID linked to a 5-ton aluminum bar is now helping scientists look for gravitational radiation. As Professor William Fairbank explains, "If a star collapses in our galaxy, the way it did recently in the supernova, then Einstein's theory says a gravitational wave goes out from the star. When it passes the earth, in this case 150,000 years later, the bar will expand and contract by about 10^{-16} centimeters for a thousandth of a second, and then the gravity wave will continue on its way. How in the world can you see the motion of an aluminum bar more than a thousand times smaller than the size of a nucleus? It turns out that if one makes a mechanical transducer on the end of the bar and attaches it to one of these SQUIDs, one can see these very small motions."

Superconducting quantum interference devices operating at liquid helium temperatures as sensitive magnetic field detectors, are already of value in many disciplines including medical diagnostics, geophysical prospecting, undersea communications, and submarine detection. SQUIDs made with the new high-temperature materials have been operated at liquid nitrogen temperatures. Relatively inexpensive SQUID-based magnetometers operating at 77 K or higher would be deployed in large numbers if electrical noise can be held to acceptably low levels.

Radiation Detectors

Superconducting microwave and far-infrared radiation detectors (quasiparticle mixers, superconducting bolometers) already exist using conventional superconductors. In spite of a loss of sensitivity due to increased electrical noise at higher temperatures, the increased energy gap of high-temperature superconductors would offer sensitive detection in a largely inaccessible frequency range, and the simplified refrigeration allows increased ease of use. Other microwave applications include high-Q waveguides, phase shifters, and antenna arrays.

ANALOG SIGNAL PROCESSORS

High-speed analog signal processors performing such functions as filtering, convolution, correlation, Fourier transformation, and analog-to-digital (A-to-D) conversion are important for many applications. Various high-speed A-to-D converters have been tested successfully at 4.2 K. If high-quality Josephson junctions can be fabricated from the new superconductors, these devices should perform comparably at 77 K. At this temperature, integration of the superconducting devices with some semiconducting devices (for example, complementary metal-oxide semi-conductors) becomes feasible, and new hybrid systems may well result in the fastest A-to-D converters available.

MAGNETIC SHIELDING

Both superconducting wires and superconducting sheets have been used for many years to create regions free from all magnetic fields or to shape magnetic fields. The advent of high-temperature superconductors may extend the range of this application. Like niobium-tin, high-temperature superconductors may be plasma-sprayed, permitting their use on surfaces of complex shape.

VOLTAGE STANDARD

Many countries now maintain a voltage standard in terms of the voltage generated across a low-temperature superconducting Josephson junction irradiated by microwaves at a precise frequency. This standard could be more cheaply maintained and more widely available with no significant loss of accuracy by operating at 77 K with the new materials.

A voltage across a Josephson junction will cause a high-frequency alternating current, according to a simple relationship (about 484 gigahertz per millivolt). Since one can measure frequencies with high precision, it is possible to exploit this property of Josephson electronics to establish a standard for a volt. Since 1972, the United States National Bureau of Standards has maintained its voltage standard using Josephson junctions. Other countries using Josephson electronics to establish their voltage standard include Japan, Australia, Canada, England, France, West Germany, and the Soviet Union.

ELECTRIC POWER AND UTILITY APPLICATIONS

Magnets have no moving parts. Technical complexities grow in electrical machinery, and in the entire range of electric utility applications. Transformers, for example, would demand more attention to AC losses than magnets, while superconducting transmission lines will almost certainly have to go underground, so long as refrigeration is required. Underground lines are costly, although already in use in many urban areas. Still, the overriding design requirement is reliability. Utilities are quite willing to trade off higher operating costs against lower probability of failures and down-time. A disabling failure, after all, can lead, not only to a blackout, but to an ongoing need to purchase power from other suppliers until repairs have been completed.

In general, high-temperature superconductor-based generators will need conductors similar to those for magnets. Dynamic forces, however, will add to static forces, while cooling also becomes more difficult. Large conventional generators already have efficiencies greater than 98 percent. Superconducting field windings can increase this to more than 99 percent. In a large machine, an improvement of 0.5 percent to 1 percent in efficiency can be significant—reducing the losses by half—while superconducting generators have the additional advantage (for utility applications) of increasing network stability as they are less sensitive to shifts in electrical load.

Worldwide, at least two dozen low-temperature superconductor generator research and development projects have been undertaken since the middle 1960s, but none has gone beyond construction and testing of a prototype. Utilities will have to be convinced that such machines offer reliable service over periods of many years before investing; high-temperature superconductors will not affect the economics much compared to low-temperature superconductors, and, lacking even the experience base of low-temperature superconductor systems, the new materials have an added hurdle to overcome. Energy storage rings, with no moving parts and tolerable failure modes, will almost certainly come first.

In recent years the electrical power industry has shown considerable interest in the promise of superconductivity to cut the costs and improve the performance of power generation equipment (see Figure 2-15), energy storage technology, and power transmission lines. The application of superconductivity in these areas may eventually help utility companies improve their profitability while reducing electrical rates.

FIGURE 2-15
Illustration of Electric Generator

Source: EPRI

Generators with superconducting wires in their rotors would offer lifetime cost savings of 40%, according to Dr. Narain Hingorani of the Electric Power Research Institute. John Hulm and his colleagues at Westinghouse built a prototype 10 megawatt superconducting generator in the early 1980s, and researchers from Japan and Germany have also built prototypes. Researchers in the Soviet Union are now building a 300 megawatt superconducting generator. (For reference, the capacity of a large nuclear power plant is about 1000 megawatts).

Superconducting magnetic energy storage (SMES) is a highly efficient technology that stores energy in the magnetic field that is formed when very large currents are made to flow in a superconducting coil.

The technology makes use of the fact that a magnetic field is created whenever an electric current flows in an electrical inductor or coil of wire. For energy storage to be practical, large currents are required and electrical resistance must be low. For this reason, the storage coil must be superconducting.

The operation of a SMES unit is conceptually simple and straightforward, as shown in Figure 2-16. The SMES unit is connected to the utility grid by way of a power conditioning system that converts the AC power on the grid to a DC current to power the superconducting coil, and then reconverts the DC current to AC power when the coil is discharged.

The actual energy storage takes place in the superconducting coil. The amount of energy stored is determined by the physical parameters of the coil and the amount of current flowing in the coil.

Because superconductivity occurs only at low temperatures, a coil refrigeration system is required. Overall SMES operation is monitored, coordinated, and controlled by a supervisory control unit.

It should be noted that a SMES unit contains no moving parts except for the refrigeration system and that the electrical energy from the utility does not go through a chemical or mechanical conversion for storage.

Energy storage technologies provide utilities with new ways to maximize the use of their most efficient generating capacity and minimize the need for "peaking" capacity which is used for only short periods and which can cost 4 or 5 times as much as baseload generation.

FIGURE 2-16

Illustration of a 20 MWh Superconducting Magnetic Energy Storage System (SMES)

Source: Virginia Power

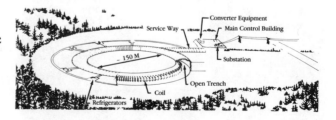

SMES is of interest because it has potential advantages over other storage technologies. In utility-scale sizes, SMES has a 90 to 95% energy efficiency as opposed to 75 to 80% for competing systems. In addition, SMES can go from full charge rate to full discharge rate in less than a second. It can also be discharged at any rate up to maximum without loss of efficiency.

In the last 15 or more years, significant effort has been devoted to developing a basic engineering approach and a conceptual design (see Figure 2-17) for a "utility sized" (1000 MW-5000 MWh) SMES coil. In addition, cost and application studies have been performed as well as laboratory component development. These for the most part have been paper studies. There has been no actual engineering, construction, and operating experience on a SMES system of any significant size.

Design and construction of an engineering test and development unit is needed to obtain experience to allow for the future development, refinement, and cost reduction of the present SMES technology.

In addition, work needs to be done to raise superconductor current density and operating temperature, or to find new structural materials and designs to allow SMES technology to be cost competitive at "modular" utility sizes in the range of 200 to 300 MW with 6 to 8 hours of storage.

The new higher temperature superconductors could help to make SMES more cost competitive at "modular" utility sizes. They will provide only modest improvement for the 5000 MWh unit.

In general, higher superconductor critical temperatures will have a major impact only if the temperatures can be raised to a point where a broader range of more-inexpensive structural materials can be used. For this to occur, temperatures significantly higher than that of liquid nitrogen must be obtained.

Superconducting power transmission lines (see Figure 2-18) would reduce operating costs for utility companies and also make it easier to gain right-of-way, since the lines would lie underground. The copper transmission lines in use today dissipate 5% of their power in resistive losses, costing U.S. utility companies a total of about $7.5 billion a year. These losses account for half the lifetime cost of a power transmission system. Superconducting power transmission lines would dramatically reduce these losses. (Power

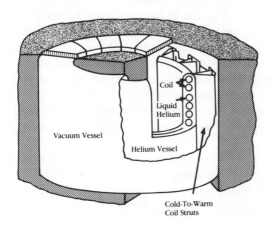

FIGURE 2-17
Conceptual Physical
Design of SMES

Source: Virginia Power

FIGURE 2-18
Photograph of Power
Transmission Cable

Source: EPRI

FIGURE 2-19

transmission lines nearly always carry alternating current, and the ac losses of supercon-ductors, while very low, are not zero.) Researchers at Brookhaven National Laboratory in Upton, New York have successfully tested a prototype superconducting power trans-mission cable. It took 10 years of experimentation to configure Nb_3Sn, a brittle alloy, into a tape that could be wound into cables without cracking.

Superconductivity is also compatible with photovoltaics. Superconducting power trans-mission lines could enhance the attractiveness of very large solar arrays in remote, sunny areas far from population centers. Small energy storage coils could enhance the outlook for small solar installations by providing an efficient way to store electricity for use in the evening.

For non-utility applications, characteristics other than efficiency and reliability come to the fore: superconducting machines promise to be smaller and lighter than conven-tional motors and generators by half and more. These are the attractions for ship propul-sion. Where a superconducting generator driving a superconducting motor could elimi-nate the gearing and shafting between turbine (or other prime mover) and propeller. With much more freedom in packaging, nuclear submarines could carry more weapons (or be smaller), as could surface ships. Submarines might also prove quieter, perhaps even faster. Moreover, with the motor/generator set(s) providing speed control (and revers-ing), efficiency during part load operation would rise (the turbine can run at its optimum speed).

More cost-sensitive, applications for motor/generator sets might also open up at some point. And of course, given high enough operating temperatures, the many large electric motors used throughout industry (ranging from pump, fan, and blower drives to machine tools and rolling mills) would be candidates for replacement.

As shown in Figure 2-19, electronics technician Richard McDaniel of Argonne National Laboratory pours liquid nitrogen into the world's first electric motor based on the unique properties of high-temperature superconductors. It turns at about 50 revolutions per minute. The new superconductors lose all resistance to electricity when cooled by liquid nitrogen to about 290 below zero Fahrenheit.

Transportation

Ambitious attempts to apply superconductivity to land and ocean transportation have been made over the years with some success in the United States, Europe, and Japan. In 1979, the Japanese National Railway, in cooperation with Fuji, Hitachi, Mitsubishi, and Toshiba, built and tested a magnetically levitating ("maglev") train (see Figure 2-20) that travelled at speeds up to 325 miles per hour. The train's wheels touch the track only during lift-off and landing. Superconducting magnets aboard the train induce eddy currents in the track, which is a normal electrical conductor. These eddy currents produce a

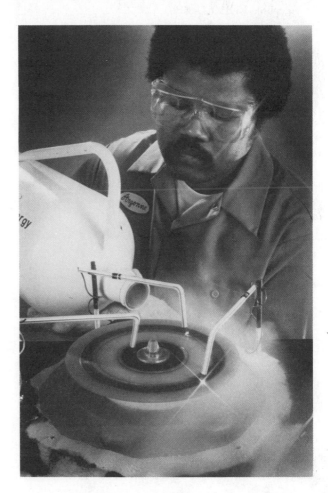

FIGURE 2-19
World's First Electric
Motor Based on
High-Temperature
Superconductors

Source: Argonne National
Laboratory

TABLE 2-4. End Use Applications by Industry

Application	Current		Emerging	
	Commercial	Government/ Defense	Prototype Demo.	R&D
Medical				
Magnetic Resonance Diagnostic Systems (MRI)	X	-	-	-
Biotechnology and Engineering	-	-	-	X
Electronics				
Semiconductors and Transistors	-	-	X	X
IC Packages and Substrates	-	-	-	X
Josephson Junction Devices	-	-	X	X
Circuit Interconnections	-	-	-	X
Particle accelerators	-	X	-	-
Sensors	X	X	-	X
Special Oscilloscopes	X	-	-	-
Aerospace and Defense				
Rail Guns	-	-	-	X
Electromagnetic Unmanned Launch Space Vehicles	-	-	-	X
Micro Wave Power Transmission	-	-	-	X
Communications	-	-	-	X
Gyroscopes	-	-	-	X
Industrial				
Separation	X	-	X	-
Processing of Materials	-	-	X	X
Sensors and Transducers	-	-	X	X
Magnetic Shielding	-	-	-	X
Magnets	X	X	-	-
Power Generation				
Motors and Generators	-	-	X	X
Energy Storage	-	-	-	X
Transmission	-	-	X	-
Fusion	-	-	-	X
Magnetic Hydrodynamics	-	-	-	X
Transportation				
Magnetic Levitated Vehicles	-	-	X	-
Marine Propulsion	-	X	X	-
High-Speed Propulsion	-	-	X	X

FIGURE 2-20
Illustration of Magnetic
Levitation Train.

Source: All American
MagnePlane Inc. and Argonne
National Laboratory

magnetic field of opposite polarity, causing the train to levitate. Technical feasibility has been proven, and further development depends on economic and cultural considerations.

Planners in government and industry are proposing commercial maglev train projects in several areas. Japan's HSST Corporation may soon build a 4.4 mile maglev train line in Las Vegas, connecting the downtown area to the casinos. California and Nevada officials are considering a 230 mile maglev line which would carry passengers from Los Angeles to Las Vegas in 70 minutes (the trip takes 5 hours by car). In Japan, the national railway hopes to build a maglev train to replace the conventional bullet train running from Tokyo to Osaka.

Superconducting ship propulsion systems were studied in England in the 1960s. The U.S. Navy successfully installed a prototype superconducting drive system (see Figure 2-21) on a small ship in 1980; the development of a 40,000-horsepower drive system continues. A second concept uses seawater as the working fluid in an MHD propulsion system; the drive scheme is known as an electromagnetic thruster (EMT). Ship models based on this propulsion principle were promoted in the United States in the 1960s and operated in Japan in the 1970s. Practical designs for a full-scale EMT ship have been proposed, and industrial collaboration is being sought.

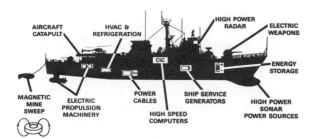

FIGURE 2-21
The Superconductivity
Revolution Potential
Shipboard Applications

Because the present worldwide systems of air, sea, and land transportation are well established and represent a substantial investment, society has not yet made use of the potential advantages that have been demonstrated in prototype transportation systems using low-temperature superconductors. In land-based systems the principal advantage is high speed. Ship-based systems are lighter, have better control, and permit the radical re-arrangement of power drive systems within ship structures. The combination of room-temperature superconductors and the low specific volume of superconducting machines could revolutionize surface transportation.

Table 2-4 provides a summary of superconductivity applications that are currently viable or are emerging technology opportunities.

CHAPTER 3

MARKET POTENTIAL

INTRODUCTION

The world superconductor industry is small, but superconducting devices are usually components of larger systems whose gross annual sales volume is many times the value of the devices themselves. Annual device sales total about $400 million (see Figure 3-1), of which magnetic resonance imaging machines and electronics instruments each account for approximately $150 million. Magnet coils represent 10 to 20 percent of device costs in magnetic resonance imaging systems, and annual sales of basic materials such as alloy rod and sheet are on the order of $20 million.

It is difficult to estimate the potential economic impact of today's high-temperature superconductors because so little is known about them and much depends on improved understanding and technological development. Assuming that satisfactory conductors can be manufactured, there are considerable advantages to operation in liquid nitrogen. Refrigeration units are simpler and cost less to operate. Conductor stability generally improves as the temperature increases because of the higher heat capacity of materials; however, the protective effect of the shunting normal conductor is reduced slightly because of its increased resistivity. Structural materials are less brittle at higher temperatures; therefore, more conventional structures can be used. Cryogenic liquids and systems, however, will still be needed. In comparing superconductor technology with present room temperature devices, the need for cooling is a serious economic and technological disadvantage. There is a great difference between switching on a machine as needed and having to supply continuous refrigeration, or having to wait for refrigeration systems to reach operating temperatures.

Assuming that some utility and heavy electric power applications can be competitively marketed using systems cooled by liquid nitrogen, the superconducting materials market may be substantially increased; the market for heavy electrical equipment, however, would be mainly a replacement one, because few new systems are being built.

For substantial business growth above that projected for low-temperature superconductors, new technology developments are needed. There is little doubt that the new materials offer technological advantages, for they promise high-magnetic-field devices and new types of electronic sensors and switches at lower refrigeration costs than before. Advances are bound to result in new applications and new economic growth. If room temperature superconductors become available, we can confidently expect a truly revolutionary expansion of superconducting applications in electrotechnology.

The worldwide market for superconductive products exceeded $1 billion in 1988. By the year 2000, the worldwide market for superconductors may approach $3 billion. Superconductive products are expanding the frontiers of science, revolutionizing the art of medical diagnosis, and developing the energy technology of the future. In general, today's customers for superconductive equipment want the highest possible performance, almost regardless of cost. The products operate within a few degrees of absolute zero, and virtually all are fabricated from niobium or niobium alloys—so far the high-temperature superconductors discovered in 1986 and 1987 have had little impact on these markets. The industry shows potential for sustained economic growth and profound societal impact.

The market for superconductors is only now emerging. Current estimates of future growth and market size are dependent upon the degree to which the technology is accepted in current and new applications. Factors affecting the consumption of raw and semi-finished superconductive materials include:

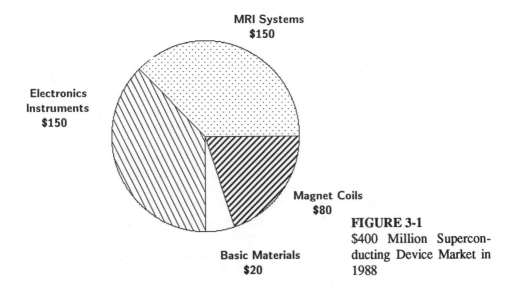

MRI Systems
$150

Electronics Instruments
$150

Magnet Coils
$80

Basic Materials
$20

FIGURE 3-1
$400 Million Superconducting Device Market in 1988

- Cost of materials
- Savings derived
- Successful implementation in known areas
- Development of new areas
- Degree of penetration of superconductive technology

Technology Research Corporation has estimated a 15% conservative growth rate for the superconductor materials market. Growth will be the largest in medical imaging, electronics, defense systems, industrial, and power generation/distribution fields.

MARKET FOR SUPERCONDUCTIVE PRODUCTS

The market for superconductive products is briefly described in this chapter. The key products are magnetic resonance imaging; production magnets; SQUIDs; oscilloscopes and electronics; wires, rods, and cables; small magnets for research; R&D niobium materials; large niobium magnets; R&D of ceramic materials; ceramic wires and tapes; ceramic coatings, thin films, and shielding; particle accelerators; and magnets for nuclear fusion research.

Magnetic Resonance Imaging

Magnetic resonance imaging (MRI) superconducting magnets, are by far the mainstay of the current market for superconductive components. Of the $200 million spent in this area in 1987, about $150 million was for magnetic resonance imaging applications. Supporting a market base of some 300 systems each year, this magnet business can be expected to continue and expand.

Magnetic Resonance Imaging is a new medical diagnostic technique—it creates high-resolution images of human tissues without harmful radiation. MRI images are so clear they can eliminate the need for exploratory surgery. The process has begun to save lives—some doctors say MRI has allowed them to remove brain tumors they would previously have considered inoperable. MRI requires intense magnetic fields (up to 2 Tesla, or 20,000 gauss) over large volumes, which can be produced economically only by superconducting magnets.

As noted in *Supercurrents*, an MRI system costs about $2 million, plus an additional $500,000 for installation. The market for these systems has grown rapidly, from $10 million in 1982 to $100 million in 1984 and nearly $1 billion (450 systems) in 1987. Over 1000 systems are now in operation worldwide. Because they operate in clinical settings where the operators are unfamiliar with superconducting magnets, the systems must be safe to operate, easy to maintain, and very reliable. The major MRI system suppliers are General Electric, Siemens, Picker International, Diasonics, and Philips.

MRI systems account for 75% of the market for superconducting magnets. MRI magnet manufacturers include Applied SuperConetics in San Diego, Intermagnetics General Corporation of Guilderland, New York, and Oxford Instruments, UK. An MRI magnet costs about $250,000.

Applied SuperConetics, Inc. (see Figure 3-2) was formed in 1983 to manufacture MRI magnets for system suppliers. It drew on the experience of its parent company, GA Technologies, Inc., which manufactured superconducting magnets for nuclear fusion research. ASC became profitable in 18 months and repaid its entire R&D investment in 26 months. It now employs 20 people, and its manufacturing facility can produce 1 magnet per week. In fiscal 1986 ASC had sales of $6.2 million. Lawrence Woolf, Staff Scientist, says the company has now captured 15% of the worldwide market for MRI magnets.

Intermagnetics General Corporation (IGC) sells MRI magnets to Philips Medical Systems Division, Picker International, and Diasonics. IGC also sells superconducting wire to General Electric for GE's MRI magnets. IGC creates its superconductive wire from niobium-titanium filaments 2 to 20 microns in diameter, embedded in a copper matrix. The company pioneered the use of mobile MRI systems housed in trucks so hospitals that are hundreds of miles apart can share the benefits as well as the high costs. IGC has formed a joint venture with Alsthom of France to manufacture and market MRI equipment in Europe. The company lost about $3 million on sales of $14 million in fiscal 1987, but showed a small profit in the first quarter of fiscal 1988.

Oxford Instruments, the world's most successful superconductivity company, earned $20 million on sales of $160 million in the year ending March 31, 1987. Oxford supplies MRI magnets for General Electric and Philips. The company also fabricates magnets for nuclear magnetic resonance systems used in analytical spectroscopy, supplying Varian, General Electric, Jeol (a Japanese company), and Bruker (a German company). The company operates 3 manufacturing facilities: Oxford Magnet Technology in Eynsham, Oxfordshire, England; Oxford Superconductor Technology in New Jersey; and Furukawa-Oxford, a joint venture in Japan.

Supercon, Inc., of Shrewsbury, Massachusetts, a manufacturer of niobium-titanium superconducting wires and cables, has supplied wires to most of the magnet manufacturers. Supercon is probably the oldest company in the world continuously devoted to the manufacture of superconducting materials. "We compete with Hitachi, Furukawa, and

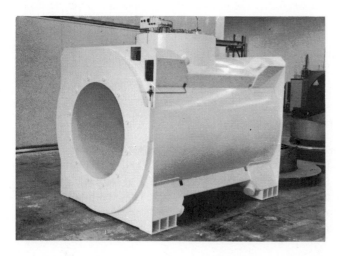

FIGURE 3-2
1-Meter Bore Magnetic
Resonance Imaging (MRI)
Magnet

Source: Applied SuperConetics, Inc.

Siemens, and we also supply Siemens and Hitachi with the more sophisticated products that they require," says Eric Gregory, Vice President and General Manager. Gregory says government has been Supercon's best customer over the years—"The Department of Energy high-energy physics program has kept this business alive."

SQUIDs

Biomagnetic Technologies, Inc. (BTI) of San Diego, the oldest commercial supplier of superconducting electronics, manufactures SQUIDs for monitoring magnetic fields generated by the human brain. The brain's magnetic fields are a billion times less intense than the earth's magnetic field—so subtle that only SQUIDs can detect them. Since the skull is transparent to magnetic fields, SQUIDs offer a non-invasive way to study normal and pathological brain activity.

One possible clinical application for SQUIDs involves the treatment of focal epilepsy. Certain types of epilepsy are caused by abnormal brain tissue confined to a small area known as the epileptic focus. By pinpointing the location of the focus, SQUIDs may improve the effectiveness of surgical removal. SQUIDs may also aid in the treatment of Alzheimer's disease, Parkinson's disease, strokes, multiple sclerosis, head injuries, and schizophrenia.

Dr. Sam Williamson of New York University believes annual brain exams involving SQUIDs will eventually be as common as annual dental exams. The exams could assess sensory and cognitive functions, sensory-motor activity, and memory. Dr. Williamson and other researchers at NYU Medical Center, the National Institute of Health, and UCLA Medical Center are pioneering the use of SQUIDs to monitor magnetic fields generated by the brain and also by muscles, limbs, the eyes, the liver, the spinal column, and the heart.

BTI's systems cost $825,000 configured for research institutions or $1.2 million configured for hospitals. BTI has installed over 50 systems during its 15-year history. Each system contains 7 SQUIDs, which simultaneously monitor magnetic fields at 7 different locations in the brain. Refrigeration devices account for most of the size of the system (the SQUIDs operate at about 4°K). The most costly component is a magnetically shielded room, which eliminates magnetic noise caused by paging systems and other hospital equipment.

The company recently raised over $4 million in a second round of venture capital financing in the expectation that a large market will develop. It has forecast new sales of $4 million.

Oscilloscopes and Electronics

In 1987 Hypres Inc. of Elmsford, New York introduced the first superconducting electronics product with a broad range of applications: the PSP-1000, the world's fastest oscilloscope. The PSP-1000 senses and displays extremely fast and faint electronic signals. Sadig Faris, President and technical leader of Hypres, claims the PSP-1000 is 5 times faster and 50 times more sensitive than any competitive oscilloscope. Faris touts his product as the beginning of an electronics revolution.

The PSP-1000 exploits at least 3 advantages of Josephson electronics: 1) Speed. Its switching time is measured in picoseconds. (A picosecond is a millionth of a millionth of a second.) 2) Lower power dissipation. Since a superconducting circuit has no resistive losses, it dissipates less than 1/1000 of the power of its semiconducting counterpart. 3) Dispersion less transmission. Superconducting circuits can transmit very high-speed signals through transmission lines without distortion.

Primary emphasis has been on the Josephson Junction, a fast electronic switch. The junction operates several orders of magnitude faster than current semiconductor switches and dissipates less power in the process. About $40 million was spent in 1987, with Hypres developing and producing the primary entry into the marketplace, an oscilloscope (see Figure 3-3) that performs measurements and has extremely high sensitivity to high frequency circuits and switching networks.

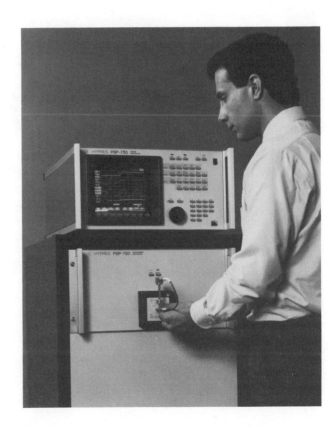

FIGURE 3-3
Hypres Inc. PSP-750
Oscilloscope. (70 Ghz
bandwidth and 50 uv
sensitivity)

Source: Hypres Inc.

Biomagnetics Technologies has developed a sensitive device that is designed to measure brain activity, promising early diagnosis of disorders. Called the neuro-magnetometer, the device recently won Food and Drug Administration approval, but its cost will still keep it from the average medical facility for several years.

Quantum Design is marketing a magnetic properties measuring system for laboratories performing advanced superconductivity research. The device is successfully being marketed with little or no competition and is backlogged with new orders. They are also marketing a rust detector for MIT, a device that senses the weak fields resulting from the chemical reaction that occurs as metals oxidize. They envisage other uses of the device, e.g., for users of steel supports (bridges) and pipelines.

Wire, Rods and Cable

Most of the materials produced in this category are used in making superconducting magnets. Teledyne makes the preponderance of niobium alloy rods. General, Oxford Instruments and Supercon use these rods as raw materials by using them to make wire to be wound into cable. A $10 million market is supporting this sector, but market dominance will depend on innovation as the niobium-tin compounds enter the marketplace. Intermagnetics General (see Figure 3-4) has already made continuous thin strands of the wire and could break away as the market leader.

FIGURE 3-4
Superconductive Wires,
Tapes, and Cables.
Developed by IGC

Source: Intermagnetics General
Corporation

Small Magnets For Research

In 1987, about $10 million was spent purchasing small niobium magnets primarily for research purposes. The work underway is focussed on finding new applications for magnets as well as improving the magnets themselves. There are three potential new uses being explored at this time:

1. As Focussing Elements for Intense Electron Beams.

The state of the art of integrated circuitry is currently limited by its ability to finely etch circuitry on substrates that are becoming more and more crowded. The conductor runs are so thin (the width of an etching beam), that new methods are needed to reduce them further. It is anticipated that x-rays focussed sharply by superconductor magnets might allow the fine etching required to further reduce the size of chips, now approaching the sub-micron level. Oxford is building a prototype system for IBM expected to cost approximately $15 million.

2. As Industrial Impurity Gatherers.

Waste impurities contained in water, coal, oil and other carriers, are potentially removeable if sufficient power can be achieved in magnet trapping systems. Eriez Magnetics, GA Technologies, and Cryogenic Consultants recently completed a $2 million system to remove impurities from clay.

3. For Improved Medical Imagers.

Magnetic imagers are the current predominant user of superconducting magnets today, and is being exploited to improve its performance. Siemens and Oxford Instruments have already produced magnets twice as powerful as available on the market today.

R&D of Niobium Materials

Niobium is the heart of today's superconductor and even though the trend is toward ceramics, the industry still intends to improve the applications for niobium. Teledyne's Wah Chang produced better compounds of niobium-titanium, the basic element in wire, cable, and therefore the heart of all magnet systems. The super collider will benefit from this technology in the interim period until the application of ceramics can be fully utilized.

Large Niobium Magnets

Both General Electric and Westinghouse were involved in power generation using superconductive magnets, but the effort lost a good bit of steam in 1987 due to the reduced costs of fuels in general. The future applications that appear most interesting are the storage applications where electricity could be maintained in reserve stored in a magnetic reservoir until peak needs required its release into the system. At low usage times, the storage device would again begin to absorb energy for the next peak period. Bechtel, GA Technologies, and General Dynamics have already demonstrated the principle for these storage devices in 1986, and when the market comes alive, they can be produced.

A more recent application of these larger superconductor magnets is in the Strategic Defense Initiative program where the need exists to provide large surges of energy in several scenarios including energy directed weapons. The Defense Nuclear Agency has been working with Bechtel to build this type of large energy storage system.

R&D of Ceramic Materials

Two major problems still exist in the applications of ceramic superconductors, inability to carry large currents and their brittleness making them hard to work with. Both IBM and Nippon are working on the current carrying problem, experimenting with thin films that have comparable current carrying capability to niobium, but the ability to develop these approaches into bulk current carriers is incomplete.

The cost of research is expensive and since research is generally geared toward potential market scenarios, the effort expended here will undoubtedly be performed by only the larger companies with sufficient resources to continue the work. IBM, AT&T, and Nippon are still working on the problems and Dupont is working on large scale production techniques.

Ceramics Wires and Tapes

Wires and cables will be the basic elements of any superconductivity oriented systems and these must be reckoned with first in the production cycle. American Superconductor is making flexible ribbon in a pilot plant. Argonne Laboratory mixes ceramic powders with a binder and creates a thread-like fiber that can be wound into cable. AT&T (see Figure 3-5) uses a similar approach while Toshiba encloses the ceramic powder in a metal tube that can be formed into wire or cable.

FIGURE 3-5
Zero resistivity in a ceramic superconducting wire coil.

Source: AT&T Bell Laboratories

The processes are all troubled by the brittle nature of the ceramic and this will require considerable effort for years to come. The rewards are apparently sufficient for the larger firms to continue this process.

Ceramic Coatings, Thin Films, and Shielding

Standard microelectronic fabrication techniques are presently being used to make coatings and thin film superconductors. The SQUID magnetometer is an example of such thin film technology. Electro-Kinetic Systems is working to develop coatings that can provide electronic shielding for computers. Having significant benefits as a security measure, i.e., preventing the radiation of signals, the work is of great interest to the intelligence community.

Particle Accelerators

Particle accelerators are high-energy physics instruments for studying the fundamental units of matter. Fermilab's Tevatron in Batavia, Illinois accelerates protons in a 3-mile loop to bombard other protons, breaking them up into smaller particles called quarks. Electronic detectors record the collisions and provide information about the new particles. The protons approach the speed of light—during 30 seconds of acceleration before the collision, they travel nearly 5 million miles. The power invested in each collision equals the instantaneous output of all the power plants in the world, concentrated for an instant in an area smaller than a proton. The magnetic fields which control the path of the protons are created by 900 superconducting magnets made from niobium-titanium flat ribbon cable, each weighing about a ton. These magnets provided a $15 million market over 5 years for IGC, Supercon, and Oxford Instruments. The Tevatron was originally constructed and operated with conventional magnets. When superconducting magnets were substituted, the power of the accelerator doubled and its energy consumption declined sharply.

In January of 1987 President Reagan endorsed construction of the Superconducting Super Collider (SSC), a much larger particle accelerator. The SSC will consist of a 52-mile underground ring (see Figure 3-6) for accelerating protons. "Basically its a microscope, and the bigger it is, the smaller they can look," says Bob Baldi of General Dynamics in San Diego. The SSC will contain 10,000 niobium-titanium superconducting magnets, each about 50 feet long. It may provide a total market of $1 billion over several years for magnet suppliers, beginning in 1989. In September 1987 the Department of Energy received 43 proposals from candidates for the SSC's location. President Reagan may name the winner before he leaves office in January 1989.

Oxford Instruments has signed a contract with IBM to produce a small particle accelerator for x-ray lithography of microcircuits. Today's standard procedure is to etch microcircuits with visible light. By etching circuits with x-rays, which have a shorter wave length, it will be possible to create smaller microchip circuits. Superconducting magnets will create the intense magnetic fields required to generate and control the x-rays. "This is a very important development—it involves new technology that will create the next generation of microchips," says Michael Cassidy, President of an Oxford Instruments sales division in Bedford, Massachusetts.

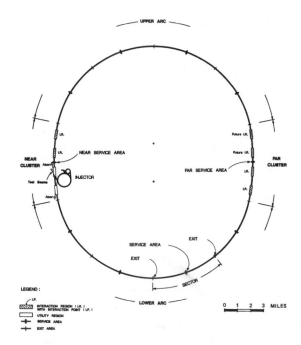

FIGURE 3-6
Schematic Layout of
Superconducting Super
Collider (SSC)

Source: U.S. Department of
Energy

Magnets for Nuclear Fusion Research

As noted in *Supercurrents*, nuclear fusion reactors, now in experimental stages, generate power the same way as the sun: by fusing two smaller hydrogen atoms into a helium atom. The helium atom contains slightly less mass than the two hydrogen atoms combined, and the leftover mass is transformed into enormous quantities of energy. The hydrogen nuclei in a gallon of water, combined in a fusion reaction, would release the same amount of energy as burning a million gallons of oil. Proponents of fusion claim the process offers a safe way to generate inexhaustible energy for the 21st century. The role of superconducting magnets in fusion is to confine the hot gases, which can reach millions of degrees centigrade. Several fusion test facilities employ superconducting magnets, including the TORE SUPRA in France, the T-15 in the USSR, and the MFTFB in the United States. Fusion magnets are huge, and they must withstand extreme mechanical forces. Bob Baldi, Manager of Accelerator Technology for General Dynamics in San Diego, says his company has built pairs of fusion magnets weighing 350 tons. "These magnets have to withstand 22 million pounds of force. A Boeing 747 weighs 800,000 pounds, fully loaded, so the mechanical forces on the magnets are the same as if thirty 747s were hanging on them." Baldi says that General Dynamics has captured 65% of the Department of Energy's superconducting magnet business for fusion and accelerators.

Oak Ridge National Lab recently completed a 4-year international research program to test 6 fusion magnets with different configurations. The magnets were built by Westinghouse, General Electric, General Dynamics, and companies from Switzerland, Germany, and Japan. The Westinghouse magnet was built from niobium-tin wire, and the other 5 from niobium-titanium. All 6 magnets exceeded the 8-tesla design criteria. The niobium-tin magnet reached 8.9 tesla, and the 5 niobium-titanium magnets exceeded 9 tesla.

BUSINESS GUIDE TO SUPERCONDUCTIVE PRODUCTS

High Technology Business in their January 1988 issue presented the following directory which listed the 48 companies active in superconductor research and sales around the world. The listings are by categories including the United States (Table 3-1), European (Table 3-2), and Japanese (Table 3-3), organizations which represent the full spectrum of superconductor activity, from small entrepreneurs to large corporations doing small-scale research to major players already profiting from this growing technology.

Tables 3-1 to 3-3 are reprinted with permission of HIGH TECHNOLOGY BUSINESS magazine, January 1988, copyright 1988 by Infotechnology Publishing Corporation, 214 Lewis Wharf, Boston, Massachusetts 02110.

SUPERCONDUCTIVITY FEDERAL FUNDING IN THE UNITED STATES

Within a few months of the initial new superconductivity discoveries, Federal agencies redirected $45 million in fiscal 1987 funds from other R&D to high-temperature super-conductors. The scientific breakthroughs prompted a dozen bills during the first session of the 100th Congress, proposals ranging from study commissions to a national program on superconductivity.

Taken together, Federal agencies spend, as shown in Table 3-4, nearly $159 million for superconductivity R&D in fiscal 1988, over half ($95 million) on the new materials (and the rest for lower-temperature superconductors). The Department of Defense (DOD) and the Department of Energy (DOE) together account for three-quarters of the lower-temperature superconductor budget, and received most of the increase. DOE, for instance, will have nearly twice as much high-temperature superconductor money as the National Science Foundation. Most of the Federal high-temperature superconductor money will go to government laboratories, contractors, and universities that are well removed from the commercial marketplace.

The President's legislative package, which reached Congress in February 1988, did not address R&D funding. Consistent with the Administration's emphasis on indirect incentives for commercialization, the package included provisions that would further liberalize U.S. antitrust policies, and extend the reach of U.S. patent protection.

SUPERCONDUCTIVITY FUNDING IN JAPAN

Japan is accelerating its funding for superconductor research in 1988 by more than 300% of the prior year. Japan has recognized the market potential, both technological and financial, of the research, development, and technology transfer to the market of super-conductors.

In 1988, Japan's superconductivity budgets provides $56.97 million for research. Table 3-5 provides the Fiscal Year 1988 budget broken down by Japanese government agencies.

Japan's organizational structure for commercialization of superconductivity is shown in Figure 3-7. The Japanese Council for Science and Technology has disseminated in Policy Recommendation Number 14, Japan's Strategy for Superconductivity Development. Their program, based on multi-core projects, has two major thrusts.

TABLE 3-1 Business Guide to U.S. Companies

Company	Officers	Financing	Staff	Superconductor Activity
1. American Magnetics Box 2509 Oak Ridge, TN 37831 (615) 482-1056	Kenneth Efferson, president Robert Jake, v.p., general manager E.T. Henson, v.p., marketing and sales	Privately owned; sales of $1 million to $5 million	20	Building custom niobium superconducting magnets and instruments for research use; pursuing the specialty-magnet market.
2. AT&T Bell Labs/Bell Communications Research (Bellcore) 600 Mountain Ave. Murray Hill, NJ 07974 (201) 582-3000	Robert Dynes, director, chemical physics research Donald Murphy, head of solid-state chemistry research Paul E. Fleury, director, physical research lab	AT&T subsidiaries; Bell Labs has an annual budget of $2.25 billion, 10 percent dedicated to basic research	40	In superconductivity research since 1950; developed niobium alloys, did early work on Josephson junctions. Made major contributions to the discovery of high-temperature ceramic superconductors. Currently pursuing basic research.
3. Bechtel National Box 3965 San Francisco, CA 94119 (415) 768-1234	Robert J. Loyd, project manager, Superconducting Magnetic Energy Storage (SMES)	Privately held; sales of $140 million. SMES contract worth more than $10 million	6 (more expected in early 1988)	Developing SMES as part of the Strategic Defense Initiative; may use experience to enter commercial utility market.
4. Biomagnetic Technologies 4174 Sorrento Valley Blvd. San Diego, CA 92121 (619) 453-6300	Stephen O. James, president, CEO William C. Black, senior v.p. Eugene Hirschkoff, v.p., operations	Privately held. Raised $5.2 million in first-round financing in 1985, $4.2 million in second round, 1987. Projected 1987 revenues, $3.7 million	85	Marketing SQUIDs and related equipment, including its Neuromagnetometer for observing brain functions. Seeking partnership with medical-equipment company to sell and support products.
5. E. I. Du Pont de Nemours 1007 Market St. Wilmington, DE 19898 (302) 774-1000	Edward Mead, manager, superconductor business development Rudolph Pariser, director, advanced materials research Arthur W. Sleight, research leader	Listed on New York Stock Exchange. 1986 earnings, $1.1 billion; sales, $29.4 billion	30	Attempting to apply its expertise in chemical processing to the large-scale production of superconductors. Wants to supply high-temperature superconductor markets as they emerge.
6. Energy Conversion Devices 1675 W. Maple Rd. Troy, MI 46084 (313) 260-1900	Stanford R. Ovshinsky, president, CEO Stephen J. Hudgens, v.p., R&D Rosa T. Young, sr. scientist, group leader	Traded on NASDAQ. Fiscal 1986 net loss of $27.9 million on revenue of $21.1 million; superconductivity work internally funded at $1 million	About 10	Studying ceramic and niobium-based superconductors. Developed a process for mixing fluoride with ceramic for higher-temperature superconductivity, but process not independently verified. Plans to license fluorination technology if a market develops.

Company	Contact	Financial	No.	Description
7. Eriez Magnetics/Eriez Manufacturing Asbury Road at Airport Erie, PA 16514 (814) 833-9881	Chester F. Giermak, president Jerry Selvaggi, consultant/engineering manager	Privately held; $40 million in sales	10	Designed and installed the first superconducting magnet for industrial use, a separator that removes impurities from clay. Plans to compete with conventional separators that remove particles from wastewater.
8. Ford Motor Box 1899 Dearborn, MI 48121 (313) 322-3000	John McTague, v.p., research Marga Roberts, director, chemistry and physical sciences Craig L. Davis, manager, physics dept.	Traded on New York Stock Exchange. 1986 earnings of $3.3 billion on sales of $62.7 billion	5	Working with Detroit's Wayne State University on high-temperature superconductors. Looking for electronic applications that would be pursued by its Aeronutronic Division in Newport Beach, Calif.
9. GA Technologies/Applied Superconetics Box 85608 San Diego, CA 92138 (619) 452-3400	Tihiro Ohkawa, vice chairman Kenneth Partain, president, Applied Superconetics John Alcorn, manager, Superconducting Magnet Group	Privately owned; 1986 sales of $154 million	100	Designing and building specialized magnets. GA's subsidiary, Applied Superconetics, sells magnets for use in magnetic-resonance imaging. Strong candidate to supply magnets for the Super Collider.
10. Garrett Box 92248 Los Angeles, CA 90009 (213) 776-1010	Anil Trivedi, assistant manager, advanced applications	Parent company, Allied-Signal, on New York Stock Exchange; Garrett had 1986 sales of $2.15 billion	10 to 20	Developing a fine-grained superconductor ceramic powder for electronics and industrial use.
11. General Dynamics Space Systems Division 5001 Kearny Villa Rd. San Diego, CA 92123 (619) 573-8000	David Walker, chief, R&D designs Robert Johnson, program manager, energy programs	Traded on New York Stock Exchange. 1986 revenues, $8.9 billion; loss of $63 million due to $420-million write-off of purchase price of Cessna Aircraft	60 at peak; now only a few preparing proposals	Built large magnets for Department of Energy research programs; may use expertise to supply magnets for the Super Collider.
12. General Electric Medical Systems Group Box 414 Milwaukee, WI 53201 (414) 544-3011	John Trani, sr. v.p., group executive Michael J. Jeffries, R&D manager, GE R&D Center	Traded on New York Stock Exchange. 1986 earnings of $2.5 billion on sales of $35.2 billion	20 at R&D Center; the Medical Systems Group employs several hundred	Supplying magnetic-resonance-imaging equipment. The R&D Center developed superconducting generators and is working on high-temperature ceramic superconductors.
13. General Motors Technical Center 30200 Mound Rd. Warren, MI 48090 (313) 575-1188	Donald J. Atwood, vice chairman Robert Frosch, v.p., GM Research Laboratories	Traded on New York Stock Exchange; 1986 earnings of $2.9 billion on earnings of $102.8 billion	5	GM Research Labs is developing ways to deposit thin films of ceramic superconductors on silicon wafers. Has demonstrated a metallo-organic deposition technique that lays down films without the use of vacuum.

TABLE 3-1 Business Guide to U.S. Companies (continued)

Company	Officers	Financing	Staff	Superconductor Activity
14. Hypres 500 Executive Blvd. Elmsford, NY 10523 (914) 592-1190	Sadeg M. Faris, president, CEO Gerald M. Haines; v.p., CFO Eric Hanson, v.p., product development	Privately held; venture funding of $2.2 million in August 1983 and $6.4 million in December 1985	75	Produces a commercial Josephson-junction microchip that it uses in electronic instruments. Plans to introduce more such devices; seeks partner to develop and market a computer.
15. IBM Watson Research Center Box 218 Yorktown Heights, NY 10598 (914) 945-3000	Prauveen Chaudhari, v.p., physical-science research Alex Malozemoff, coordinator, superconductivity program	Traded on New York Stock Exchange. 1986 earnings of $4.8 billion on revenues of $51.2 billion; 1986 R&D and engineering budget, $5.2 billion	Not available	Studying high-temperature materials to achieve superconductivity at room temperature.
16. Intermagnetics General Charles Industrial Park Box 566 Guilderland, NY 12084 (518) 456-5456	Carl Rosner, chairman, president C. Richard Mullen, sr. v.p., operations Bruce A. Zeitlin, v.p., materials technology	Traded on NASDAQ. Lost $3.9 million on revenues of $14.3 million in 1987; 1986 profit of $1.6 million on revenues of $21.2 million	More than 300, including production workers	The leading U.S. maker of wire and cable, and magnets for commercial and research markets. Saw 1987 loss after its largest customer, Johnson & Johnson, discontinued product line. Positioned to be leading supplier of magnets for the Super Collider.
17. Microelectronics and Computer Technology 3500 W. Balcones Center Dr. Austin, TX 78759 (512) 343-0978	Grant A. Dove, chairman, CEO Barry Whalen, v.p. Harry Kroger, technical director, packaging and interconnects	Owned by consortium; $75-million operating budget	7	Coordinates research efforts of electronics companies that own it. Developing high-temperature superconductors for electronics packaging and interconnects. Seeking new participants.
18. Quantum Design 11578 Sorrento Valley Rd. San Diego, CA 92121 (619) 481-4400	William B. Lindgren, president, general manager Michael B. Simmonds, v.p.	Privately held; recently topped $1 million in annual sales	22	Making instruments that measure magnetic properties, using SQUIDs from Biomagnetics Technologies. A subsidiary, Quantum Magnetics, will market additional SQUID-based instruments, including a rust detector.
19. Supercon 830 Boston Turnpike Rd. Shrewsbury, MA 01545 (617) 842-0174	James Wong, president Eric Gregory, v.p., general manager	Privately held; annual sales of $1 million to $5 million	30	Manufacturing niobium-alloy wire and cable. Supplies research labs, GE, and GA Technologies. Maneuvering to supply the Super Collider.

Company	People	Ownership/Financial	Staff	Activity
20. Teledyne Wah Chang Albany Division Box 460 Albany, OR 97321 (503) 926-4211	Al Riesen, president Chet Leroy, v.p., technology	Traded on New York Stock Exchange. Earned $129 million on sales of $1.6 billion for the first half of 1987.	10 in R&D; many more in production	Leading supplier of niobium-alloy wire for magnets made by companies including Oxford, Intermagnetics, and Supercon. Plans to be major supplier of wire for magnets in the Super Collider.
21. TRW 1 Space Park Redondo Beach, CA 92077 (213) 535-4321	William Simmons, director, group research Arnold Silver, head, superconducting electronics	Traded on New York Stock Exchange. 1986 earnings of $217 million on sales of $6.4 billion	About 20	Researching Josephson-junction circuits for the Defense Department. Will develop products for military and aerospace markets.
22. Westinghouse Electric Research and Development Center 1310 Beulah Rd. Pittsburgh, PA 15235 (412) 256-1352	John Hulm, director, research Richard D. Blaugher, manager, cryogenic technology and electronics Alex Braginski, manager, superconducting materials	Traded on New York Stock Exchange. 1986 earnings of $671 million on revenues of $10.7 billion	7	Researching high-temperature ceramic superconductors; developing Josephson-junction technology for the Air Force. Well positioned to be a major magnet supplier for the Super Collider. Superconducting generator technology may interest the Navy.
23. American Superconductor 21 Erie St. Cambridge, MA 02139 (617) 499-2600	George McKinney, president Terry Loucks, v.p., technology Francis Hughes, treasurer	Privately held; $4.35 million from American Research & Development, Rothschild Ventures, and Venrock	4	Holds a license on an MIT process for making ceramic wire and tape; plans to open pilot plant in 1988 to produce wires and ribbons, windings for magnets, and possibly thin wires for electronics. Significant profits not expected for 7 to 10 years.
24. AppliTech of Indiana 8150 Zionsville Rd. Indianapolis, IN 46200 (317) 872-6109	N. Quick, founder	Privately held; financial data unavailable	4	Developing a process for making very high-quality ceramic; testing a laser process for eliminating flaws. Pilot plant expected in two years.
25. Arch Development 1115-25 E. 56th St. Chicago, IL 60637 (312) 702-7417	Steven Lazarus, president, CEO Brian R.T. Frost, director, Technology Transfer Center, Argonne National Laboratory Janett Truhatch, associate v.p. for Research, University of Chicago	Nonprofit; funded by Argonne National Laboratory and University of Chicago	50 scientists and technicians at Argonne	Setting up a company to develop a way to make ceramic wire. Plans to license patents, form cooperative R&D partnerships, and create new companies.

TABLE 3-1 Business Guide to U.S. Companies (continued)

Company	Officers	Financing	Staff	Superconductor Activity
26. Ceramics Process Systems 840 Memorial Dr. Cambridge, MA 02139 (617) 354-2020	H. Kent Bowen, chairman Clayton M. Christensen, president, director George A. Neil Jr., exec. v.p., operations	Common stock traded on NASDAQ. 1986 revenue, $2.6 million; net loss, $4.1 million	As many as 12	Developing metal-ceramic layered packages for integrated circuits. Wants to link marketing with other companies. Focusing on developing products that can be made using micro-smooth sheet forming, metal-ceramic laminates, and molding.
27. Conductor Technologies 1001 Connecticut Ave. N.W. Washington, DC 20036 (202) 452-0900	Stephen J. Lawrence, president Laurence Storch, v.p.	Privately held; undisclosed amount from private sources	None full-time	Supports the work of MIT researchers who are developing electronic devices made from ceramic superconductors; seeking priority in licensing resulting patents.
28. Conductus 2275 E. Bayshore Rd. Palo Alto, CA 94303 (415) 494-7836	John Shoch, president, CEO Tony Sun, CFO	$6 million in first-round financing	None full-time	Developing fabrication methods using thin-film techniques similar to those used to produce semiconductors. Exploring very high-speed digital devices, magnetic field detectors (SQUIDs) and other sensors, and high-speed electronic interconnections.
29. Electro-Kinetic Systems 701 Chestnut St. Tramer, PA 19013 (215) 497-4660	Jack Reilly, chairman, CEO Burton Lederman, director, R&D	Listed on NASDAQ. Sales of $2 million, earnings of $65,000	2	Developing ceramic-based materials that can be applied as coatings and which superconduct at liquid-nitrogen temperature. Working with MIT.
30. Guernsey Coating Labs 4464 McGrath St., Unit 106 Ventura, CA 93003 (805) 642-1508	Peter Guernsey, president Sam Pellicori, consulting physicist	Privately held; sales of $500,000 in optical coatings; seeking $500,000 in venture capital	1 part-time	An established optical-coating lab, diversifying into custom coating with ceramic superconductors.
31. Monolithic Superconductors Box 1654 Lake Oswego, OR 97035 (503) 684-2974	Lawrence E. Murr, owner, founder Alan Hare, owner, founder	Privately held; financial data unavailable	6	Developing a way to produce bulk ceramic material by using shock waves to bond particles. Seeking venture capital and cooperative research to help commercialize the technique.

TABLE 3-2 Business Guide to European Companies

Company	Superconductor Activity
. ASEA-Brown Boveri S-721 83 Vafteraf, Sweden 021-10 00 00	Researching high-temperature superconductors for use in high-powered magnets and generators.
Cryogenic Consultants Metrostore Building 231 The Vale London W3 7QS, England 01-743-6049	Designing and manufacturing superconductor and superconductor-cooling equipment, including magnets for research, mineral separators, SQUIDs, and related electronic devices.
General Electric 1 Stanhope Gate London W1A 1EH, England 01-493-8484	Makes magnetic-resonance-imaging equipment with magnets from Oxford Instruments. Researching high-temperature superconductors for use in large magnets, electrical machiners, and electronics.
Oxford Instruments Group Eynsham Oxford OX8 1TL, England (0865) 881-437	Has about 50 percent market share in magnets for magnetic-resonance imaging; also makes niobium-alloy wire and cable. Expected to become major supplier of wire and cable for magnets used in the Super Collider.
Plessey Vicarage Lane Ilford, Essex IG1 4AQ, England 01-478-3040	Developing ceramic superconductors; studying applications in electric-power cables, Josephson-junction circuits, SQUIDs, and thin films.
Siemens Wittelsbacherplatz 2, Muenchen Postfach 103, D-8000 1 Federal Republic of Germany (089) 234-10	A leading builder of superconductor magnet systems and magnetic-resonance imaging equipment. Developing test magnets for medical market.

TABLE 3-3 Business Guide to Japanese Companies

Company	Superconductor Activity	Company	Superconductor Activity
Fujitsu Marunouchi Building 6-1 Marunouchi, 1-chome Chiyoda-ku, Tokyo 100, Japan (03) 216-3211	Researching high-temperature ceramics, with particular interest in thin films. Working on Josephson junctions to develop a superconducting computer.	**NEC** 5-33-1, Shiba Minato-ku, Tokyo 108, Japan (03) 454-1111	Researching Josephson-junction technology for computers and other electronic applications.
Furukawa Electric 2-6-1, Marunouchi Chiyoda-ku, Tokyo 100, Japan (03) 286-3001	Major electrical cable maker; developing a ceramic-based, ring-shaped superconducting magnet.	**Nippon Steel** 2-6-3, Otemachi Chiyoda-ku, Tokyo 100, Japan (03) 242-4111	Developing ceramic-based, superconducting wire.
Hitachi Central Research Laboratory Kokubunji, Tokyo 185, Japan 0423-23-1111 x 3217	Leading developer of niobium-based Josephson junctions. Also developing ceramic-based superconductors for electronics. Hitachi Cable division makes niobium-based wire.	**Nippon Telegraph & Telephone** 1-6-1, Uchisaiwaicho Chiyoda-ku, Tokyo 100, Japan (03) 509-5035	Pursuing Japan's largest development effort in Josephson-junction technology; also experimenting with techniques for producing ceramic crystals and films.
Kawasaki Steel 2-2-3, Uchisaiwaicho Chiyoda-ku, Tokyo 100, Japan (03) 597-3111	Developing an experimental superconducting wire made of ceramic.	**Sumitomo Electric Industries** 3-12-1 Moto-Akafaka Minato-ku, Tokyo 107, Japan (03) 423-5111	Has more than 400 Japanese patent applications in ceramic superconductors. Affiliate Sumitomo Heavy Industries is building a superconducting synchotron, expected by 1989, to etch chips.
Matsushita Electric Industrial 1-1-2, Shibakoen Minato-ku, Tokyo 105, Japan (03) 437-1121	Working on ceramic thin films for silicon wafers, possibly leading to a process that uses superconductors in integrated circuits.	**Toshiba** 1-1-1, Shibaura Minato-ku, Kawasaki 105, Japan (03) 597-7111	Developing experimental ceramic wire and tape by bonding superconducting powders inside a metal capillary.
Mitsubishi Electric 2-2-3, Marunouchi Chiyoda-ku, Tokyo 100, Japan (03) 218-2111	Researching superconductivity since 1958; experimenting with ceramic-based, high-temperature materials. Makes superconducting tape.		

TABLE 3-4 Summary of Fiscal 1988 Superconductivity Budgets of United States Government Agencies

	Million Dollars		
	FY 1987	Estimate FY 1988	Estimate FY 1989
I. Department of Energy	$ 41.0	$ 66.9	$ 94.8
II. Department of Defense	26.0	53.0	71.0
III. National Science Foundation	13.7	17.5	20.1
IV. National Aeronautics and Space Administration	2.6	6.9	9.8
V. Department of Commerce (National Bureau of Standards)	1.1	2.8	9.3
VI. Department of Interior	0.1	0.1	0.1
VII. Department of Transportation	0.1	0.1	0.3
Total:	**$ 84.6**	**$147.5[3]**	**$205.4[2]**

1. Both high temperature and low temperature.
2. Not including $46.9 million for the procurement of superconducting wire and designed systems.
3. Not including $11 million for procurement of designed systems. ($147.5 + 11 = 158.5$)

Source: Budget Figures from Committee on Materials (COMAT): U.S. Office of Science and Technology Policy.

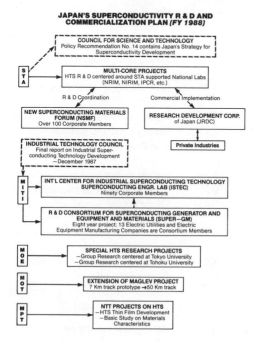

FIGURE 3-7
Japan's Superconductivity R & D and Commercialization Plan (FY 1988)

Source: CSAC

TABLE 3-5 Fiscal 1988 Superconductivity Budgets of Japanese Government Agencies

I. Ministry of International Trade and Industry

Project Title	Million* Dollars	Remarks
Superconducting Materials	$ 2.36	To national research institutes
	3.30	To private industry and the International Superconductivity Research Center
Organic Superconductivity	0.48	To national research institutes
High-Energy, High-Flux Materials	2.50	Wire for electric power
Josephson Computer Technology	0.82	To national research institutes
	2.55	To consortium of private companies
Cable, Power, Generators	0.52	To national research institutes
	10.88	To consortium of private companies
	1.32	Administrative expenses
Joint International Research	0.12	Research with U.S. NBS on SQUID, etc.
Ultra-low Temperature Electronics	0.13	Electrical properties of metals
Energy Storage	0.35	Agency of National Resources and Energy
Rare Earth Technologies	1.05	Agency of National Research and Energy
Total	**$ 26.38**	**620% increase over FY 87**

II. Science and Technology Agency

Project Title	Million* Dollars	Remarks
New Projects		
National Research Institute for Metals	$ 8.38	Theory, thin films, evaluation
National Institute for Research in Inorganic Materials	3.75	New materials, crystals, analysis of crystals
Institute of Physical and Chemical Research	0.95	Micro process technology, structural analysis
Japan Atomic Energy Research Institute	2.25	Theory database, neutron-beam analytic system
Power Reactor and Nuclear Fuel Development Corporation	0.20	Survey on applications for nuclear power
National Space Development Agency	0.25	Survey on space applications
Research and Development Department	0.03	Overall coordination
Continuing Projects		
National Research Institute for Metals	0.53	Ultra-low temperature equipment
Research Development Corporation of Japan	3.04	Promotion of innovative research
Japan Atomic Energy Research Institute	5.22	Superconducting coils
Special Atomic Power Research Grant	0.05	Superconducting wire, to National Research Institute for Metals
Total	**$ 24.65**	**201% increase over FY 87**

III. Ministry of Education

Project Title	Million* Dollars	Remarks
Scientific Research Grants		
Intensive Area Grants	NP**	Mechanisms of superconductivity Grants to 20 researchers
Special Project Grant	$ 0.69	Tanaka's group at Univ. of Tokyo (total $1.98 million)
Special Research Projects	0.50	Superconductivity and the New Electronics
Special Project	NP**	Nuclear fission, magnets, high density current
Scientific Research Grants		
& Appropriations for Education	3.31	
Total	**$ 4.50**	

Special National Appropriations for Education

1. Laboratory in material properties at Materials Science Department, Tohoku University.
2. Ceramics center at Tokyo Institute of Technology.

IV. Ministry of Transportation

$1 million for superconductivity-related part of Maglev railway project.

V. Ministry of Posts and Telecommunications

$440 thousand for research in high-speed telecommunications. The goal is development of a mixer that can handle THz frequencies. Half of funds to Radio Research Laboratory, half to industry. A six-year project.

*Does not fully reflect salary expenses.
** NP = Data Not Provided

Source: Nikkei Electronics, Feb. 22, 1988-Exchange rate: Y130/$1.

The first is R&D coordination with the more than 100 corporate members of the New Superconducting Materials Forum and the second element is the commercial implementation efforts, through the Japan Research Development Corporation (JRDC), working with private industries.

The Industrial Technology Council provided, in late 1987, their final report on Industrial Superconducting Technology Development which has been used by MITI in coordinating the International Center for Industrial Superconducting Technology Superconducting Engineering Lab (ISTEC) and the R&D Consortium for Superconducting Generator and Equipment and Materials (SUPER-GM) efforts.

The Ministry of Education (MOE) is funding research programs at Tokyo University and new electronics projects at Tohoku University. The Ministry of Transportation (MOT) is investing $1 million for superconductivity-related parts for the MAGLEV railway project. The Ministry of Posts and Telecommunications is supporting research on thin film development and basic studies of material characteristics.

The Japanese have developed an effective overall plan for research and development in superconductivity and could provide transfer of the technology to their industrial base faster than the United States or Europe.

In comparing the United States and Japanese Government budgets for superconductivity, one finds that the fiscal 1988 total for the United States is $147.5 million (estimated) compared to a 1988 Japanese superconductivity budget of nearly $57 million.

Does this mean that the United States is far outspending Japan? Not necessarily, as the Japanese budget figures usually do not reflect salary expenses.

Informed sources estimate that the Japanese figures could be doubled, or perhaps tripled, in order to perform a valid U.S./Japanese funding comparison.

Furthermore, a significant portion of the budget for Japan's Ministry of International Trade and Industry is going toward commercial applications research, unlike the federal U.S. superconductivity effort which is geared more toward basic research, with a fair amount earmarked for national defense programs.

The $57 million total depicted for Japan's 1988 fiscal year reflects an accurate face value figure. The $57 million represents a 300 percent increase in funding relative to Japan's total FY 1987 government superconductivity outlay.

Of extreme interest in the future is how the Japanese and U.S. government expenditures ramp up in coming years. For a preview, note the proposed U.S. FY 1989 spending levels, will reflect a 234 percent increase over FY 1987 if approved by congressional appropriations processes.

WORLDWIDE COMPETITION

On a global scale, today's world superconductor industry is small but mature and principally confined to the developed countries. Basic research capabilities are more widespread.

Although much of the early impetus for research and development came from the United States, technology transfer has not been unidirectional. National and international conferences on all aspects of low-temperature physics have become routine.

Over the past 25 years, in several countries a wide variety of applications of superconducting electrotechnology have been examined in prototype development programs. No replacements for conventional applications have reached the market, however. As a result, the demand for superconducting materials has been relatively small and has lacked continuity, being largely oriented toward development. Nevertheless, in most countries, government programs have supported a fledgling industry.

In the United States, magnet development for high-energy physics machines has been carried out in the national laboratories. Fusion and magnetohydrodynamics magnets have been built both in the national laboratories and in private industry. There is also a rapidly growing commercial market based mainly on new medical imaging systems. A small materials and wire industry serves magnet development efforts. Many U.S. firms have supported their own research and development efforts in superconductor technology, both for power and electronic applications. A few small, continuing ventures have succeeded in superconducting electronics; a large market for superconducting electronic devices or systems has not yet developed.

Most American firms—viewing payoffs from high-temperature superconductor R&D as uncertain and distant—have declined to invest heavily. A few major corporations—e.g., DuPont, IBM, AT&T—are mounting substantial efforts. A number of small firms and venture startups have also been pursuing the new technology. By and large, however, American companies have taken a wait-and-see attitude. They plan to take advantage of developments as they emerge from the laboratory—someone else's laboratory—or buy into emerging markets when the time is right. Unfortunately, reactive strategies such as these have seldom worked in industries like electronics over the past 10 to 15 years, while many American firms seem to have forgotten how to adapt technologies originating elsewhere.

Corporations in Europe and Japan (see Table 3-6) have also fostered and maintained an expertise in superconductivity. In those nations, foreign governments have to some degree protected their superconductor industries by ensuring that equipment for government laboratories is built by domestic private industry; foreign bids are not accepted, a policy that ensures national industrial expertise. By comparison, much of this work in the United States is carried out in the federal laboratories from which there is little transfer to industry. In addition, foreign superconductor firms are allowed to bid on equipment needed by the United States government.

Corporate executives in Japan see high-temperature superconductors as a major new opportunity, one that could set the pattern of international competition for the 21st century. Japanese companies have made substantial commitments of people and funds, pursuing research and applications-related work (see Table 3-7) in parallel. Japanese firms are at work in more lines of business than United States firms. Steel companies and glassmakers, as well as chemical producers and electronics manufacturers, are seeking new businesses, ways to diversify. Japanese managers see in high-temperature superconductors a road to continued expansion and exporting and are willing to take the risks that follow from such a view.

In the past, the United States has provided world leadership in superconducting science and technology, and has generously shared its own technology with other nations. Current collaborative efforts include the international Large Coil Fusion Project at Oak Ridge National Laboratory and Japanese development of a very large detector magnet for the Fermilab collider interaction area. In the evaluation of conductors for the physics collider magnets (the Superconducting Super Collider and the heavy-ion collider), material has been purchased from Japanese and European firms. The United States, through Brookhaven National Laboratory, also provides test evaluations of cables and conductors for the Hadron Electron Ring Accelerator (HERA) collider under construction in Hamburg, West Germany. Two prototype magnets for the Relativistic Heavy-Ion Collider (RHIC) have been purchased from the Brown Boveri Corporation, in West Germany, because this was the cheapest and quickest way to obtain them (Brown Boveri had the necessary tooling because of their work for HERA).

TABLE 3-6 European and Japanese Superconductor Competition

Company	Superconductive Product		
	Raw Material	Semi-Finished Material	End Product and Equipment
Europe			
Ciba-Geigy	X	-	-
Enichem	X	-	-
Hoechst	X	-	-
Johnson Matthey	X	-	-
Lonza	X	-	-
Philips	-	X	X
Pirelli	-	X	-
Rhone-Poulenc	X	-	-
Siemens	-	-	X
Solvay	X	X	-
Japan			
Furukowa	X	-	-
Hitachi	X	X	X
Kobe	X	-	-
Kyocera	X	-	-
Mitsubishi Electric	-	X	X
NEC	-	-	X
Nippon Shukubai America	X	-	-
NTK	X	X	-
NTT	-	-	X
Sanyo	-	-	X
Showa Denka	X	-	-
Sony	-	-	X
Sumitomo Electric	X	X	-
TDK	X	-	-
Toshiba	-	X	X

Source: Chemical Marketing & Management—Fall 1987

TABLE 3-7 Projected Implementation Times for Various Superconducting Devices

Application	K. Tachikawa (Tokai U.)	S. Tanaka (U. Tokyo)	N. Makino (Mitsubishi Res.)
Magnetic levitated train	~2000	1995-2000	1995-2000
S.C.-driven ship	~2000	~1997	1990-1995
Power line	1995-2000	~2000	~2000
Power generator	~2000	~2000	1995-2000
Power storage	~2000	~2000	1990-1995
Josephson device	~1990	1990-1995	~1990
IC package/substrate	~1990	1990-1995	1990-1995
S.C. semicon. IC	~1990	1990-1995	~1990
Magnetic sensors	~1990	1995-2000	~1990
IR sensors	~1990	1995-2000	~1990
NMR-CT	1990-1995	1990-1995	~1990
Magnetic shielding	1990-1995	Now	1990-1995
Toys or kits	1990-1995	~1990	1990-1995
S.C. magnet (fusion)	~2000	~1995	1990-1995

Source: From Nippon Keizai Shimbun (18 May 1987), based on interviews with three leading Japanese researchers.

What effect will high-temperature superconductivity have on the marketplace of the future? At the 1987 International Superconductivity Electronics conference in Tokyo, one researcher quoted a Japanese market research firm which predicted the market for superconducting products would reach $100 billion in annual sales by the year 2010. Approximately 33% of the sales would come from the power industry, and the other 67% from the information and medical industries.

CHAPTER 4

COMMERCIALIZING SUPERCONDUCTIVITY

This chapter is based on "Commercializing High-Temperature Superconductivity" published in June 1988, by the Office of Technology Assessment.

U.S. GOVERNMENT AND INDUSTRY

The United States invents and Japan commercializes. So say some. Is it true? If so, this would suggest not only that American companies fail to capitalize on technologies developed here, but that Japanese firms get a free ride on American research and development. Furthermore, if this has happened in other industries (steel) and with other technologies (video tape recorders, TV), it could happen with high-temperature superconductors (HTS).

Has American industry really had that much difficulty in commercialization—in designing, developing, manufacturing, and marketing products based on new technologies? Yes—in some industries and with some kinds of technologies. In other cases—for example, biotechnology or computer software—American firms continue to do better at commercialization than their world class competitors. Nonetheless, the competitive difficulties of American semiconductor firms have long since shown that continuing U.S. advantages in high technology cannot be assumed. And sectors like consumer electronics demonstrate that, when it comes to engineering, if not science, Japan has been a formidable presence since the 1960s.

Commercialization is the job of the private sector. Government plays a critical role in two respects:

1. Research and development funding. Federal agencies will spend some $60 billion on research and development in 1988. Government dollars create much of the technology base that companies throughout the economy draw on. In 1988, the U.S. Government will spend some $95 million on high-temperature superconductor research and development. This is about as much as the American firms surveyed by OTA say they will spend on superconductivity research and development (low-temperature superconductors as well as high-temperature superconductors) in 1988.
2. The environment for innovation and technology development. A host of policies—ranging from regulation of financial markets, to protection for intellectual property, and education and training—affect commercialization by companies large and small.

Private firms use scientific and technical results—more or less freely available, including knowledge originating overseas—in their efforts to establish proprietary advantages. Universities and national laboratories create much of the science base. Some industrial research contributes to the storehouse of scientific knowledge. All three groups—universities, government laboratories, industry—contribute to the larger technology base.

A lot of technical knowledge remains closely held—protected by patents, by secrecy (classification for reasons of national security, trade secrets), or simply as proprietary expertise. Much proprietary information resides in peoples' heads, in organizational routines, management styles, as tacit know-how. Companies also write down some of their organizational knowledge: in product drawings and specifications; in process sheets, manuals, and computer programs for running production lines and entire factories. The manufacturing skills that helped Japanese semiconductor manufacturers outstrip their American competitors depend heavily on proprietary know-how, much of it embodied in the skills of their employees—skills that people often cannot fully articulate or explain.

Commercialization of high-temperature superconductors will depend on scientific knowledge, much of this available to anyone who can understand it. It will also depend on know-how, hard-won learning and experience—making good thin films, orienting the grains in superconducting ceramics to increase current-carrying capacity. The knowledge of markets will also be very important. Government contributes directly through support for the technology and science base. Federal agencies may spend their high-temperature superconductor research and development budgets wisely, or not. National laboratories may transfer technologies to the private sectors quickly, or only after long bureaucratic delays. Government policies also affect commercialization indirectly. Patents and legal protection of trade secrets help firms stake out proprietary technical positions.

No one anticipated superconductivity at 90 or 125 degrees K. No one can predict what will come next. More likely than not, 5 to 10 years of research and development—much of it supported by Federal agencies—lie ahead before high-temperature superconductor markets will have much size or begin to grow rapidly. A few niche products could come sooner. So could some military applications. New discoveries could change the picture radically. The ways in which the Federal Government spends its research and development dollars matter right now. Policymakers may have a bit more leisure to review the other channels of policy influence on commercialization of high-temperature superconductors. The stakes are high—for the private sector, and for government decisionmakers.

Potential for dramatic breakthroughs, coupled with great uncertainty, makes for difficult decisions. OTA sees no reason to rule out the possibility of room-temperature superconductivity (next month, next year). Room-temperature superconductivity—in a cheap material, easy to work with—has almost unimaginable implications. Companies with proprietary technical positions could reap huge rewards. The risks of inaction are high; on the other hand, progress could stall. High expectations and media hype could be followed by difficulty in raising capital and inaction on the policy front. Robotics, artificial intelligence, and biotechnology have already lived through several such waves. High-temperature superconductors could follow the same course.

Early applications of new technologies tend to be relatively specialized, of modest economic significance. The public may lose interest, financial markets downgrade the prospects. No one can know, at this point, whether high-temperature superconductors could turn out to be a solution in search of a problem.

OTA's analysis suggests that commercialization of high-temperature superconductors will proceed somewhat faster than many American companies expect, though not as fast as the Japanese companies that have been making heavy investments seem to anticipate. As American companies move down the learning curves that mark accumulated knowledge and experience in high-temperature superconductors, federally funded research and development will provide critical support for the technology base that all firms—but particularly smaller companies—draw from. The analysis in this chapter leads to the following conclusions:

- Small, entrepreneurial firms will be well placed to develop commercial applications of high-temperature superconductors. The conditions are right: a new science-based technology; synergistic links with existing industries, including low-temperature superconductors (LTS) and electronics; and venture capital for good ideas. But while small companies have been a major source of U.S. strength in high technology, few can assemble the financing, the technological breadth, or the production and marketing capabilities to grow as fast as their markets.

- Larger American corporations may find that they are starting out behind some of their Japanese rivals. The new high-temperature superconductor materials are ceramics. Japanese firms have a substantial lead in both structural and electronic ceramics. Some of this expertise will transfer to high-temperature superconductors. So will a good deal of know-how developed for fabricating microelectronic devices—another field where Japanese firms have demonstrated themselves to be at least as good and sometimes better than American companies.

- Processing and fabrication techniques will be critical for commercialization. American companies have fallen down in manufacturing skills across the board; the more heavily process-dependent high-temperature superconductor applications turn out to be, the more difficult it will be for U.S. firms to keep up with the Japanese.

- Product as well as process technologies will demand much trial-and-error development. Japanese engineers and Japanese corporations are good at this. American companies are not. To the extent that commercialization of high-temperature superconductors depends on step-by-step, incremental improvements—brute-force engineering—U.S. companies will be in relatively poor positions to compete.

- Research and development funded by the U.S. Government will help American companies in commercializing high-temperature superconductors, but the spinoffs from defense-related research and development may not be large or long-lasting if military requirements become highly specialized and diverge from commercial needs.

- Indirect policy measures—intended to remove the roadblocks to commercialization and increase the rewards for innovators and entrepreneurs—can also help. But the indirect approach alone will not be an adequate response to the coming international competition in high-temperature superconductors.

What about U.S. commercialization in general—the backdrop for the statements above?

- Mobility among scientists, engineers, and managers has spurred rapid growth and technological innovation in postwar U.S. high-technology industries ranging from computers and semiconductors (starting in the 1940s and 1950s) to biotechnology (beginning in the late 1970s). Venture capital for small, high-technology firms, likewise, has been a consistent source of competitive strength, one that will continue to work to U.S. advantage in high-temperature superconductors.

- Many larger American companies have pulled back from basic research and risky technology development projects. Ease in establishing new small firms compensates in part for these relatively conservative investment decisions; indeed, negative decisions on proposed research and development projects sometimes spawn startups that go on to commercialize new technologies. Some of this will probably happen in high-temperature superconductors.

- With few American firms self-sufficient in technology, a lack of long-term research and development in the private sector, and managements that look for home-run opportunities rather than building technologies and markets step-by-step, the Federal Government has, by default, become a primary source of support for technology development. As yet, government agency missions do not reflect this new role.

- Despite the onslaught of foreign firms since the late 1960s, many American companies have not yet made the changes in their own organizations necessary to compete more effectively. Paying little more than lip service to well-known engineering methods such as simultaneous product and process design, they fail to give manufacturing high priority. Neither managers nor engineers in the United States have learned to take advantage of technologies originating overseas.

The indirect policy approach the U.S. Government has traditionally relied on to stimulate innovation and commercialization worked well for many years. Today, with foreign competition stronger than ever before, it seems time to explore new directions. The Federal research and development budget has grown rapidly over the postwar period. Management practices in government agencies, mechanisms for setting priorities, for ensuring an adequate technology base, have not kept pace.

The climate for innovation can always be improved and the barriers reduced. Short-term perspectives of U.S. corporations, many of which have been unwilling to keep pace with foreign investments in new technologies, stem in part from the removal of another set of barriers—deregulation in U.S. financial markets.

Unless the United States learns to match the kinds of supports for commercialization that have proven effective elsewhere, only small improvements can be expected. U.S. industry could fall behind in high-temperature superconductors, and in the applications that this new technology will provide.

THE GOVERNMENT ROLE

High-temperature superconductors are fresh from scientific laboratories, but many commercial innovations begin with existing knowledge, gleaned from textbooks, design manuals, and the schoolhouse of experience. The work of commercialization centers on engineering: development of new products and new manufacturing processes. Companies support their development groups with marketing people, and in some cases with research. Sometimes new science is part of commercialization, but not always.

The process may begin with an idea that is old, but has never been reduced to practice because of gaps in the technology base. The automobile, the airplane, and the liquid-fueled rocket all had to await needed pieces of technical knowledge. The Wright brothers learned to steer and stabilize their flying machine.

Despite years of trial and error (and centuries of speculation), they were the first to find a way around these technical barriers.

Superconductivity itself has a long history as a specialized field of physics, and a shorter history—beginning about 1960—as a technology that private firms sought to exploit.

Support for Industry: Direct and Indirect

What does this have to do with government? Today, governments finance much of the research and development that provides the starting point for commercialization. Companies everywhere start with this publicly available pool of technical knowledge in their search for proprietary know-how and competitive advantage. Second, public policies influence the choices companies make in financing their own research and development, and in using the knowledge available to them. Tax and regulatory policies encourage or discourage investments in commercial technology development. Patents create incentives, high capital gains taxes disincentives.

Smaller companies depend heavily on externally generated knowledge; many manufacturing firms with hundreds of employees have few if any engineers on their payrolls. But if smaller companies have the greatest need, science and technology move so fast today that big companies also rely heavily on government research and development. Moreover, pressures for near-term profits have forced many larger U.S. corporations away from basic research. In the United States, a few hundred larger companies account for the lion's share of industry-funded research and development—three firms (IBM, AT&T, General Motors) for more than 15 percent.

Half of all U.S. research and development dollars come from the Federal treasury. The fraction is smaller in most other countries, but in all industrial economies public funds pay for a substantial share of national research and development. The reasons begin with health and national defense, but competitiveness has been one of the rationales: the first government research laboratories, established in the early years of this century in Britain,

Germany, and the United States, were intended to help domestic industries meet foreign competition.

Foreign firms have access to many of the results of federally funded research and development, just as U.S. firms can tap some of the technical knowledge generated with foreign government support. Governments seek to use technology policy to help domestic firms compete, while commercial enterprises seek to take advantage of the world store of technical knowledge. Technology policy begins with research and development spending—setting broad priorities, making funding decisions at the project level, agency management. Other tools include intellectual property protection, which can help domestic firms establish and protect a technological edge. Of course, many countries also provide direct funding for commercially oriented research and development.

The U.S. Position in Technology

Past OTA assessments have examined U.S. competitiveness in a number of industries, and linked technological position with competitiveness; the most recent found signs of slowdown in U.S. research and development productivity, as well as evidence that newly industrializing countries have made surprising gains in technology. Principal findings from these earlier assessments include:

- Technology is vital for competitive success in some industries. In others, it may be secondary. But in all or nearly all sectors, the technological advantages of American firms have been shrinking for years. The United States may be able to retain narrow margins in some technologies. Parity will be the goal in others. Regaining the advantages of the 1960s will, in the ordinary course of events, be impossible.
- In newer technologies, those that have developed since the 1960s, the Japanese have been able to enter on a par with American firms, and to keep up or move ahead. Examples include optical communications, and both structural and electronic ceramics. European firms, in contrast with the Japanese, have had trouble turning technical knowledge into competitive advantage.
- Today, U.S. military and space expenditures yield fewer and less dramatic spinoffs than two decades ago. The U.S. economy is vast and diverse. Defense research and development—increasingly specialized when not truly exotic—cannot provide the breadth and depth of support needed for a competitive set of industries.
- Japan and several European countries place higher priorities on commercial technology development than does the United States. research and development spending by Japanese industry reached 2.1 percent of gross domestic product in 1986, compared with 1.4 percent here in the United States.

Productivity, Innovation, Competitiveness, Commercialization

Import penetration in steel and consumer electronics, going back two decades, marks the beginnings of the wave of concern over lagging U.S. productivity growth and competitiveness. Commercialization is simply the latest catch phrase for problems that are part of the puzzle. The ongoing political debate has centered on the proper mix of policies in the United States, where government has been reluctant to intervene as directly or as deeply in the affairs of industry as, say, in Japan or France.

Research and Development Funding and Objectives

The weight of specific changes in U.S. technology policy during the 1980s has been on the indirect side, the direct role of the Federal Government has also changed—though not in the direction of support for commercial technology development. Government research and development has grown under the Reagan Administration, but much of the expansion has been for defense. Support for commercially oriented research and development has lagged, and in many cases been cut back.

Department of Defense (DOD) research and development went from $20.1 billion in fiscal 1982 to $37.9 billion in 1988. DOD research and development, plus the defense-related portion of DOE spending (about half the Department's research and development), account for nearly 70 percent of all Federal research and development; the great majority consists of applied research and the engineering of weapons systems.

The U.S. Government has not paid much attention, relatively speaking, to research and development of interest to companies outside the defense, aerospace, and health sectors. And in the 1980s, Federal agencies have backed away even further (e.g., from energy research and development). The Reagan Administration has held that government has no business supporting commercial technology development. Fundamental research, yes, but anything more would be a subsidy—unjustified and likely to create harmful economic distortions.

Research carried on in industrial laboratories, almost by definition, has a practical orientation. So does engineering research in universities and nonprofit laboratories. Plainly, distinctions such as that between untargeted and directed research will always be arbitrary. Nonetheless, such distinctions help in thinking about research and development and how it supports commercialization.

Within directed research, further distinctions can be made. Incremental work, for example, takes a step-by-step approach toward reasonably well-defined goals. The problems may be technically difficult, but the territory has been at least partially explored. Much of the work on synthesis of new materials that laid the groundwork for the discovery of high-temperature superconductors falls in this category.

Most research serves the needs of government or industry. Military needs, social objectives such as health care, and industrial competition have driven the scientific enterprise at least since the end of the 19th century.

COMMERCIALIZATION

Both industry and government support directed research. Promising results lead naturally into development. Research and development then go on in parallel, with research outcomes suggesting new avenues for development, and problems encountered in development defining new research problems.

While the U.S. Government has a long tradition of support for basic research and mission-oriented research and development, it usually leaves pursuit of commercial technologies to the private sector. This policy worked well for many years. For instance, continued development of fiber-reinforced composite materials— lighter and with greater stiffness, strength, and toughness than many metals—builds on a technology base that has been expanding at a rapid rate since the 1950s. The primary stimulus came from the military, where composites found their first applications in missiles, later in manned aircraft. Penetration into commercial aircraft followed.

When it comes to technologies where Federal agencies have been less active, U.S. firms have often fallen behind. Although the U.S. Government has spent several hundred million dollars for research and development on structural ceramics since the early 1970s, the effort has been a small one compared with fiber composites. Japan, meanwhile, has established a useful lead in structural (as well as electronic) ceramics. In semiconductors, American firms established a commanding lead during the 1950s and 1960s, when military procurement provided much of the demand. In later years, as production swung towards civilian markets, Japanese firms closed the gap.

Inputs to Commercialization: Technology and the Marketplace

Product or process development—whether adapting magnet technology for medical imaging systems, or generic techniques for computerized process control to steelmaking—depends on at least two inputs from outside the development group itself. The first of these is knowledge drawn from the technology base, including science, engineering, and shop floor know-how (see Figure 4-1). The second input is knowledge of markets—what potential customers want and need.

Research and Development and Innovation

Innovations follow their own paths. Figure 4-2 summarizes the later stages for magnetic resonance imaging (MRI)—those after initial research and feasibility demonstration. Science came first, the complete chronology beginning in 1936 with theoretical predictions of the underlying phenomenon of nuclear magnetic resonance. Experimental demonstrations followed a decade later, with the first two dimensional images in 1973.

Heavy continuing involvement by physicians and scientists made MRI something of an exception. Normally, commercialization is a job for engineers, supported on the one side by knowledge flowing from the technology and science base, and on the other by information on customer wants, needs, and perceptions.

Much of the early work in high-temperature superconductors will be undertaken by multidisciplinary groups including physicists, chemists, materials, scientists, and ceramists, along with electrical, chemical, and mechanical engineers. The known high-temperature superconductor materials are oxide ceramics—brittle and difficult to work with. Learning to use them means drawing where possible on past research and development—work undertaken earlier and for other purposes on structural and electronic ceramics, as well as processing, fabrication, and design techniques from microelectronics.

As applications come into view, companies will call on marketing tools ranging from feasibility studies (which may include detailed projections of manufacturing costs) to consumer surveys. Technical objectives shift as prospective markets emerge; some firms use "technology gatekeepers" to help match research results and market needs. This is an area where U.S. and Japanese strategies in high-temperature superconductors contrast markedly, with Japanese companies much quicker to begin thinking about applications and the marketplace.

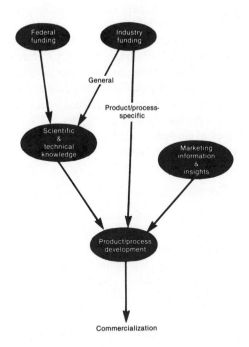

FIGURE 4-1
The Process of
Commercialization

Source: Office of Technology
Assessment, 1988

FIGURE 4-2
Development Stages for
Magnetic Resonance
Imaging
(MRI) Systems
* Primarily aimed at
developing the technology
of MRI and, learning to
use it (establishing
efficacy, subject safety,
etc.)
** Primarily aimed   at
testing prototype MRI
system.

Source: Based on Health
Technology Case Study 27:
Nuclear Magnetic Resonance
Imaging Technology, Washing-
ton, D.C.: Office of Technology
Assessment, September 1984,
Ch. 4

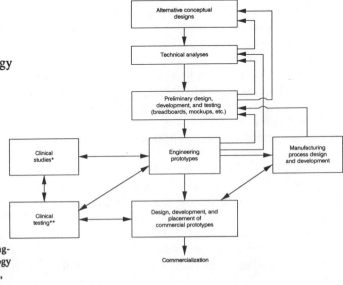

What makes for success or failure in commercialization? Product and process engineering, marketing skills, luck, sometimes research results. No one has a recipe, any more than a recipe for room-temperature superconductivity.

Costs are central for some products, but for others—MRI is one—competition revolves around non-price features. Many hospitals will readily pay a premium of several hundred thousand dollars for an MRI system with superior imaging performance. At the same time, small private clinics or rural hospitals make up a niche market for which a number of manufacturers have designed low-cost systems.

Products may come out too late or too early. A company may fall behind its competitors and never get much market penetration. Early innovators in the semiconductor industry have sometimes failed and sometimes succeeded. The pioneer minicomputer manufacturer, Digital Equipment Corp., whose PDP-8 established this part of the market, went on to become the second largest computer firm in the world. On the other hand, the microcomputer pioneers—Altair, Imsai, Polymorphic Systems—disappeared.

Cost and Risk

As firms move further along the development path, mistakes become more costly. Only one of ten projects launched at the research and development stage ever brings in profits. Before reaching the marketplace, half of all research and development projects fail for technical reasons; poor management or financial stringencies kill two or three more. Of those that do enter production, some never earn enough to cover development costs.

The vast majority of project budgets go for product engineering, process design and development, tooling and production start-up, and test marketing. Introducing an MRI system means investments of $15 million and up for research and development alone; pilot production and field trials require much larger financial commitments. Seldom does research account for more than 10 percent of total project outlays, although the distribution of costs varies a good deal from project to project and industry to industry. The distribution also varies between the United States and Japan.

As Table 4-1 shows, Japanese companies (for the industries and time period examined) spent a bit less on research and development than the average American firm, and much less on manufacturing startup and product introduction. They budgeted more in gearing up for production—on facilities, tooling, and special manufacturing equipment (a difference that may also reflect higher projected volumes). Japanese firms no doubt have lower startup costs because they invest more in front-end process development. Yet a substantial difference remains. Adding the percentages for tooling and equipment to those for manufacturing startup gives a total of 40 percent for the U.S. companies, 54 percent for the Japanese. The greater proportion of total project expenses for tooling and equipment reflects the higher priorities Japanese managers place on manufacturing as an element in competitive strategy.

Such priorities will make a difference in commercialization of high-temperature superconductors, which will depend critically on process know-how. U.S. firms have underinvested in process technology for years—one reason for competitive slippage in industries ranging from steel to automobiles to electronics.

TABLE 4-1 **Distribution of Costs for Development and Introduction of New Products and Processes**[a]

	Percentage of Total Project Cost				
	Research Develop. & & Design	Prototype or Pilot plant[b]	Tooling & Equipment	Mfg. Start-up	Market Start-up
U.S. Companies	26%	17%	23%	17%	17%
Japanese Companies	21%	16%	44%	10%	8%

[a] Survey figures from 1985 for 50 matched pairs of U.S. and Japanese firms. The total of 100 included 36 chemical companies, 30 machinery, 20 electrical and electronics, and 14 from the rubber and metals industries.

[b] For cases of product development, the costs are for prototyping; for process development, they include investments in pilot plants.

NOTE: Totals may not add because of rounding.

Source: E. Mansfield, The Process of Industrial Innovation in the United States and Japan: An Empirical Study, unpublished seminar paper presented March 1, 1988 in Washington, D.C.

COMPETITIVE ADVANTAGE

What does it take to use technology effectively? The examples mentioned above, and others, point to the following common factors:

- Appropriate use of technology and science, new and old—whether a company generates the knowledge internally, or gets it elsewhere. Much of the science base for high-temperature superconductors will be available to everyone. To establish a competitive advantage, companies will have to develop proprietary know-how, and do it ahead of their competitors.
- Effective linkages between engineering and marketing. Customers for many of the early applications of high-temperature superconductors—in military systems, electronics, or perhaps energy storage—will be technically astute. Marketing will count, but not so heavily as for consumer products.
- Effective linkages between product development groups and manufacturing—a point already stressed for high-temperature superconductors.
- Managerial commitment to risky and uncertain research and development projects.

What are the conditions under which American firms have trouble in commercialization—in the effective utilization of technical knowledge, new and old? Under what circumstances do American firms perform best? The most problems occur when it comes to shop-floor manufacturing technologies. In the earlier years of high technology, the United States had potent competitive advantages; entrepreneurship and venture capital; a decentralized science infrastructure with many centers of excellence both inside and outside the Nation's universities; flexible labor markets, with high mobility among engineering scientists, and managers. These strengths have begun to wane. In many industries, Japanese companies are out-engineering American firms. Even in high technology, the Japanese have been able to move quickly from the laboratory to the marketplace. The days when U.S. companies could take their time in commercializing research and development are gone.

What does the discussion above imply for U.S. abilities in commercialization? The first point is simply that taking a new product into the marketplace is always difficult. In their efforts to penetrate the U.S. market, Japanese automakers suffered from many problems they failed to anticipate. Powertrains wore out quickly in long-distance driving. Companies like Honda found themselves trying to sell cars with fenders that would rust through after one or two northern winters.

The high-fashion, product differentiation strategy is new for Japanese companies only in the automobile industry. It is one the Japanese have used in the past in cases like consumer electronics and motorcycles. Successful targeting of markets—whether for consumer goods, for capital equipment (machine tools), or for intermediate products (semiconductor chips)—has been a hallmark of Japan's competitive success.

Japanese companies have already put a good deal of effort into thinking about new applications of superconductivity; they may well locate some of the possible market niches before American firms. The Japanese have often carved out substantial markets by starting from small niches; large, integrated Japanese firms have been more aggressive than their American counterparts in pursuing specialized products, including advanced materials. Japanese companies are willing to start with small-volume production and grow with their markets—a strategy likely to prove successful in high-temperature superconductors, indeed one that may prove necessary.

How do companies based in Japan do so well at defining and attacking market segments, particularly in countries foreign to them? Most Japanese companies do use market research techniques, although Table 4-1 showed they spend less on this than American companies. As some U.S. firms also realize, the best marketing research often remains as informal today as it was 50 years ago—a matter of good judgment from within the company more so than from consulting firms, focus groups, and consumer surveys.

Japanese firms in many industries have also capitalized on the quality of their goods. Lagging quality not only leaves customers unhappy, it raises manufacturing costs. Quality and reliability problems have plagued American industries ranging from automobiles to semiconductors. Careful control of the production process will be necessary for fabricating the new high-temperature superconductor materials, as it is for high-technology electronic and structural ceramics, or for integrated circuits.

The primary point is this: by the 1960s, American firms had come to think of their skills in engineering and marketing as far and away the best in the world. If this was true then, it is true no longer. Many U.S. companies have not yet faced up to the need to do better. Others periodically rediscover such well-known management and engineering practices as mechatronics, simultaneous engineering, design for production, or quality engineering, but fail to follow through with actions that institutionalize them. Some still look to techniques like quality circles for miracle cures.

MECHATRONICS

Mechatronics and simultaneous engineering means nothing more than tackling product and process development in parallel, with overlapping responsibilities in design and manufacturing groups, if not a fully integrated approach. Simultaneous engineering may be hard to achieve in a modern American corporation, but in principle is nothing but common sense. A hundred years ago, technology was simpler and no one had discovered any need to separate design and manufacturing.

The chain can be extended back to research. But given the uncertainties that accompany the search for new knowledge, and the high costs of downstream development, many U.S. executives have come to view research, development, and product planning as sequential processes. Only when consistent, verifiable, and potentially useful results begin to emerge from the laboratory do American companies think about incorporating engineers into the effort. Even at this point, research may remain separate from development: the scientists pass along their findings, but the two groups continue to work independently. Under these circumstances, the entire process can become almost purely sequential—running from applied research to product planning and development to manufacturing engineering, with little overlap.

Technology-based Japanese companies, in contrast, have developed mechatronics and simultaneous/parallel processes to a high level. Many are now busy integrating backward into research—a task they see as necessary for commercializing high technologies like high-temperature superconductors. Already, they do a better job of responding to design and marketing requirements through incremental, applied research.

Japanese managers, moreover, tend to be optimistic about research in general and about high-temperature superconductors specifically. Perhaps because they mix development and engineering personnel into project groups at an early stage, the belief seems pervasive that useful results of one sort or another will inevitably emerge from high-temperature superconductor research and development. Japanese managers have strong convictions on these matters. They believe it is wrong to think about technical developments as proceeding more-or-less linearly from basic research to applied research, then to development and product design, and finally to process engineering. More to the point, they are acting on these beliefs in high-temperature superconductors.

American managers know just as well that many of the steps should take place in parallel. But for reasons ranging from trouble in learning to manage parallel processes effectively (one reason for longer product development cycles), to the characteristics of U.S. financial markets, they do not always act on this knowledge. When it comes to high-temperature superconductors, American managers have been relatively cautious; they want to see results from the laboratory before taking the next step.

COMMERCIALIZING HIGH-TEMPERATURE SUPERCONDUCTORS

There is a bright side. The United States retains major sources of strength in commercializing new technologies. Table 4-2—which draws heavily on past OTA assessments of competitiveness—summarizes advantages and disadvantages of U.S. firms. Table 4-3 outlines the implications for high-temperature superconductors.

TABLE 4-2 U.S. Strengths and Weaknesses in Commercialization

U.S. strengths	U.S. weaknesses	Comments
Factor 1. Industry and market structure; market dynamics. In the past, U.S. firms performed well in rapidly growing industries and markets, especially during the early stages in R&D-intensive industries.	American companies have had trouble coping with slow growth or contraction. Although new technologies promising greater productivity might improve competitive performance in industries like steel, corporate executives frequently choose to invest in unrelated businesses. Where foreign firms might take a more active approach to managing contraction, American companies sometimes let troubled divisions struggle along, without new investment, until profits disappear. Then they shut the doors.	Other countries frequently look to public policies to help companies and their employees adjust to decline.
Factor 2. Blue and gray collar labor force. High labor mobility helps American companies attract the people they need.	Many development projects depend on craftsmen who can fabricate prototypes and modify them quickly based on test results and field experience. In some U.S. industries, shortages of skilled labor—e.g., technicians, modelmakers —have begun to slow commercialization. U.S. apprenticeship programs have been in decline. Vocational training reaches greater fractions of the labor force in nations like West Germany; large Japanese companies invest more heavily in job-related training for blue- and gray-collar employees than do American firms.	In the past, U.S. wage rates worked to the disadvantage of American firms, while creating incentives for investments in R&D and new manufacturing technologies that could raise productivity. Today, international differences in labor costs are less of a factor than in the 1970s.

TABLE 4-2 U.S. Strengths and Weaknesses in Commercialization (continued)

U.S. strengths	U.S. weaknesses	Comments
Factor 3. Professional and managerial work force. Mobility among managers and technical professionals has stimulated early commercialization in high-technology industries. New products have reached the marketplace more quickly because people have left one company and started another to pursue their own ideas. Deep and well-integrated financial markets—e.g., for venture capital—have helped.	American companies underinvest in process (as opposed to product) technologies. This is part of a bigger problem: too many managers and engineers in the United States avoid the factory floor: • for managers, marketing or finance has been the road to the top. • engineers—schooled according to an applied science model—have been insensitive, not only to role of manufacturing, but to the significance of design and marketing. Put simply, the engineering profession has divorced itself from the marketplace, and the needs and desires of potential customers (particularly when it comes to consumer products). Compounding these problems, many American companies underutilize their engineers. Finally, many U.S. firms provide little support for continuing education of their technical employees. Managers and professionals in the United States sometimes place individual ambition over company goals. Competition among individuals may make cooperation within the organization more difficult (e.g., between product engineering and manufacturing).	More upper level managers in Japanese and West German firms have technical backgrounds than in the United States; they appear more sensitive to the strategic significance of manufacturing, and in at least some cases to new technological opportunities.
Factor 4. Industrial infrastructure (also see Factor 6 below). American companies can call on a vast array of vendors, suppliers, subcontractors, and service firms for needs ranging from fabrication of prototypes to financing, legal services, and marketing research. Few other countries have a comparable range of capabilities so easily available.[a]	U.S. competitiveness in capital goods like machine tools has slipped, compounding the problems in manufacturing technology. Arms-length relationships between American firms and their vendors and suppliers may not be as conducive to commercialization as the relationships found in Japan (relations which might be classified as close and cooperative, or perhaps with equal accuracy as coercive and dependent).	At present, the independent computer software and services industry is perhaps the preeminent illustration of U.S. infrastructural strength.

Factor 5. Technology and science base: (also see Factor 7 below).

U.S. strength in basic research—both science and engineering—has been a cornerstone of commercialization.

The national laboratory system is a major resource, although one that has not been turned to the needs of industry.

Multidisciplinary R&D—essential in industrial (and government) laboratories—has been the exception rather than the rule in American universities. Foreign university systems, however, have probably been even worse at multidisciplinary research.

U.S. strength in basic research has not always been matched by strength in applied research, nor in the application of technical knowledge. The Nation depends heavily on a relatively small number of large corporations for industrial R&D and the development of new commercial technologies. When R&D is not close enough to anyone's interests, gaps open in the technology base. Moreover, U.S. firms seem to be falling behind in their ability to move swiftly from the R&D laboratory to the marketplace. Diffusion of technology within the U.S. economy has been a persistent and serious problem.

American engineers and their employers have often remained unfamiliar with technologies developed elsewhere, reluctant to adopt them. This reluctance is evident, for instance, when it comes to rules of thumb and informal procedures—sanctioned by experience if not by scientific knowledge. Examples include shop-floor practices for job scheduling and quality control.

The science base and technology base are not identical. The latter spreads much more broadly, encompassing, for instance, the intuitive rules and methods—many of them tacit rather than formally codified—that lie at the heart of technological practice. The semiconductor and biotechnology industries have both sprung from scientific advances. But the theoretical foundations for each remain relatively weak. As a result, progress depends heavily on experience and empirical know-how—again, part of the technology base but not the science base.

Japanese and German firms give commercial technology development higher priorities. Governments in these countries also give more consistent support to generic, pre-competitive R&D.

Factor 6. The business environment for innovation and technology diffusion (also see Factor 7 below).

Clusters of knowledge and skills such as found in the Boston area, or Silicon Valley, help speed commercialization. While some of this entrepreneurial vitality can be linked to major research universities, other regions have become centers of high-technology development even though lacking well-known schools like MIT or Stanford.

The size and wealth of the U.S. market, and the sophistication of customers—especially business customers—work to the advantage of innovators; indeed, foreign companies sometimes come to the United States simply to try out new ideas.

Many American firms seem preoccupied with home runs—major breakthroughs in the marketplace—unwilling to begin with niche products and grow gradually.

Poor labor relations sometimes slow adoption of new technology. Reluctance among American engineers and managers to learn from shop-floor employees hurts productivity and competitiveness.

Companies in other parts of the world may be somewhat more willing to cooperate in R&D.

Linkages between universities and industry could be stronger, but nonetheless probably function better in the United States than elsewhere.

Business and consumer confidence encourage innovation and rapid commercialization. Over the past few years, business confidence appears to have ebbed somewhat—a casualty of Federal budget deficits, trade imbalances, rapid exchange rate swings, and the evident inability of the Government to address these issues. At the same time, the political stability of the United States remains a major strength.

[a]On the importance of specialty firms for the U.S. microelectronics industry, particularly those supplying semiconductor manufacturing equipment, see *International Competitiveness in Electronics* (Washington, DC: November 1983), pp. 144-145. On service inputs, see *International Competition in Services* (Washington, DC: July 1987), pp. 32-34 and 55-57.

TABLE 4-2 U.S. Strengths and Weaknesses in Commercialization (continued)

Factor 7. The policy environment for innovation and technology development.
The United States has a deeply rooted commitment to open markets and vigorous competition. (So does Japan, when it comes to domestic markets and domestic competition.) With widespread economic deregulation since the early 1970s—plus a tax system and financial markets that reward entrepreneurs—startups and smaller companies have often been leaders in commercializing new technologies.

Purchases by the Federal Government have stimulated some industries, particularly in their early years. Examples range from aircraft and computers to lasers and semiconductors.

A broad range of other U.S. policies—e.g., strong legal protections for intellectual property—helps companies stake out and exploit proprietary technological positions.

Deregulated U.S. financial markets bear some of the blame for the risk aversion and short-term decisions common in American business.

Sometimes U.S. regulatory policies delay commercialization. Examples include approvals for drugs and pharmaceutical products.

Many government policies act on commercialization indirectly. Industries have evolved in different ways in different countries, in part because of these influences:

• Along with antitrust, financial market regulations—e.g., rules covering holdings of stock in one company by others—affect the extent of vertical and horizontal integration.

• Tax policies—treatment of capital gains, R&D and investment tax credits—influence corporate decisions on investments in new products and processes.

• Antitrust enforcement helps set the environment for inter-firm cooperation in R&D.

• Trade protection can reduce the risks of new investment, thereby stimulating commercialization. On the other hand, protected firms may grow complacent and decline to invest in new technologies.

• Technical standards sometimes act to speed the adoption of new technologies. If premature or poorly conceived, however, they can impede commercialization.

• Education and training have enormous long run impacts on commercialization and competitiveness.

Source: Office of Technology Assessment, 1988

TABLE 4-3 U.S. Advantages and Disadvantages in Commercializing High-Temperature Superconductors

U.S. advantages	U.S. disadvantages	Comments
Factor 1. Industry and market structure; market dynamics. New science and technology make for conditions under which American firms should be able to commercialize quickly and compete effectively.	At some point, financing constraints may make it difficult for startups and smaller U.S. companies to continue in HTS on an independent basis. Mergers may be necessary for growth.	Past U.S. successes in high technology came when international competition was a minor factor. Foreign firms have now proven they can move quickly from the R&D stage to the marketplace. Mergers or other arrangements driven by financing needs sometimes help, sometimes hurt. Ties with larger companies may stifle innovation. In biotechnology, linkages between small firms and larger companies have helped with regulatory approvals and process scale-ups. American semiconductor firms, however, have seldom been willing to sacrifice their independence for new capital—one reason they have fallen behind large, integrated Japanese competitors.
Factor 2. Blue and gray collar labor force. Some American companies start with a core of employees having experience in low-temperature superconductivity (LTS). A portion of these skills will translate to HTS. At the same time, given that the new HTS superconductors are fundamentally different materials—ceramics rather than metals—a wide array of quite different skills will be needed. Some of the skilled employees may come from related industries, including electronics.	Japanese companies with ceramics businesses can draw on larger numbers of people with relevant skills. These employees will help give Japan a head start in certain kinds of HTS R&D—e.g., mechanical behavior, processing and fabrication. Japanese firms also have many workers with extensive and transferable experience in microelectronics.	So far, few American ceramics firms have been prominent in HTS R&D.

95

TABLE 4-3 U.S. Advantages and Disadvantages in Commercializing High-Temperature Superconductors (continued)

U.S. advantages	U.S. disadvantages	Comments
Factor 3. Professional and managerial work force. Managers, engineers, and scientists moving into U.S. HTS companies from industries like microelectronics will bring new insights and new ideas.	Decisionmakers in American companies, large and small, may not be willing or able to make long-term commitments to HTS-related R&D, particularly more basic work.	At least initially, HTS startups will have managerial staffs with strong technical backgrounds. Some larger U.S. firms with the resources to compete in HTS-related markets may chose other investments because managers fail to understand the technology or recognize the opportunities.
	Processing and fabrication will pose difficult technical problems, of a sort that American companies have not been very good at solving.	Managers with previous experience in LTS may tend to err on the side of conservatism. On the other hand, HTS has had more than its share of exaggerated publicity already. A cautious view of HTS, born of past experience in LTS, could prove realistic.
	Much of the R&D needed to develop HTS will be empirically-based engineering, with heavy doses of trial and error. Japanese companies do very well at this kind of development, often better than their American counterparts.	
Factor 4. Industrial infrastructure (also see Factor 6 below). The generally strong U.S. infrastructure for high technology should be an advantage in HTS.	When it comes to the science and technology of ceramics, specifically, the U.S. infrastructure is weak. American HTS companies with ceramics-related technical problems may have trouble finding help.	Japan's HTS infrastructure exists mostly inside large, integrated companies. In the United States, startups will have to rely heavily on help from outside. The U.S. approach has advantages in flexibility and creative problem-solving, while Japan's reliance on internal resources creates reservoirs of skills and expertise that will be very effective over the longer term.

Factor 5. Technology and science base (also see Factor 7 below). Despite lack of attention to ceramics compared with Japan, the United States has a relatively strong base in materials R&D.

In the early years of HTS development, the defense emphasis of federally supported R&D will work in some ways to the U.S. advantage. Funding from the Department of Defense (DoD) will help train engineers and scientists, and may support the development of some dual-use HTS technologies (e.g., powerful magnets). DoD support for processing R&D could be especially important.

A number of national laboratories have the resources, including specialized equipment, to help with the technical problems of HTS.

Military and civilian applications of HTS will diverge rapidly, limiting the spillover effects from DoD R&D spending.

In 1986, U.S. engineering schools granted 3700 PhDs—but only 14 in ceramics.

Without major policy shifts, Federal agencies will fund little R&D that directly supports commercialization. Nonetheless, the United States is beginning to address the problems of transferring federally funded R&D to industry.

American companies will probably be at a disadvantage for years to come in solving the manufacturing-related problems of HTS. To make progress here, American scientists and engineers—including those engaged in university research—must be willing to spend more of their time working on industrial problems (even if the scientific and university communities continue to view practical work as less than fully respectable). Without substantial efforts in manufacturing R&D, some American companies could be forced into partnerships with Japanese firms simply to get access to processing know-how.

Factor 6. The business environment for innovation and technology diffusion (also see Factor 7 below). U.S. markets should prove receptive to new products based on HTS. Some foreign companies could find they need an R&D presence here simply to keep up.

With Japanese firms starting on a par with American companies, know-how from abroad may prove essential for keeping pace. Many American companies have been unable or unwilling to reach useful technology transfer agreements with Japanese firms. Lack of experience in doing business with the Japanese could become a significant handicap in HTS.

University-industry relations in the United States seem to be following patterns similar to those in biotechnology, with strong and productive linkages developing.

Small U.S. firms have begun devising strategies for commercializing HTS. Many larger American firms with the resources to compete in HTS, however, seem to be adopting a wait-and-see attitude.

TABLE 4-3 U.S. Advantages and Disadvantages in Commercializing High-Temperature Superconductors (continued)

U.S. advantages	U.S. disadvantages	Comments
Factor 7. The policy environment for innovation and technology development.		
So far, the U.S. policy approach seems conducive to entrepreneurial startups in HTS. There is little indication that the 1986 changes in U.S. tax law—which increased rates on capital gains—have choked off funds for HTS startups.	After the initial announcement of the Administration's 11-point superconductivity initiative, little was heard for 7 months—a long time in such a fast-moving field. Budgetary uncertainties, moreover, delayed decisions on Federal R&D funding well into the 1988 fiscal year, hampering progress in universities, industry, and the national laboratories.	While Federal procurements helped the U.S. semiconductor industry get off the ground in the 1960s, poor experience with demonstration projects in energy and transportation has soured prospects for some kinds of policy options that otherwise might provide stability and support for HTS during a long period of gestation.

Some companies continue to express concern that U.S. antitrust policies will limit opportunities for consortia and other forms of joint R&D. However, OTA has not learned of any case in which U.S. antitrust enforcement has in fact stopped firms from cooperating in R&D. |

Table 4-2 has a simple message: the United States has a number of areas of advantage, coupled with several serious handicaps. Those handicaps—emphasis on short-term financial paybacks, low priorities for commercial technology development and for manufacturing—have put U.S. firms at a severe disadvantage in competing with Japan. Some of the consequences can already be seen in high-temperature superconductors.

On the other hand, American firms have often been successful—at least in the past—when new science has led to new products and new industries, especially where fast-growing and volatile markets promise rich awards (Table 4-2, factor 1). American companies perform less well, and often poorly, at incremental innovation—more-or-less routine improvements to existing products and processes. These kinds of problems have been much more prevalent in steel than in chemicals, in machine tools than in computer software, in automobiles than in commercial aircraft.

Most of the success stories came in the years before U.S. industry had much to worry about from international competition. Table 4-3 summarizes the lessons that past performance and events thus far hold for high-temperature superconductors, and compares the strengths and weaknesses of American companies with those in Japan. Some of the U.S. entrants will be new companies, started specifically to exploit high-temperature superconductors and staffed by people with strong credentials in related fields of science and technology. Other firms will move in from a base in low-temperature superconductors. Both kinds of companies should be able to respond effectively to the problems and opportunities that emerge in the early years of high-temperature superconductors—with good ideas and a strong science base, together with venture capital and entrepreneurial drive, leading to success in specialized products and niche markets.

The picture could change as the technology stabilizes and financial strength becomes more important. When production volumes increase, manufacturing capabilities will grow more important. Companies will have to carefully tailor products to emerging markets, and find capital for expansion. U.S. industries that flourished as infants have run into difficulty as competitors—primarily the Japanese—caught up and pulled ahead in the race to capitalize on new approaches to factory production or new knowledge concerning electron devices; in the years ahead, the biotechnology industry could stumble, just like the semiconductor industry.

CHAPTER 5

SUPERCONDUCTIVITY DEFINITIONS A–Z

A

absolute   The basic temperature scale used in scientific and engineering research, in which the theoretical zero point is –273.15°C (–459.69°F); also called the Kelvin scale.

absolute expansion   The true expansion of a liquid with temperature, as calculated when the expansion of the container in which the volume of the liquid is measured is taken into account.

absolute pressure   The pressure above the absolute zero value of pressure that is theoretically obtained in empty space or at the absolute zero temperature.

absolute specific gravity   The ratio of the weight of a given volume of a substance in a vacuum at a given temperature to the weight of an equal volume of water in a vacuum at a given temperature.

absolute system of units   A set of units for measuring physical quantities, defined by interrelated equations in terms of arbitrary fundamental quantities of length, mass, time, and charge or current.

absolute temperature   The temperature measurable in theory on the thermodynamic temperature scale. The temperature in Celsius degrees relative to absolute zero at –273.15°C (the Kelvin scale) or in Fahrenheit degrees relative to absolute zero at –459.69°F.

absolute vacuum   A void completely empty of matter. Also known as perfect vacuum.

absolute zero   The lowest temperature theoretically possible; the temperature at which the thermal energy of random motion of the particles of a system in thermal equilibrium is zero. It is, therefore, the zero of thermodynamic temperature: $0K = -273.15°C = -459.69°F$.

absorption   The action of a solid or liquid in taking up and retaining another substance uniformly throughout its internal structure. This action may be only mechanical, such as the absorption of a liquid by a solid, the liquid being recovered by controlled pressure or heat. When the absorbed material is a gas, it may dissolve in a liquid to form a solution, or enter into a chemical reaction with it to form a new compound.

accelerator   A device that increases the speed, and thus the energy, of charged particles such as electrons and protons.

acceptor   An impurity element that increases the number of holes in a semiconductor crystal such as germanium or silicon; aluminum, gallium, and indium are examples.

acceptor atom   An atom of a substance added to a semiconductor crystal to increase the number of holes in the conduction band.

active current   The component of an electric current in a branch of an alternating-current circuit that is in phase with the voltage. Also known as watt current.

active substrate   A semiconductor or ferrite material in which active elements are formed; also a mechanical support for the other elements of a semiconductor device or integrated circuit.

adhesion   Any mutually attractive force holding together two magnetic bodies, or two oppositely charged nonconducting bodies. The force of static friction between two bodies, or the effects of this force.

adiabatic   Referring to any change in which there is no gain or loss of heat.

adiabatic compression   A reduction in volume of a substance without heat flow, in or out.

adiabatic cooling   A process in which the temperature of a system is reduced without any heat being exchanged between the system and its surroundings.

adiabatic envelope   A surface enclosing a thermodynamic system in an equilibrium which can be disturbed only by long-range forces or by motion of part of the envelope; intuitively, this means that no heat can flow through the surface.

adiabatic process   Any thermodynamic procedure which takes place in a system without the exchange of heat with the surroundings.

adiabatic vaporization   Vaporization of a liquid with virtually no heat exchange between it and its surroundings.

adsorption   Attachment of the molecules of a gas or liquid to the surface of another substance (usually a solid); these molecules form a closely adherent film or layer held in place by electrostatic forces that are considerably weaker than chemical bonds. The finer the particle size of the solid, or the greater its porosity, the more efficient an absorbent it will be because of the increase in surface area thus provided.

allotropy   The existence of a solid, liquid, or gaseous substance in two or more forms that differ in physical rather than chemical properties.

allowed energy bands   The restricted regions of possible electron energy levels in a solid.

allowed transition   A transition between two states which is permitted by the selection rules and which consequently has a relatively high priority.

alloy   A mixture or solution of metals, either solid or liquid, which may or may not include a nonmetal. In certain types the components are not completely miscible as liquids or tend to separate on solidification; others, in which the components are miscible, can be considered as solid solutions. An alloying element is used for a constructive and beneficial purpose rather than as an adulterant.

alternating current   Electric current that reverses direction periodically, usually many times per second. Abbreviated ac.

alternating current losses   Conventional superconductors exhibit losses in alternating current applications, such as in 60 Hertz power transmission or in microwave devices. Although little is known about the alternating current characteristics of the new high-temperature superconductors, there is no reason to expect that the new materials will exhibit lower alternating current losses than other superconducting materials. Recent measurements on thin films in parallel applied fields show the presence of a large surface barrier for the entry of flux, which indicates that hysteresis losses would be small. More extensive measurements of such losses are required.

alternating-current motor   A machine that converts alternating-current electrical energy into mechanical energy by utilizing forces exerted by magnetic fields produced by the current flow through conductors.

alternating gradient   A magnetic field in which successive magnets have gradients of opposite sign, so that the field increases with radius in one magnet and decreases with radius in the next; used in synchrotons and cyclotrons.

ambient temperature   The temperature of the surrounding medium, such as gas or liquid, which comes into contact with the apparatus.

amorphous   Lacking a crystal structure (literally, without shape); having random atomic arrangement without a three-dimensional structure. This property is characteristic of liquids, in which the molecules are not densely ordered but, rather, are comparatively free to move. A few solids also may have amorphous forms (carbon, for example) when their crystal orientation is sufficiently disordered to permit random movement of the atoms. When a solid melts, it changes from the crystalline to the amorphous state in which it remains until recrystallization takes place.

amorphous semiconductor   A semiconductor material which is not entirely crystalline, having only short-range order in its structure.

amorphous solid   Solids may be amorphous, which means they do not have long-range formation patterns.

ampere   The unit of electric current in the rationalized meter-kilogram-second system of units; defined in terms of the force of attraction between two parallel current-carrying conductors. Abbreviated a; A; amp.

Ampere currents   A postulated "molecular-ring" current used to explain the phenomena of magnetism, as well as the apparent nonexistence of isolated magnetic poles.

Ampere law   A law giving the magnetic induction at a point due to given currents in terms of the current elements and their positions relative to the point. Also known as Laplace law. A law giving the line integral over a closed path of the magnetic induction due to given currents in terms of the total current linking the path.

ampere meter squared   The Standard International unit of electromagnetic moment. Abbreviated Am2.

ampere-minute   A unit of electrical charge, equal to the charge transported in 1 minute by a current of 1 ampere, or to 60 coulombs. Abbreviated A min.

ampere per square inch   A unit of current density, equal to the uniform current density of a current of 1 ampere flowing through an area of 1 square inch. Abbreviated A/in^2.

Ampere rule   The rule which states that the direction of the magnetic field surrounding a conductor will be clockwise when viewed from the conductor if the direction of current flow is away from the observer.

ampere square meter per joule second   The Standard International unit of gyromagnetic ratio. Abbreviated Am^2/Js.

Ampere theorem   The theorem which states that an electric current flowing in a circuit produces a magnetic field at external points equivalent to that due to a magnetic shell whose bounding edge is the conductor and whose strength is equal to the strength of the current.

ampere-turn   A unit of magnetomotive force in the meter-kilogram-second system defined as the force of a closed loop of one turn when there is a current of 1 ampere flowing in the loop. Abbreviated amp-turn.

analog signal processor   High-speed analog signal processors performing such functions as filtering, convolution, correlation, Fourier transformation, and analog-to-digital (A-to-D) conversion are important for many applications. Various high-speed A-to-D converters have been tested successfully at 4.2 K. If high-quality Josephson junctions can be fabricated from the new superconductors, these devices should perform comparably at 77 K. At this temperature, integration of the superconducting devices with some semiconducting devices (for example, complementary metal-oxide semiconductors) becomes feasible, and new hybrid systems may well result in the fastest A-to-D converters available.

angstrom   A linear dimensional unit, equal to one-ten thousandth micron or one-100 millionth centimeter (one-250 millionth inch). It is used to measure the spacing of atoms in crystals, the size of atoms and molecules, and the length of light waves. The diameter of a single atom is from 0.64 A (hydrogen) and 1.54 A (carbon) up to 5 A for heavier atoms; the length of a carbon-to-hydrogen bond is 1.09 A. Molecules range from about 3 A for hydrogen to over 1000 A for polymers. The shortest wavelength of visible light is about 4000 A. Ten thousand angstroms equal one micron. The abbreviation is either A or A.U. Named for Anders Jons Angstrom (1814-1874), a Swedish physicist.

anisotropic   A term used in crystallography to indicate a major group of crystalling structures in which the optical properties (e.g., refractive index) vary with the direction of the axes, that is, with the direction of the vibrations of crystal. In isotopic crystals there is no such variation, as light passes through them at the same rate in all directions. Cubic crystals are isotropic, but other structures are generally anisotropic. The word literally means "not turning the same way."

anistropy   The variation of physical properties with direction.

annihilation   A process in which an antiparticle and a particle combine and release their rest energies in other particles.

annular conductor   A number of wires stranded in three reverse concentric layers around a saturated core.

antiatom   An atom made up of antiprotons, antineutrons, and positrons in the same way that an ordinary atom is made up of protons, neutrons, and electrons.

antiferroelectric crystal   A crystalline substance characterized by a state of lower symmetry consisting of two interpenetrating sublattices with equal but opposite electric polarization, and a state of higher symmetry in which the sublattices are unpolarized and indistinguishable.

antiferromagnetic domain   A region in a solid within which equal groups of elementary atomic or molecular magnetic moments are aligned antiparallel.

antiferromagnetism   A property possessed by some metals, alloys, and salts of transition elements by which the atomic magnetic moments form an ordered array which alternates or spirals so as to give no net total moment in zero applied magnetic field.

antiparticle   An antiparticle is identical to the particle with which it is associated, except that certain of its elementary properties, notably electric charge, are exactly opposite.

antiproton   The negatively charged antiparticle of the proton.

apparent expansion   The expansion of a liquid with temperature, as measured in a graduated container without taking into account the container's expansion.

armature   That part of an electric rotating machine that includes the main current-carrying winding in which the electromotive force produced by magnetic flux rotation is induced; it may be rotating or stationary. The movable part of an electromagnetic device, such as the movable iron part of a relay.

armature reactance   The inductive reactance due to the flux produced by the armature current and enclosed by the conductors in the armature slots and the end connections.

armature reaction   Interaction between the magnetic flux produced by armature current and that of the main magnetic field in an electric motor or generator.

armature resistance   The ohmic resistance in the main current-carrying windings of an electric generator or motor.

atmosphere   The unit of pressure equal to 1.013250×10^6 dynes/cm^2, which is the air pressure measured at mean sea level. Abbreviated atm. Also known as standard atmosphere.

atom   The smallest unit of a chemical element. Every atom has a central nucleus of protons and neutrons, around which orbit electrons. Atoms are held together by the electromagnetic force. Derived from the Greek, meaning "indivisible," an atom is the smallest unit of an element that can exist. Atoms have long been known to be made up of particles of matter called protons, neutrons, and electrons. The nucleus of an atom contains one or more protons, which are positively charged, and two or more neutrons (except in hydrogen), which have no electric charge. Each proton and neutron has a mass number of 1; they are bound together in nucleus by exceedingly strong forces. The electrons move around the nucleus in shells (groups of orbitals) which are determined by their energy levels or wave functions. An electron is a negatively charged unit which behaves as both a particle and a wave, i.e., it is partly mass and partly energy. Every atom has the same number of electrons as protons and is thus electrically neutral. Atoms which have gained or lost one or more electrons are called ions.

atomic mass unit   An arbitrarily defined unit in terms of which the masses of individual atoms are expressed; the standard is the unit of mass equal to 1/2 the mass of the carbon atom, having as nucleus the isotope with mass number 12. Abbreviated amu.

atomic number   The number of protons, or positively charged mass units, in the nucleus of an atom, or the number of electrons revolving around the nucleus, upon which its structure and properties depend. This number represents the location of an element in the Periodic Table; it is always the same as the number of negatively charged electrons in the shells. All the isotopes of an element have the same atomic number although different isotopes have different mass numbers. Thus an atom is electrically neutral, except in an ionized state, that is, when one or more electrons have been gained or lost. Atomic numbers range from 1, for hydrogen, to 106 for the most recently discovered element. Syn. proton number. Symbol: Z.

atomic weight   The average weight (or mass) of all the isotopes of an element, as determined from the proportions in which they are present in a given element, compared with the mass of the 12 isotope of carbon (taken as precisely 12.000), which is now the official international standard. The true atomic weight of carbon, when the masses of its isotopes are averaged, is 12.01115. The total mass of any atom is the sum of the masses of all its constituents (protons, neutrons, and electrons). The atomic weight of an element expressed in grams is called the gram atomic weight.

atomic volume   The volume in the solid state of one mole of an element. Thus, atomic volume equals atomic weight divided by the density of the solid.

B

back bond   A chemical bond between an atom in the surface layer of a solid and an atom in the second layer.

back electromotive force   An electromotive force that opposes the normal flow of current in an electric circuit.

band   A restricted range in which the energies of electrons in solids lie, or from which they are excluded, as understood in quantum-mechanical terms.

band gap   An energy difference between two allowed bands of electron energy in a metal.

band scheme   The identification of energy bands of a solid with the levels of independent atoms from which they arise as the atoms are brought together to form the solid, together with the width and spacing of the bands.

band theory of solids   A quantum-mechanical theory of the motion of electrons in solids that predicts certain restricted ranges or bands for the energies of these electrons.

Bardeen-Cooper-Schrieffer theory   A theory of superconductivity that describes quantum-mechanically those states of the system in which conduction electrons cooperate in their motion so as to reduce the total energy appreciably below that of other states by exploiting their effective mutual attraction; these states predominate in a superconducting material. Bardeen, Cooper, and Schrieffer published the theory, now commonly known as the BCS Theory, that earned them the Nobel Prize in 1972. The BCS Theory is based on the existence of an energy gap in what would normally be a continuum of electron energy states in a partly filled band. This energy gap is created by the collective motions of pairs of electrons with opposite spins and momenta. Once excited above the energy gap, electrons cannot decay to their normal states. These electrons become free to move through the lattice without scattering by the ion cores.

barium   The element Barium, m.p. 729°C (1344°F), is a heavy alkaline-earth metal derived from barytes (barium sulfate) and witherite (barium carbonate) found in Missouri, Georgia, Mexico, and Canada. It can be extruded and machined and is used in special alloys, as a "getter" in vacuum tubes, and as a thin-film lubricant in electronic instruments. It exhibits a green color when exposed to flame. Barium is quite active chemically, reacting with halogens, water, oxygen, and acids. In powder form it takes fire at room temperature and thus must be stored and handled in an atmosphere of inert gas. Symbol - Ba, Atomic No. 56, State - Solid, Atomic Wt. 137.34, Group - IIA, Valence 2, Isotopes 7.

Barkhausen effect   The succession of abrupt changes in magnetization occurring when the magnetizing force acting on a piece of iron or other magnetic material is varied.

bar magnet   A bar of hard steel that has been strongly magnetized and holds its magnetism, thereby serving as a permanent magnet.

Barnett effect   The development of a slight magnetization in an initially unmagnetized iron rod when it is rotated at high speed about its axis.

base load   Minimum load over a given period of time. (In addition, the terms base voltage, base current, etc., are used to define the rated values for a system.)

batch process   Any industrial procedure in which raw materials are mixed in individual lots or batches.

bead   A glass, ceramic, or plastic insulator through which passes the inner conductor of a coaxial transmission line and by means of which the inner conductor is supported in a position coaxial with the outer conductor.

B-H curve   A graphical curve showing the relation between magnetic induction B and magnetizing force H for a magnetic material. Also known as magnetization curve.

bismuth   The element Bismuth, m.p. 271°C (518°F), is a hard, brittle metal with a pronounced rhombohedral structure; it has only one stable form. It has two unusual properties: it expands 3.3% when it solidifies, and it is strongly repelled by magnetic fields, assuming a position at right angles to the lines of magnetic force. It forms compounds with halogens, oxygen, nitrogen, and sulfur; it also occurs in a few organic compounds. The nitrate and subnitrate are derived from concentrated solutions of bismuth in nitric acid. Symbol - Bi, Atomic No. 83, State - Solid, Atomic Wt. 208,9806, Group - VA, Valence 2,3,4,5.

Bloch wall   The transition layer, with finite thickness of a few hundred lattice constants, between adjacent ferromagnetic domains magnetized in different directions. It allows the spin directions to change slowly from one orientation to another, rather than abruptly.

FIGURE 5-1
Illustration of spin direction in a Bloch wall with 180° rotation.

boiling point   For an enclosed liquid or mixture of liquids, the temperature at which the upward pressure of molecules escaping from the surface (vapor pressure) equals the downward pressure of the atmosphere (14.7 pounds per square inch at sea level). When the liquid is in a closed container, the boiling point may be defined as the temperature at which the liquid and its vapor are in equilibrium, that is, when evaporation and condensation are occurring at the same rate at constant atmospheric pressure. The temperature is the same in both cases. The boiling point is increased somewhat by the presence of molecules of dissolved substances and substantially by the presence of ions. It decreases with decreasing atmospheric pressure.

boil-off   The vaporization of a liquid, such as liquid oxygen or liquid hydrogen, as its temperature reaches its boiling point under conditions of exposure.

bond   An electrostatic attraction acting between atoms of the same or different elements, enabling them to unite to form molecules or metallic crystals. The number of bonds that each element can form, that is, the number of combining forces it can exert, is called its valence and is determined by the electronic configuration of the particular element.

boson   Every particle is either a boson or a fermion, according to whether the value of its spin, a quantum mechanical property, is a whole number (0,1,2...) or a half number (1/2,3/2...). The particles that mediate the fundamental forces are all bosons, and the quarks and the leptons are all fermions.

bulk lifetime   The average time that elapses between the formation and recombination of minority charge carriers in the bulk material of a semiconductor.

C

capture   A process in which an atomic or nuclear system acquires an additional particle; for example, the capture of electrons by positive ions, or capture of neutrons by nuclei.

carbon   The element carbon is classed as a nonmetal, carbon characterizes all organic compounds and is essential to the photosynthetic reaction on which all living organisms depend. It also occurs in a few inorganic compounds (carbon oxides and metallic carbonates). Though inactive at room temperature, it is highly reactive above 540°C (1000°F). Carbon occurs in compounds that number in the hundreds of thousands, e.g., hydrocarbons (petroleum) and carbohydrates (plants). It is the only element that can form four covalent bonds, resulting in stable chains in which carbon combines both with itself and with other nonmetals (hydrogen, oxygen, sulfur, nitrogen). It also forms binary compounds called carbides with many metals and some nonmetals. Its electronegative nature makes it a reducing agent for metallic oxides; and its high electrical conductivity accounts for its use as electrodes, brushes, contacts, and other electrical equipment. Symbol - C, Atomic No. 6, State - Solid, Atomic Wt. 12.0111, Group - IVA, Valence (2),4, Isotopes 2 stable and 4 radioactive (natural).

Celsius degree   Unit of temperature interval or difference equal to the kelvin.

Celsius temperature scale   Temperature scale in which the temperature 0_c in degrees Celsius (°C) is related to the temperature T_k in kelvins by the formula $0_c = T_k - -273.15$; the freezing point of water at standard atmospheric pressure is very nearly 0°C and the corresponding boiling point is very nearly 100°C.

center of pressure   The point on a plane surface, immersed in a fluid, at which the resultant pressure on the surface may be taken to act. If the surface is horizontal the center of pressure coincides with the center of gravity; otherwise it is below the center of gravity but gets nearer to it as the liquid depth increases.

central-field approximation   In the central-field approximation, each electron in an atom moves under the action of a spherically symmetric electric field caused by the nucleus and all the other electrons. In this approximation, the quantum state of each electron is identified by the four quantum numbers n,l,m, and s.

ceramic   Any product made from earth-derived materials such as clays, silicates, sand, and the like, usually requiring the application of high temperature in a kiln or oven at some stage in the process.

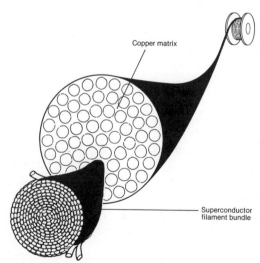

FIGURE 5-2
Illustration of making superconducting wire from ceramics.

Source: EPRI

characterization   Necessary knowledge to relate superconducting properties to composition, microstructure, and defects.

charge   A property of some elementary particles that causes them to exert forces on one another. The natural unit of negative charge is that possessed by the electron and the proton has an equal amount of positive charge. The use of the terms negative and positive are purely conventional and are used to differentiate the types of forces that charged particles exert on each other. Like charges repel and unlike charges attract each other. The force is thought to result from the exchange of photons between the charged particles. The charge of a body or region arises as a result of an excess or defect of electrons with respect to protons. Charge is the integral of electric current with time and is measured in coulombs. The electron has a charge of $1.602\ 192 \times 10^{-19}$. Symbol: Q.

charge carrier   A mobile conduction electron or mobile hole in a semiconductor. Also known as carrier.

charging current   Current that flows into the capacitance of a transmission line when voltage is applied at its terminals.

chemical elements   See table for chemical element symbol, atomic number, and atomic weight. (Page 110)

chemical stability   The 1-2-3 compounds readily react with the ambient atmosphere at typical ambient temperatures. These problems seem to be less severe, however, as the purity and density of the materials are improved. Both water and carbon dioxide participate in the degradation through the formation of hydroxides and carbonates. Further study of the nature of this degradation is needed to develop handling procedures or protective coatings that will ensure against impairment of superconducting properties by atmospheric attack. Chemical stability is also limited because oxygen leaves the structure under vacuum, even at room temperature. Surface protection techniques need to be developed to allow satisfactory performance and lifetime of the materials under various conditions of storage and operation. These concerns are heightened in thin films, in which, for some applications, the chemical composition of the outer atomic layers near the surface must be maintained through many processing steps, and in which diffusion into the substrate interface could degrade superconducting properties.

chemical vapor deposition   A new chemical vapor deposition (CVD) process able to create superconducting thin films 500 to 1,000 times faster than conventional CVD methods. The process is a variation on the high-frequency plasma method used in fabricating metal alloys. In it, materials for the ceramic are conbined in powder form with argon gas in a reactive chamber that has been heated to 10,000°C by high-frequency (about 4 mHz) coils. The powders are vaporized and deposited on a cooled magnesium oxide substrate in the center of the chamber. Fabrication speed is about 10 microns per minute. The resulting ceramic is uniform in texture. There is no need to combine the materials for the ceramic before insertion into the reactive chamber. In the normal high-frequency plasma process, electrodes are placed inside the chamber, which means that temperatures cannot be raised as high as 10,000°C, and also leads to

TABLE 5-1 Chemical Elements

Element	Symbol	Atomic number	Atomic weight	Element	Symbol	Atomic number	Atomic weight
Actinium	Ac	89	227	Mendelevium	Md	101	256
Aluminum	A	13	26.98154	Mercury	Hg	80	200.59
Americium	Am	95	241	Molybdenum	Mo	42	95.94
Antimony	Sb	51	121.75	Neodymium	Nd	60	144.24
Argon	Ar	18	39.948	Neon	Ne	10	20.179
Arsenic	As	33	74.9216	Neptunium	Np	93	237.0482
Astatine	At	85	211	Nickel	Ni	28	58.70
Barium	Ba	56	137.34	Niobium	Nb	41	92.9064
Berkelium	Bk	97	249	Nitrogen	N	7	14.0067
Beryllium	Be	4	9.01218	Nobelium	No	02	254
Bismuth	Bi	83	208.9804	Osmium	Os	76	190.2
Boron	B	5	10.81	Oxygen	O	8	15.9994
Bromine	Br	35	79.904	Palladium	Pd	46	106.4
Cadmium	Cd	48	112.40	Phosphorus	P	15	30.97376
Calcium	Ca	20	40.08	Platinum	Pt	78	195.09
Californium	Cf	98	252	Plutonium	Pu	94	239.11
Carbon	C	6	12.011	Polonium	Po	84	210
Cerium	Ce	58	140.12	Potassium	K	19	39.098
Cesium	Cs	55	132.9054	Praseodynium	Pr	59	140.9077
Chlorine	Cl	17	35.453	Promethium	Pm	61	147
Chromium	Cr	24	51.996	Protactinium	Pa	91	231.0359
Cobalt	Co	27	58.9332	Radium	Ra	88	226.0254
Copper	Cu	29	63.546	Radon	Rn	86	222
Curium	Cm	96	244	Rhenium	Re	75	186.207
Dysprosium	Dy	66	162.50	Rhodium	Rh	45	102.9055
Einsteinium	Es	99	253	Rubidium	Rb	37	85.4678
Erbium	Er	68	167.26	Ruthenium	Ru	44	101.07
Europium	Eu	63	151.96	Rutherfordium	Rf	104	257
Fermium	Fm	100	254	Samarium	Sm	62	150.4
Fluorine	F	9	19.99840	Scandium	Sc	21	44.9559
Francium	Fr	87	223	Selenium	Se	34	78.96
Gadolinium	Gd	64	157.25	Silicon	Si	14	28.086
Gallium	Ga	31	69.72	Silver	Ag	47	107.868
Germanium	Ge	32	72.59	Sodium	Na	11	22.98977
Gold	Au	79	196.9665	Strontium	Sr	38	87.62
Hafnium	Hf	72	178.49	Sulfur	S	16	32.06
Hahnium	Ha	105	260	Tantalum	Ta	73	180.9479
Helium	He	2	4.00260	Technetium	Tc	43	98.9062
Holmium	Ho	67	164.9304	Tellurium	Te	52	127.60
Hydrogen	H	1	1.0079	Terbium	Tb	65	158.9254
Indium	In	49	114.82	Thallium	Tl	81	204.37
Iodine	I	53	126.9045	Thorium	Th	90	232.0381
Iridium	Ir	77	192.22	Thulium	Tm	69	168.9342
Iron	Fe	26	55.847	Tin	Sn	50	118.69
Krypton	Kr	36	83.80	Titanium	Ti	22	47.90
Lanthanum	La	57	138.9055	Tungsten	W	74	183.85
Lawrencium	Lr	103	257	Uranium	U	92	238.029
Lead	Pb	82	207.2	Vanadium	V	23	50.9414
Lithium	Li	3	6.941	Xenon	Xe	54	131.30
Lutetium	Lu	71	174.97	Ytterbium	Yb	70	173.04
Magnesium	Mg	12	24.305	Yttrium	Y	39	88.9059
Manganese	Mn	25	54.9380	Zinc	Zn	30	65.38
				Zirconium	Zr	40	91.22

oxidation if the reactive gas contains oxygen. The University of Tokyo group solved this problem by placing the high-frequency coils outside the chamber.

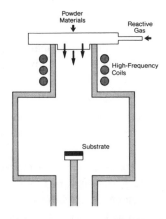

FIGURE 5-3
Illustration of High-Speed Chemical Vapor Deposition (CVD) Fabrication Process

Source: Nikkei Sangyo Shinbun, March 15, 1988, page 3

circuit breaker   Device for interrupting a circuit between separable contacts under normal or abnormal conditions; they are ordinarily required to operate only infrequently, although some classes of breakers are suitable for frequent operation.

closed cycle   A thermodynamic cycle in which the thermodynamic fluid does not enter or leave the system, but is used over and over again.

closed system   A system which is isolated so that it cannot exchange matter or energy with its surroundings and can therefore attain a state of thermodynamic equilibrium.

Cockcroft-Walton generator or accelerator   A high voltage direct-current accelerator especially for the acceleration of protons. The d.c. voltage is produced from a circuit of rectifiers and capacitances to which a low a.c. voltage is applied.

coherence length   Correlation distance of the superconducting electrons.

coil/rail gun   Uses a rapidly changing magnetic field in a spiral coil (coil gun) or a linear conductor (rail gun) to accelerate a projectile via magnetic forces. Much greater velocities can be reached than are possible with gas expansion (as in a conventional gun).

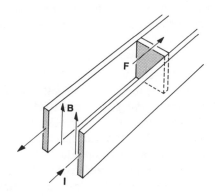

FIGURE 5-4
Illustration of Coil/Rail Gun Concept Force (F) drives projectile out of rail gun.

collider An accelerator in which two opposed beams of particles collide head-on, designed primarily to smash particles.

commercialization Commercial introduction of a technology requires the following:

- All technical and engineering problems associated with the applications of the technology must have been resolved.

- The economics of the technology must be such that manufacturers can obtain a fair return on their investment at prices that offer users an attractive alternative to conventional technologies.

- Users must accept the technology as effective and reliable.

- The public must be satisfied that the socioeconomic and environmental effects of the new technology have been properly analyzed, and that actions are being taken to eliminate any adverse effects.

  When these conditions are satisfied and commercialization proceeds, an institutional infrastructure (producers, distributors, suppliers of raw materials, industry associations, etc.) will evolve. The development of this infrastructure can be accelerated by certain Government actions such as the early establishment of industry performance standards, tax incentives, etc.

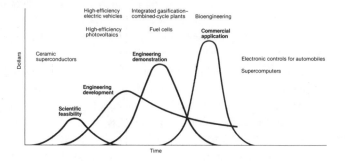

FIGURE 5-5
Phases of Typical
Commercialization
Development

Source: EPRI

conduction of heat The transference of heat through a body, without visible motion of any part of the body, due to a temperature gradient. The heat energy diffuses through the body by the action of molecules possessing greater kinetic energy on those possessing less. In the case of liquids and gases this is achieved by molecular collisions, in the case of solid electric insulators it is dependent on the elastic binding forces between the atoms. For solid electric conductors the former mechanism applied to the free electrons in the material predominates.

conductivity The reciprocal of resistivity. The current density divided by the electric field strength: this definition is often more useful when considering solutions; it is then known as the electrolytic conductivity. The property of a substance of transmitting an electric current by flow of electrons through a dense solid (metal) or by movement of the ions of a dissolved electrolyte to the electrodes when a potential difference exists between them. In the case of solids, no chemical change occurs when the current is passed, but in electrolytic solutions, decomposition of the electrolyte takes place. The term "conductance" is often used in preference to "conductivity" when referring to electrolytic solutions. Conductivity is measured in siemens per meter.

conductor   A substance or body that offers a relatively small resistance to the passage of an electric current.

conductor phenomena   One of the most crucial factors in determining the performance of a magnet is the design of the conductor. This design affects the ultimate field achieved by the magnet, the rate at which the magnet can be energized and the drift rate in the persistent mode of operation. Several phenomena observed in magnets are caused by the conductor itself. One of the earliest phenomena observed in supercon- ducting magnets wound with single filament conductors was flux jumping.

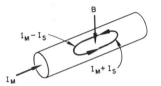

FIGURE 5-6
Illustration of the
Conductor Phenomena

connected load   Sum of the continuous ratings of the load-consuming apparatus con- nected to the system, or part of the system, under consideration.

Cooper Pairs   The term given to pairs of bound electrons which occur in a supercon- ducting medium according to the Bardeen-Cooper- Schrieffer (BCS) theory.

copper   The element copper is a reddish metal, m.p. 1083°C (1981°F), occurring in the form of sulfide and oxide ores in the western states, as well as in Chile, Canada, Mexico, and Peru. It is a relatively soft metal but is distinctive in its high electrical conductivity and in its resistance to corrosion; both these properties contribute to its premier use in wiring and in many other electrical applications. It is also widely used as an alloying metal (brass and bronze). Copper is essentially non-toxic. Symbol - Cu, Atomic Wt. 63.546, State - Solid, Valence 1,2, Group -IB, Isotopes 2 stable, Atomic No. 29.

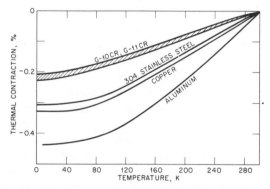

FIGURE 5-7
Illustration of Copper
Wire Thermal Contraction

Source: American Magnetics,
Inc., "Selection Guide."

conservation of charge   The principle that the total net charge of any system is constant.

covalent bond   A type of chemical bond in which atoms of the same or different ele- ments combine to form a molecule (or in some cases a crystal) by sharing pairs of electrons; the molecule so formed is stable and does not ionize. In a covalent bond, one or more pairs of electrons occupy the space between atoms and cause an attractive interaction. To satisfy the Pauli Exclusion principle, the two electrons in a pair must have opposite spins.

convection current   A stream of fluid, warmer or colder than the surrounding fluid and in motion because of the buoyancy forces arising from the consequent differences in density.

copper loss   The power loss in watts due to the flow of electric current in the windings of an electrical machine or transformer. It is equal to the product of the square of the current and the resistance of the winding.

core   In general, it is the part of the magnetic circuit which is situated within the winding.

core loss   The total power loss in the iron core of a magnetic circuit when subjected to cyclic changes of magnetization such as that occurring in the core of a transformer. The loss is due to magnetic hysteresis and eddy currents. It is usually expressed in watts at a given frequency and the value of the maximum flux density.

core-type transformer   A transformer in which the windings enclose the greater part of the core.

coulomb   The Standard International unit of electric charge, defined as the charge transported in one second by an electric current of one ampere. Symbol: C.

coulomb force   A force of attraction or repulsion resulting from the interaction of the electric fields surrounding two charged particles. The magnitude of the force is inversely proportional to the square of the distance between the particles.

critical current   Cooled material experiences superconductivity properties up to a critical current, above which the material would revert to its normal state and exhibit resistance.

critical current density   The maximum value of the electrical current per unit of cross-sectional area that a superconductor can carry without reverting to the normal (non-superconducting) state. The critical current density drops as the temperature rises toward the transition temperature, and as the magnetic field increases. A material can superconduct only if the current passing through a given area is below a certain threshold, known as the critical current density (J_c). For practical applications, J_c values in excess of 10^3 amperes per square millimeter (A/mm^2), are desirable both in bulk conductors for power applications and in thin film superconductors for microelectronics. Bulk ceramic conductors of $YBa_2Cu_3O_7$ have achieved about 10^2 A/mm^2 at 4.2 K and 6 T. However, J_c falls off very steeply to levels around 1-10 A/mm^2 at 77 K and 6 T. These J_c values are determined from magnetization measurements; J_c values derived from transport measurements are usually lower. There is no clear understanding of these reduced J_c levels at the present time, but achieving acceptable values for J_c in bulk high-temperature superconductors is of critical importance and should be a principal focus of research on fabrication processes. Based on current experience, a reasonable target specification for a commercial magnet conductor would be J_c of 10^3 A/mm^2 at 77 K and 5 T, measured at an effective conductor resistivity of less than 10^{-14} ohm-m, with strain tolerance of 0.5 percent, and availability at prices comparable to or less than those of conventional low-temperature superconductors. Preliminary measurements on epitaxially grown single-crystal thin films indicate J_c values in excess of 10^4 A/mm^2 at 77 K and zero magnetic field. These values seem adequate for microelectronic applications.

critical field   The maximum magnetic field strength that a superconductor can emit.

critical length   That length of a cable for which the charging current equals the rated current at the input terminals.

critical magnetic field   Above this value of externally applied magnetic field a superconductor becomes nonsuperconducting (normal). The two types of superconductors define their magnetic field boundaries in different ways. Type I superconductor, material with perfect electrical conductivity for direct current that also possesses perfect diamagnetism (i.e., magnetic flux is totally excluded from the material): most metal element superconductors are Type I. When an external magnetic field is applied on this superconductor, the transition from superconducting to normal state is sharp. Type II superconductor, material with perfect electrical conductivity for direct current (at least for low currents) that does not possess perfect diamagnetism, (i.e., flux penetration of the material is possible); most metal alloys and compounds that are superconductors are Type II. When an external magnetic field is applied, the transition from superconducting to normal state occurs after going through a broad "mixed state" region.

critical point   The point at which a significant and notable change occurs in the properties of a substance or in a state of matter, for example, the glass transition temperature. The point may be that of temperature, pressure, concentration, electrical charging, etc.

critical pressure   The saturated vapor pressure of a liquid at its critical temperature.

critical state   The state of a substance when it is at its critical temperature, pressure, and volume. Under these conditions the density of the liquid is the same as that of the vapor. The point corresponding to this state on an isotherm is the critical point.

critical temperature   Superconducting metals and alloys have characteristic transition temperatures called its critical temperature (T_c). Below the superconducting transition temperature, the resistivity of a material is believed to be exactly zero. A rule of thumb for general applications is that materials must be operated at a temperature of $3/4\ T_c$ or below. At about $3/4\ T_c$ critical fields have reached roughly half their low-temperature limit, and critical current densities roughly a quarter of their limit. Thus, to operate at liquid nitrogen temperature (77 K), one would like T_c near 100 K, making the 95 K material just sufficient. To operate at room temperature (293 K) one requires a material with T_c greater than 400 K, well above the highest demonstrated value. Higher T_c materials would be superior across the board for applications, and materials with T_cs above 400 K would have a truly revolutionary impact on technology. In this temperature domain one could consider mass market applications.

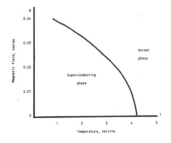

FIGURE 5-8
Critical Temperature
Phase (magnetic field vs.
temperature)

critical volume   The volume of a certain mass of substance measured at the critical pressure and temperature.

cryogenics   A branch of physics dealing with the properties of matter at extremely low temperatures. The study of the production and effects of very low temperatures. A cryogen is a refrigerant used for obtaining very low temperatures.

cryometer   A thermometer designed for the measurement of very low temperatures.

cryostabilization   Stabilization obtained if, after a temperature rise caused by either internal or external perturbations, the ohmic dissipative energy caused by the electrical current can be removed by heat transfer to a cryogen more rapidly than it is generated. The system will not then suffer total thermal runaway, but instead will ultimately decrease in temperature until the superconductive material again operates at the design point.

cryostat   A vessel that can be maintained at a specified low temperature; a low-temperature thermostat.

cryothermometer   A cryothermometer is a highly recommended temperature measuring device used in large superconducting magnet systems. It is an aid in ascertaining the proper rate for transferring liquid nitrogen for precooling and liquid helium for final cooling of the superconducting magnet.

FIGURE 5-9
Illustration of
Cryothermometer

Source: American Magnetics
Inc.

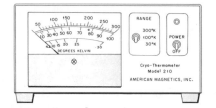

cryotron   A type of switch that depends on superconductivity. It consists of a wire surrounded by a coil in a liquid helium bath. Both the wire and the coil are superconducting and a low voltage can produce a current in the wire. If a current is also passed through the coil its magnetic field alters the superconducting properties of the wire and switches off the current, thus the presence or absence of a current in the coil determines the ability of the wire to conduct.

crystal   The fundamental unit of a solid substance. Crystals characterize the solid state of matter, from the molecular to the visible size ranges. They represent a highly ordered state of molecular structure in three dimensions, as opposed to the disordered amorphous or liquid state. Crystals have a limited degree of vibration, and when enough energy is supplied in the form of heat to overcome the restraining forces, the crystals melt and the substance changes to the liquid phase. The shape of a crystal is an indication of its orderly internal structure, called a lattice; the atoms comprising the lattice act as a diffraction grating which permits determination of the structure by means of x-rays. Solids are made up of crystals of various forms, or "habits," united in aggregates which can be split apart along their cleavage planes (adjacent crystal surfaces) The shapes of crystals always involve flat surfaces and straight lines and may be cubes, rhomboids, lozenges, needles, plates, etc.

FIGURE 5-10
Photograph of 1-2-3 Superconductor Single Crystal

Source: IBM Research

crystal structure The specification both of the geometric framework to which the crystal may be referred, and of the arrangement of atoms or electron density distribution relative to that framework.

crystalline solid Solids may be crystalline, which means they have a definite formation pattern extending over many atoms or molecules.

curie point The magentic transition temperature; ferromagnetic curie temperature.

current The rate of flow of electricity; the unit is the ampere. A conduction current is a current flowing in a conductor, the electricity being conveyed by the motion of electrons or ions through the material of the conductor. A conduction current of 1 ampere is equivalent to the flow of about 10^{18} electrons per second.

current density The maximum pinning force and the maximum current density sustainable by the conductor increase with reduced temperature. Similarly, as the magnetic induction B increases, the sustainable current density decreases. Consequently the current density, magnetic field, and critical temperature are interdependent, j_c being a function of B and T, and B_c being a function of j and T. By increasing any of these parameters to a sufficiently high value, superconductivity can be destroyed and the conductor will revert to the normally conducting state.

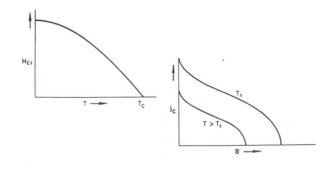

FIGURE 5-11
Illustration of
Superconductivity Current
Density Factors

current meter   A current meter is used to determine the magnetic field being generated in the dewar since the field and current are directly related.

cyclotron   A cyclic accelerator in which the charged particles spiral outward from the center of the machine as they gain energy. Charged particles describe a spiral path of many turns at right angles to a constant magnetic field, and are given an acceleration, always in the same sense, from an alternating electric field, each time they cross the gap between the two conductors.

cylindrical winding   A type of winding used in transformers. The coil is helically wound and may be single-layer or multilayer. Its axial length is usually several times its diameter.

D

DARPA   Defense Advanced Research Project Agency of the U.S. Department of Defense.

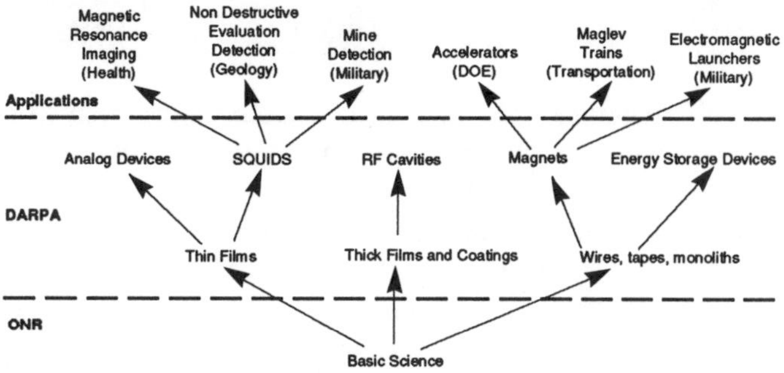

FIGURE 5-12
DARPA High Tempera-
ture Superconducting
Technology Research
Program Integration

Source: DARPA

defect   All crystalline solids consist of regular periodic arrangements of atoms or molecules. Departures from regularity are known as defects.

defect condition   In a semiconductor, conduction due to the presence of holes in the valence band.

degree   A unit of temperature difference. The Celsius and Fahrenheit degrees are defined as 1/100th and 1/180th respectively of the temperature difference between the ice and steam points, so that $1°C = 9/5°F$. The unit of thermodynamic temperature, no longer called a degree, is the kelvin.

demonstrations   Simple samples or devices (motor, electronic, etc.) illustrating the use of superconducting components.

depletion layer   A region in a semiconductor in which the mobile charge carrier density is not sufficient to neutralize the net fixed charge density of donors and acceptors.

detector   Any device that can detect the presence of a particle and measure one or more of its properties.

dewar   Named after the Scottish chemist and physicist. A double-walled flask with a vacuum between the walls that are silvered on the inside, used especially for storage of liquefied gases. Superconducting magnets must be operated at liquid helium temperature. Consequently, they require the use of some cryogenic apparatus in addition to the magnet itself. A liquid helium dewar, a means of mounting the magnet in the dewar, a power supply, a liquid helium level meter/sensor, and a helium transfer tube are basic and necessary. Most dewars are constructed from non-magnetic stainless steel with copper and aluminum radiation shields. All non-demountable joints are inert gas welded providing a rigid, mechanically strong dewar for long and reliable use. All joints should be checked with a helium leak detector at every stage of construction, and vacuum integrity must be guaranteed. The inner and outer surfaces of the dewar are highly polished to insure minimum emissivities. Many liquid helium dewars utilize liquid nitrogen shielding for reduced liquid helium consumption. In these cases, the vacuum jackets of the nitrogen and helium reservoirs are connected via an optically dense common vacuum pumping port through the radiation shield. Where liquid nitrogen shielding is undesirable, a vapor shielded dewar can be provided with multilayer superinsulation, reducing the radiation heat load into the helium reservoir to a very low value. Dewars are generally supplied with a reliable bellows sealed evacuation valve, a safety pressure relief valve and provisions for mounting a thermocouple gauge for monitoring the dewar vacuum.

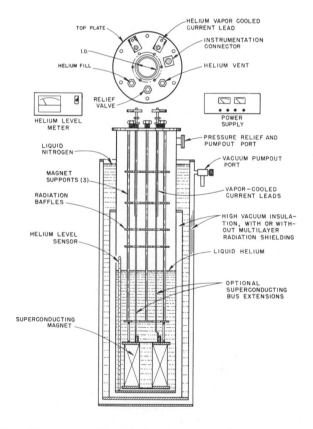

FIGURE 5-13
Illustration of Liquid
Helium Dewar

Source: American Magnetics
Inc.

diatomic molecules   Diatomic molecules have sets of energy levels associated with vibrational and rotational motion; the rotational levels are usually more closely spaced than the vibrational ones. Transitions between these levels result in molecular spectra.

dielectric   Any material which has strong electrical insulating properties. Among the solids of this type are those containing a high percentage of silica, such as glass, diatomaceous earth, mica, etc.; cellulose-containing materials (wood, cotton); and most high-polymers, both natural and synthetic (rubber, plastics). Liquid dielectrics, often called transformer oils, include silicone oils, mineral oils, and chlorinated hydrocarbons (askarels). Values used, to define this property of materials, are dielectric constant and dielectric strength.

dielectric constant   A value that serves as an index of the ability of a substance to resist the transmission of an electrostatic force from one charged body to another, as in a condenser. The lower the value, the greater the resistance. The standard apparatus utilizes a vacuum, whose dielectric constant is 1; in reference to this, various materials interposed between the charged terminals have the following value: air, 1.00058; glass, 3; benzene, 2.3; acetic acid, 6.2; ammonia, 15.5; ethyl alcohol, 25; glycerol, 56; and water, 81. The exceptionally high value for water accounts for its unique behavior as a solvent and in electrolytic solutions. Most hydrocarbons have high resistance (low conductivity).

direct current   Unidirectional current in which the changes in value are either zero or so small that they may be neglected. As ordinarily used, the term designates a practically nonpulsating current. Power transmitted by dc transmission lines is measured in watts (W), compared to volts—amperes (VA) for ac; these are numerically the same for unity power factor in ac circuits.

donor   A nucleophilic atom or ion which supplies an electron pair to another atom, thus creating a covalent bond. The electron-deficient atom, called the acceptor, is said to be electrophilic.

E

eddy current   A current induced in a conductor when subjected to a varying magnetic field. Such currents are a source of energy dissipation (eddy-current loss) in a.c. machinery. The reaction between the eddy currents in a moving conductor and the magnetic field in which it is moving is such as to retard the motion, and can be used to produce electromagnetic damping.

effective resistance   The resistance of a conductor or other element of an electric circuit when used with alternating current. It is measured in ohms and is the power in watts dissipated as heat divided by the square of the current in amperes. It may differ from the normal value of resistance as measured with direct current since it includes the effects of eddy currents within the conducting material, skin effect.

electric field   The space surrounding an electric charge within which it is capable of exerting a perceptible force on another electric charge.

electric field strength   Formerly called electric intensity. Symbol: E. The strength of an electric field at a given point in terms of the force exerted by the field on unit charge at that point. It is measured in volts per meter.

electric flux   The quantity of electricity displaced across a given area in a dielectric. It is defined as the scalar product of the electric displacement and the area; it is measured in coulombs.

electric generator   A source of electricity, especially one that transforms mechanical or heat energy into electric energy.

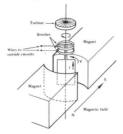

FIGURE 5-14
Illustration of an Electric Generator

Source: Handbook of Energy Technology, V. Daniel Hunt

electric power substation   An assemblage of equipment for purposes other than generation or utilization through which electric energy in bulk is passed for the purpose of switching or modifying its characteristics. A substation is of such size or complexity that it incorporates one or more buses and a multiplicity of circuit breakers and usually either is the sole receiving point of commonly more than one supply circuit or sectionalizes the transmission circuit passing through it by means of circuit breakers.

electromagnet   An iron core encircled by coils of wire that becomes magnetic when current flows through the wire.

electromagnetic force   A long-range force which acts between electric charges, currents, and magnets. It is mediated by photons.

Electrotechnical Laboratory (ETL)   This Japanese laboratory, administered by the Ministry of International Trade and Industry, has been involved in superconductivity research since the mid-1960s.

electron   A particle of negative electricity having a mass represented by the fraction 1/1837. Its properties combine those of radiation (light waves) and of particles. This double nature of the electron has been rationalized by the application of quantum or wave mechanics, in which the electron is regarded as a standing wave of energy moving around the atomic nucleus in a so-called orbital. Electrons may occupy from one to seven energy levels (depending on the element involved) which are at varying distances from the nucleus. These are called shells, or groups of orbitals; each shell contains a specific number of electrons ranging from 1 to 32, though in many cases the shells are not filled. Those in the outer shell are known as valence electrons, as they play a predominant part in chemical bonding; for example, carbon has two shells, the inner one being occupied by two electrons and the outer one by four electrons, which represents its valence. Thus, there is a definite relationship between the number of outermost electrons and the chemical activity of an element. No orbital can contain more than two electrons; two electrons in the same orbital are known as paired electrons. The negative charge on, or represented by, an electron is regarded as being diffuse rather than concentrated and thus has the appearance of a cloud around the nucleus which becomes gradually less dense as it extends outward, the density being indicated by the square of the wave function. Electrons were discovered in 1897 by Sir Joseph John Thomson, a British physicist (1856-1940), Nobel Prize 1906, and their behavior in atoms was described in 1913 by Niels Bohr, a Danish physicist (1885-1962), Nobel Prize 1922.

electron volt   A unit of electrical energy used by nuclear scientists in measuring electronic forces. It is defined as the energy an electron receives when it falls through a potential difference of 1 volt, i.e., less than a billionth of an erg. The rupture of a chemical bond yields from 5 to 10 electron volts, whereas the splitting of an atomic nucleus released about 200 million electron volts. Approved abbreviation is eV for electron volt and MeV for million electron volts. According to the equivalence of mass and energy expressed in Einstein's theory of relativity, the mass of particles can also be measured in electron volts.

electroweak force   A fundamental force comprising the unified electromagnetic and weak forces.

element   A unique arrangement of fundamental units of matter having characteristic properties; 106 elements are presently known, of which 92 occur in nature, the others being synthetic. The smallest amount of an element that can exist is the atom. The number of protons in the atomic nucleus (atomic number) and the arrangement of electrons around them determine the properties of an element, especially in relation to chemical bonding and reactivity. When arranged in sequence by increasing atomic number, the elements are seen to have recurring similarity of certain properties; it was on the basis of this observation, in 1869 by Dimitri I. Mendeleeff (1834-1907), that the Periodic Table was developed. About 75% of the elements are metals, and all those beyond lead are radioactive. All atoms of a given element are identical in every respect except mass; the presence of one or more extra neutrons in the nucleus accounts for the existence of isotopes. Besides the major groups of elements of the Periodic Table, there are several classifications or series based on type or properties; some of these line within a major group, while others straddle several groups.

elementary particle   A particle that cannot be split up into other particles. According to present knowledge, quarks, leptons, and the force-carrying bosons are elementary particles.

energy   The fundamental active entity of the universe. Among its more important manifestations are (1) electromagnetic radiation (photons, or radiant energy); (2) the energy of combustion (thermal energy); (3) electrical energy; (4) kinetic energy (the energy of motion); (5) nuclear energy, the binding force that holds protons and neutrons together in the atomic nucleus; and (6) free energy, a thermodynamic function describing the energy available to a substance for reaction with other substances. The three laws of thermodynamics state the energy conditions which determine the nature and extent of chemical reactions. The first and best-known of these (conservation of energy) is that the total of energy in the universe is constant and cannot be increased or diminished. Einstein was the first to formulate the concept of the equivalence of energy and mass, expressed by the equation $E = mc^2$.

energy bands   Electron states can be understood better based on energy bands; which are the ranges of allowed electron energies separated by ranges of forbidden energies.

FIGURE 5-15
Illustration of Energy
Bands

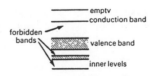

energy gap   Gap in the low-energy excitations of a superconductor.

excitation   The production of magnetic flux in an electromagnet by means of a current in a winding. The current is referred to as the exciting current and the ampere-turns produced as the exciting ampere-turns.

excitation energy   The energy required to change the energy level of an atom or molecule to a higher energy level. It is equal to the difference in energy of the levels and is usually the difference in energy between the ground state of the atom and a specified excited state.

excited state   A higher than normal energy level of the electrons of an atom, group, or molecule, resulting from absorption of photons (quanta) from a radiation source. When the energizing source is removed or discontinued, the atom or molecule returns to its normal or stable state either by emitting the absorbed photons or by transferring the energy to other atoms or molecules. The increased vibrational activity of the atom or molecule yields line or band spectra characteristic of its structure, thus permitting identification.

exciter   A small generator that provides current for the field structure of a large generator.

exciton   An electron in combination with a hole in a crystalline solid. The electron has gained sufficient energy to be in an excited state and is bound by electrostatic attraction to the positive hole. The exciton may migrate through the solid and eventually the hole and electron recombine with emission of a photon.

exclusion principle   See Pauli exclusion principle.

F

fabrication   Shaping of superconducting materials into wires/filaments, coils, tapes, thin films, coatings, single crystals, dense monoliths, and composites.

Fahrenheit   The temperature scale commonly used in the U.S. though centigrade is preferred by scientists. On the Fahrenheit scale, the freezing point of water is 32° and the boiling point 212° at sea level (1 atmosphere; or 760 mm of mercury). To convert °F to °C, first subtract 32 from the Fahrenheit temperature and then multiply the balance by 5/9 or 0.55. The temperature value for both Fahrenheit and centigrade cases is the same at –40°. This scale was devised by Gabriel Daniel Fahrenheit, a German physicist (1686–1736).

faraday balance magnets   Magnetic susceptibility measurements can be made with great precision by measuring the force exerted on a sample placed in a magnetic field that is shaped to maximize BdB/dx. Magnets of this type are comprised of two parts, a main coil that is relatively homogeneous and a second coil section that generates a linear field gradient over a specified volume.

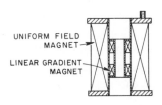

UNIFORM FIELD MAGNET

LINEAR GRADIENT MAGNET

FIGURE 5-16
Illustration of Faraday Balance Magnets

fault   In a wire or cable a partial or total local failure in the insulation or continuity of a conductor.

ferromagnetism   A property, exhibited by certain metals, alloys, and compounds of the transition (iron group) rare-earth and actinide elements, in which the internal magnetic moments spontaneously organize in a common direction; gives rise to a permeability considerably greater than that of vacuum, and to magnetic hysteresis.

fiber-optics   Use of glass fibers to transmit light (produced by lasers) for telecommunications and computer networking. Optical fibers can carry much more information than electrical wires.

field structure   The device that creates the magnetic field through which the armature moves in an electrical generator.

Fine Ceramics Center (FCC)   A laboratory in Nagoya, Japan jointly funded by industry and the Ministry of International Trade and Industry. Participating firms send researchers to work at the facility, which opened in the spring of 1987.

fission   The joining of the nuclei of two lightweight atoms into a single nucleus. The resulting nucleus weighs less than the sum of the components. The missing mass is converted to energy.

fixed-target accelerator   An accelerator in which a particle beam hits a stationary target.

flux   The literal meaning of this term is "flow," thus, it is used by physicists in the sense of a neutron flux or a light flux, referring to the rate of flow or emanation of radiation from a given source.

flux flow   At very high fields and high current densities, fluxoids will migrate comparatively rapidly, giving rise to a phenomenon called flux flow. These phenomena are crucial to the operation of magnets in the persistent mode.

flux jumping   The flux jumping phenomenon arises from current induced in the conductor by the presence of the transverse field generated by the magnet. If a superconductor is placed transverse to the magnetic field, currents are set up in the conductor which shield the bulk of the conductor from the external magnetic field. These circulating currents extend for only a finite length along the conductor, flowing in one direction on one side of the conductor and returning on the other to complete the circuit. Since the conductor carries the basic magnet current in addition to the shielding current, the current increases on one side of the conductor and decreases on the other. Consequently, at fields and currents that are quite low in comparison with their critical values, the conductor is driven into the resistive state. In this state, heat is dissipated in the small normal zone and the attendant increase in temperature causes the normal zone to expand and propagate both along the length of the conductor and transverse to it. This results in the magnet being discharged as the energy in the magnet is dissipated in the resistive portion of the conductor.

fluxoid   These quanta are comprised of circulating vortices of current and flux contained in the vortices. The total flux in a vortex is 2×10^{-7} Gauss-cm^2. Great numbers of these vortices or fluxoids can exist in a superconductor.

flux pinning   Superconducting material properties are altered locally by the presence of defects in the material. A fluxoid encompassing or adjacent to such a defect in the material has its energy altered and its free motion through the superconductor is inhibited. This phenomenon is called flux pinning. Flux pinning causes a field gradient in the superconductor and gives rise to a net current in the material.

forbidden transition   A transition of an atom or molecule between two energy levels, involving a change of quantum number that is not allowed by the selection rules.

free magnetism   An imaginary magnetic fluid to which the magnetic effects of a magnet are conventionally ascribed. In a bar magnet, the free magnetism is often regarded as being concentrated in the poles but the actual distribution of free magnetism along the bar can be studied. The algebraical sum of the free magnetism on any specimen is always zero.

freezing mixture   A mixture of two or more substances that absorb heat when they mix and thus produce a lower temperature than that of the original constituents.

freezing point   The temperature at which the solid and liquid phases of a substance can exist in equilibrium together at a defined pressure, normally standard pressure of 101 325 Pa.

fusion   The change of the state of a substance from solid to liquid which occurs at a definite temperature (melting point) at a given applied pressure.

G

gas   A substance that continues to occupy in a continuous manner the whole of the space in which it is placed, however large or small this space is made, the temperature remaining constant.

gauss   The CGS-electromagnetic unit of magnetic flux density. $1 \text{ G} = 10^{-4}$ tesla. Symbol: G.

Gauss's theorem   Total electric flux acting normal to any closed surface drawn in an electric field is equal to the total charge of electricity inside the closed surface. If the surface encloses no charge the total flux over it is equal to zero. Gauss's theorem applies also to surfaces drawn in a magnetic field.

gluon   Any of eight massless bosons that mediate the strong force.

gram   A standard unit of weight equivalent to 1/453.59 pound, or 15.4 grains. It is the weight of 1 milliliter (ml) of water at 4°C and 1 atmosphere pressure. A kilogram is 1000 grams; a milligram (mg) is 1/1000 gram; and a microgram (ug) one-millionth gram. One milliliter so closely approximates 1 cubic centimeter that, for all but the most accurate purposes, it may be said that 1 cc of water weighs 1 gram.

Grand Unified Theory   A theory that combines the strong and electroweak forces into a single unified scheme. The correct Grand Unified Theory has not yet been found.

gravitation   A long-range, purely attractive force influencing all particles. It is so weak that its effects are seen only in interactions between macroscopic objects.

graviton   The hypothetical massless boson that carries the gravitational force.

ground   A conducting connection, whether intentional or accidental, between an electric circuit or equipment and the earth or to some conducting body that serves in place of the earth.

H

heat   That form of energy transferred between two bodies as a result of a difference in their temperatures, and governed by the laws of thermodynamics. A mode of energy associated directly with and proportional to the random molecular activity (motion) of a material system. It maintains or raises the temperature of the system and can be transferred to another body (gas, liquid, or solid) by radiation, convection, or conduction. It can also be converted to other forms of energy, e.g., electricity and motion. Its absence is associated with the temperature of absolute zero. Heat is also defined as energy which is transferred under a temperature gradient or difference from one body to another. Heat is variously generated, e.g., by chemical reaction, flow of electricity, friction, nuclear fission and fusion, etc.

heat capacity   Formerly, thermal capacity. The quantity of heat required to raise the temperature of a body through one degree. It is measured in joules per kelvin.

heat exchanger   A device for transferring heat from one fluid to another without the fluids coming in contact. Its purpose is either to regulate the temperatures of the fluids for optimum efficiency of some process, or to make use of heat that would otherwise be wasted. The simplest form consists of two concentric pipes, the inner one finned on the outside to maximize the contact area, with the fluids moving through the pipes in opposite directions.

heating effect of a current   When an electric current (I) passes through a resistor a rise of temperature is observed. If all the electrical energy is turned into heat energy Q, $Q = I^2Rt$, where R is the resistance at the temperature involved, and t is the time for which the current passes. Q does not increase indefinitely with t since a temperature is reached at which the rate of emission of heat from the surface of the resistor is equal to its rate of generation.

heat of transition   The heat evolved or absorbed when a unit mass of a given substance is converted from one crystalline form to another crystalline form. It can be calculated from the respective heats of formation of each crystalline form and is usually expressed in kilocalories per gram-formula weight of the substance.

heat of vaporization   The quantity of energy required to evaporate 1 mole, or a unit mass, of a liquid, at constant pressure and temperature. Also known as enthalpy of vaporization; heat of vaporation; latent heat of vaporization.

helium   The element helium (from Greek, meaning "the sun") was so named because of its original discovery in the sun by spectrographic analysis (1868); it was not found on earth until 1895. It is a component of natural gas, from which it is separated by compressing the gas at low temperatures until all the other components have been liquefied. Helium does not become liquid until a temperature of 4.2 Kelvin (about –265°C) is reached and is thus a cryogenic liquid, which remains fluid at temperatures near absolute zero. The nuclei of helium atoms are called alpha particles and are emitted during radioactive decay of unstable nuclei. Both gaseous and liquid forms can penetrate even dense solids three times as rapidly as air. Though helium cannot enter into chemical combination and thus forms no compounds, it has unusual properties as a liquid which have been the subject of much physical research. The chief uses of helium are as a coolant for reactors, leak detection in high-vacuum equipment, and as inert medium for semiconductor crystal growth. It is noncombustible and nontoxic. Symbol - He, Atomic Wt. 4.00260, State - Gas, Valence 0, Group - Noble Gas, Isotopes 2 stable, Atomic No. 2.

T(K)	Volume (cc/gm)	Enthalpy (J/gm)
3	6.99	5.74
4	7.70	8.87
L4.214	8.01	9.98
V4.214	59.83	30.88
5	82.88	36.24
10	198.1	64.77
20	408.6	117.8
30	615.4	170.1
40	821.3	222.2
50	1026	274.3
60	1232	326.3
70	1437	378.2
80	1643	430.2
90	1848	482.2
100	2053	534.1
110	2258	586.0
120	2463	638.0
130	2668	689.9
140	2873	741.9
150	3078	793.8
160	3283	845.7
170	3488	897.7
180	3693	949.6
190	3898	1002
200	4104	1053
210	4309	1105
220	4514	1157
230	4718	1209
240	4924	1261
250	5129	1313
260	5334	1365
270	5539	1417
280	5744	1469
290	5949	1521
300	6154	1573

FIGURE 5-17
Properties of Helium at
1.0 Atmosphere

Source: Douglas B. Mann,
National Bureau of Standards,
Tech. Note 154

TABLE 5-2 Heat of Vaporization for Liquid Helium

Temperature (K)	H.V. J/g
2.20	22.8
2.60	23.3
3.00	23.7
3.60	23.2
4.20	20.9
4.60	18.0
5.00	12.0

Superfluid transition occurs at 2.17 K.

Density of liquid helium at 4.2 K; 125 g/l.

1 Watt vaporizes 1.38 liters/hr. @ 4.2 K.

Source: American Magnetics, Inc. "Selection Guide."

helium bore split coil magnets   Magnets of this type are made with two coil sections separated by a mid-plane spacer. Radial access to the field is usually provided through four holes drilled radially through the spacer. Less frequently, the radial holes are required to be square or rectangular. In neutron scattering experiments, the radial holes can take the form of sectors of a circle or "pie slice" shapes. All these latter forms of radial access are more difficult to construct than a simple circular hole and can require that the bobbin be split and rejoined for fabrication. When the dimensions of the radial ports become large, the bobbin may not be capable of supporting the forces between the two coil sections and the epoxy in the winding might break.

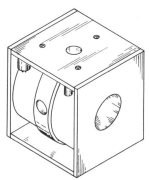

FIGURE 5-18
Rectangular Split Coil
Magnet with Helium Bore

helium vapor cooled current leads   The introduction of current into a helium dewar with minimal heat input is a crucial factor in the economics of operating a superconducting magnet system. The heat generated in the leads can be extracted to a great extent by cooling them with the helium gas exhausted from the dewar. In some cases where the magnet is expected to operate for periods of weeks or months at a fixed field, the current leads can be removed from the dewar after the magnet is in the persistent mode. This removes the heat input associated with the leads and allows the system to be operated for long periods of time without refilling the helium reservoir.

hertz   The Standard International unit of frequency, defined as the frequency of a periodic phenomenon that has a period of one second.

higgs boson   A hypothetical particle arising in unified theories of the electroweak interaction. The search for this particle will be one of the goals of the experimental program at the Superconducting Super Collider (SSC).

high energy physics   The study of the interactions and properties of energetic particles. Also known as particle physics. Concerned with obtaining and understanding the nature of elementary particles.

high-temperature superconductor (HTS)   Refers to materials—four classes of which have been discovered since 1986—with much higher transition temperatures than previously known superconductors.

hole   A term used in semiconductor technology to refer to an energy deficit in a crystal lattice due to electrons ejected from unsatisfied covalent bonds at sites where an atom is missing (vacancy) or to electrons supplied by atoms of impurities present in the crystal, e.g., arsenic or boron. The free electrons from these two sources move through the crystal leaving energy deficits which are positively charged; these deficit sites, or holes, are also considered to move as they become alternately filled and vacated by electrons; thus, a flow of positive electricity results. In a solid, an unfilled vacancy in an electronic energy level. These vacancies behave as if they were positive electronic charges with positive mass and are mathematically equivalent to the positron.

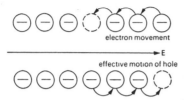

FIGURE 5-19
Illustration of Hole
Conduction

homogeneous magnets   The magnetic field in a simple solenoid decreases approximately quadratically with the axial distance from the center of the magnet. The homogeneity at the center of the magnet can be improved significantly by adding windings to each end of the solenoid to compensate for this decrease in the field. Mathematically, this technique results in a very uniform magnetic field at the center of the coil. In practice, however, winding imperfections limit the homogeneity attained and additional series connected trim windings are required to achieve the maximum homogeneity. Since the trim coils are connected in series, this type of magnet has only two current leads and one persistent switch is used for the complete magnet. The homogeneity of such a magnet is about $+ 1$ part of 10^5 of the central field in a spherical volume having a diameter that is approximately 1/5 the bore of the magnet in small magnets.

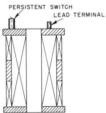

FIGURE 5-20
Illustration of
Homogeneous Magnets

hydrogen   The word hydrogen literally means "water-maker," since water is an oxide of hydrogen. Hydrogen is the lightest of all elements and is extremely flammable over a concentration range of 4 to 75% by volume. It is formed readily by the action of steam on iron, on hydrocarbons, or on carbon monoxide, as well as by many other chemical reactions including electrolysis of water. Hydrogen is the characteristic element of acids and is a constituent of water, hydrocarbons, and carbohydrates. Hydrogen is used as a low-temperature cryogenics element. Symbol - H, Atomic Wt. 1.0079, State -Gas, Valence 1, Group - IA, Isotopes - H^1 H^2 (stable) (deuterium) (radioactive) H^3 (tritium), Atomic No. 1.

I

impedance   In a two-terminal circuit the ratio of applied voltage to input current with ac excitation, taking into account the phase relationship of the two quantities. Impedance can be expressed as a complex number with real and imaginary parts.

impulse   Surge of unidirectional polarity.

impurity An extremely low percentage of an extraneous substance either naturally present in a material or added to it accidentally or intentionally. Often it is impossible to completely eliminate naturally occurring impurities by any separation method, but there are many processes that are effective in reducing them to an acceptable minimum. So small are the amounts involved that impurities are usually indicated in parts per million (ppm). Of special importance is the fact that not all impurities are undesirable; their presence may have a beneficial effect, and they are often added intentionally. Examples are arsenic or boron in semiconductors, and doping agents in crystals. In a semiconductor, foreign atoms, either naturally occurring or deliberately placed in the semiconductor. They have a fundamental effect on the amount and type of conductivity.

inert Having little or no chemical affinity or activity. The six elements comprising the noble-gas group of the Periodic Table (sometimes called Group 0) are either wholly or relatively inert. The first three (helium, neon, argon) have a valence of 0 and do not enter into any chemical combination. The last three (krypton, xenon, radon) have several valences but are capable of only limited compound formation. Thus, these gases are collectively called "noble." Less specifically, nitrogen and carbon dioxide are often referred to as "inert atmospheres," as they are unreactive under normal conditions; they are used to protect certain highly reactive substances from contact with atmospheric oxygen.

injector An accelerator whose beam is injected into another, higher energy accelerator.

insulator A substance that does not conduct electricity.

integrated circuit A complete circuit manufactured in a single package. In hybrid integrated circuits separate components and/or circuits are attached to a ceramic substrate and interconnected. The complete circuit, bearing some resemblance to a circuit made from discrete components, is enclosed in a single package. It cannot be modified without destroying the whole circuit. In monolithic integrated circuits all the components are manufactured into or on top of a chip of silicon. They are formed by selective diffusion of the appropriate type of impurity into a layer of semiconductor. Interconnections between individual components are made by means of metallization patterns and the individual components are not separable from the complete circuit.

intensity of magnetization For a uniformly magnetized body, the ratio of its magnetic moment to its volume.

interconnection tie Feeder interconnecting two electric supply systems. The normal flow of energy in such a feeder may be in either direction.

internal resistance Of a cell, accumulator, or dynamo. The resistance in ohms obtained by dividing the difference in volts between the generated electromotive force and the potential difference between the terminals of the cell by the current in amperes.

International Superconductivity Technology Center (ISTEC) An organization for superconductivity R&D set up by Japan's Ministry of International Trade and Industry.

intermetallic compound Chemical compounds of nominally fixed composition, one or more elements of which are metals. Most intermetallic compounds—e.g., the superconductor niobium-tin (Nb_3Sn)—are brittle and therefore hard to work with.

interrupting time Interval existing between the energizing of the trip coil at rated voltage and the interruption of the circuit.

ion An atom (H, Ag, Cl), group (OH, SO_4), or molecule (O_2) that has either lost one or more electrons and thus become positively charged (cation) or gained one or more electrons and thus become negatively charged (anion). The properties of ions are quite different from those of the neutral units from which they are derived, because of the presence of the electric charge. In a solution, they are closely associated with molecules of the solvent; they also may exist in solids and gases. Positive ions are formed in some gases by electric discharge or electromagnetic radiation. Strong ionizing radiation (x- and gamma rays) can remove electrons from atoms and molecules of organic substances, with lethal effects on living organisms. Short-lived, highly reactive organic ions can be formed with carbon; these act as initiators in organic synthesis. The word ion is derived from the Latin for go, referring to the rapid motion of these particles.

ionic bond A type of chemical bond, often called electrostatic, in which atoms of different elements unite by transferring one or more electrons from one atom to the other to form an ionizing or polar compound. For example, an element such as sodium which has but one electron in its outer shell readily links with chlorine, which has seven; this provides a total of eight electrons (characteristic of argon), which is energetically ideal. The sodium loses its electron and thus becomes positively charged, while the chlorine gains the electron and becomes negatively charged. Bonds of this kind are less stable than covalent bonds; the compounds formed will dissociate into ions when dissolved in water, as a result of the high dielectric constant of the water, which reduces the strength of the bond to the point of rupture.

ionic conduction The movement of charges within a semiconductor due to the displacement of ions within the crystal lattice. An external contribution of energy is required to maintain such movement.

ionic semiconductor A solid in which the electrical conductivity due to the flow of ions predominates over that due to the movement of electrons and holes.

isotope Any of two or more forms of an element whose weights differ by one or more mass units as a result of variation in the number of neutrons in the nuclei. The literal meaning of this term is "the same place" (in the Periodic Table). The basic work on isotopes was done by the British chemists, Frederick Soddy (1877-1956), Nobel Prize 1921, who suggested the word in 1913, and Francis W. Aston (1877-1945), Nobel Prize 1922, who invented the mass spectrograph (1920).

J

Josephson effect An effect that occurs when a sufficiently thin layer of insulating material is introduced into a superconducting material. A superconducting current can flow across the junction, known as a Josephson junction, in the absence of an applied voltage. This is the direct-current Josephson effect. If the value of the current exceeds a critical value, Ic, determined by the properties of the insulating barrier, current can only flow when a finite voltage is applied. The current-voltage characteristic is shown in the following Josephson junction diagram, in which the dashed curve is the current-voltage characteristic in the nonsuperconducting state. The alternating-current Josephson effect occurs when a small direct voltage, V, is applied across a Josephson junction. The superconducting current across the junction becomes an alternating current. The direct-current Josephson effect is utilized in several devices, particularly the Josephson memory. The alternating-current Josephson effect is utilized for radiofrequency detection, the determination of h/e, for accurate measurement of frequency, and as a monitor of voltage changes in standard cells or for the comparison of cells at different standards laboratories.

Josephson Junction   A Josephson junction (JJ) consists of two superconductors separated by a thin insulating barrier. Pairs of superconducting electrons will tunnel through the barrier. As long as the current is below the critical current of the junction, it will flow with zero resistance, and there will be no voltage drop across the junction. This strange phenomenon is known as the DC Josephson effect. The critical current is not an actual current—it is a threshold value which serves as a boundary for superconductivity. Below the threshold, the junction has no resistance; above it, the junction has some resistance. A Josephson junction could carry a zero resistance supercurrent. We can place it next to a wire. If we pass a current through the wire, it will generate a magnetic field. This magnetic field lowers the critical current of the junction. The actual current passing through the junction, which has not changed, becomes higher than the critical current, which was lowered. So the junction develops resistance. This resistance can cause the current to branch off in another direction—to switch away from the junction to a path of least resistance. A Josephson junction used as a switch in this way offers two powerful advantages. First, it is very fast, it switches from a zero resistance to a positive resistance in a few trillionths of a second. Second, it can be controlled by a simple low-power device, a control wire generating a subtle magnetic field. Similar Josephson junction configurations can serve as memory cells in computers, where the existence of a zero or non-zero voltage across the junction is a form of data storage. We could place two control wires near the Josephson junction, and use it as a logic element, for example, an "OR" gate or an "AND" gate. (These are two of the primary building blocks for circuitry in digital computers, video games, and pocket calculators.) The device is an "OR" gate if each control wire generates a strong enough magnetic field to switch the junction to a resistive state without help from the other. The device is an "AND" gate if neither control wire can generate a strong enough magnetic field to switch the junction to a resistive state by itself, but both control wires can switch the junction when acting together.

FIGURE 5-21

Illustration of Current and Voltage Characteristics of Josephson Junction

Source: Dictionary of Electronics

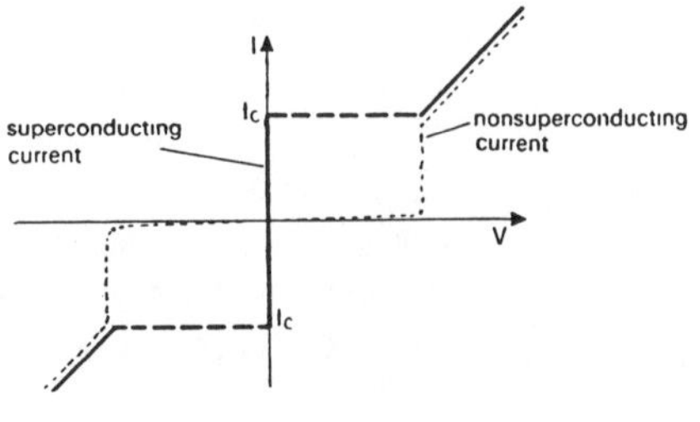

Josephson memory A cryogenic memory that consists of an array of Josephson cells, i.e., memory cells containing Josephson junctions, held at a temperature very close to critical temperature. In the absence of an external magnetic field the cell is superconducting but the presence of a magnetic field destroys the superconductivity and hence the voltage across the device changes. Information is stored in the form of local variations of a magnetic field; the data is sensed by the voltage across the appropriate cell.

joule The Standard International unit of all forms of energy (mechanical, thermal, and electrical), defined as the energy equivalent to the work performed as the point of application of a force of one newton moves through one meter distance in the direction of the force. In electrical theory one joule equals one watt second.

K

Kelvin The absolute scale of temperatures, with $0°K$ ($-494°F$) equal to absolute zero (a temperature that can be approached but never reached). A temperature scale that was named after Lord Kelvin, an English physicist (1824-1907) of the late nineteenth century. The Kelvin is the Standard International unit of thermodynamic temperature, defined as $1/273.16$ of the thermodynamic temperature of the triple point of water. The kelvin is also used as a unit of temperature difference on the Kelvin and Celsius scales, where $1 K = 1$ deg C. Absolute temperatures are denoted in degrees Kelvin by adding 273 to the centigrade temperature; thus, 25 degrees centigrade would be expressed as 298 degrees Kelvin.

kinetic energy The energy of motion of any object.

L

lanthanum The first of the rare-earth metal elements comprising the lanthanide series of the Periodic Table. It is recovered from monazite sands and the ore bastnasite by action of sulfuric acid. It undergoes considerable surface attack in moist air, and in finely divided form it will catch fire at room temperature, i.e., it is pyrophoric. Lanthanum melts at $920°C$ ($1688°F$). Lanthanum compounds are used in glass and ceramics. Symbol - La, Atomic Wt. 138.9055, State - Solid, Valence 3, Group - IIIB, Isotopes 2 stable, Atomic No. 57.

lattice A regular periodically repeated three-dimensional array of points that specifies the positions of atoms, molecules, or ions in a crystal. The stable geometric arrangement of atoms in a crystal, metals, nonmetals, and their compounds. The lattice structure constitutes a submicroscopic three- dimensional grating or network of atoms, which refracts incident light of short wavelength. Measurement of the angles of this refraction has made it possible to determine with x-rays the arrangement of atoms in a crystal with a high degree of accuracy. The science of crystallography is largely based on such measurements. Imperfections or dislocations may occur in lattices due to lack of enough atoms to satisfy all the covalent bonds; such imperfections have a marked effect on the electrical and optical properties of the crystal.

lattice constant The length of edge or the angle between the axes of the unit cell of a crystal. It is usually the edge length of a cubic unit cell.

lead Lead is a soft, heavy metal element, m.p. 327°C (621°F), obtained chiefly from the sulfide ore galena. Lead has such high density that it is almost impermeable to high-energy radiation. Lead is not toxic in bulk form and is safe to handle and process. However, fine powders and fumes are damaging if inhaled, and the presence of lead, in any form, in foods is hazardous. Symbol - Pb, Atomic Wt. 207.2, State - Solid, Valence 2,4, Group - IVA, Isotopes 4 stable, Atomic No. 82.

lepton Any of six elementary particles that feel the weak but not the strong force. The known leptons are the electron, the muon, and the tau lepton, along with the associated electron, muon, and tau neutrinos. Each of these six also has an antiparticle.

linear accelerator A particle accelerator in which electrons or protons are accelerated along a straight evacuated chamber by an electric field of radiofrequency (r.f.) produced by a klystron or magnetron. In older machines, cylindrical electrodes (or drift tubes) of the r.f. supply are aligned coaxially with the chamber. Keeping in phase with the r.f. supply, the charged particles are accelerated in the gaps between the electrodes. Modern high-energy linear accelerators are usually travelling-wave accelerators in which particles are accelerated by the electric component of a travelling wave set up in a waveguide. No drift tubes are used, the r.f. being boosted at regular intervals along the chamber length by means of klystrons. Only a small magnetic field, supplied by magnetic lenses between the r.f. cavities, is required to focus the particles and maintain them in a straight line.

line of flux An imaginary line whose direction at all points along its length is that of the electric or magnetic field at those points.

liquid helium level meter A liquid helium level meter gives an instantaneous and reliable readout of the level of the liquid helium in the dewar; used in transferring, and to avoid operation of a superconducting magnet with too little helium. During standard operation, a small current is conducted through the wire, causing it to be resistive in the helium gas and superconducting in the liquid. The liquid helium level, expressed as a percentage of the sensor length, is displayed on a front panel meter. Output terminals on the meter provide a corresponding 0-100 mV signal for a recorder. Each instrument is calibrated for a sensor of a specific length, but can be recalibrated for use with other sensors. All level sensors are unaffected by magnetic fields to at least 10 tesla.

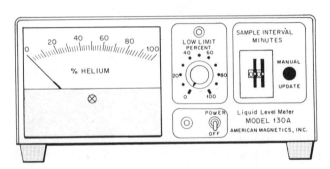

FIGURE 5-22
Illustration of Liquid
Helium Level Meter

Source: American Magnetics, Inc.

liquid helium level sensors   Liquid helium level sensors are typically constructed in 0.25 inch diameter non-conductive tubes of various lengths and can be used with a liquid helium level meter. Standard active lengths from 1 to 60 inches can be provided. The overall length typically exceeds the active length by one inch. Liquid helium level measurements are made by measuring the resistance of a fine superconductive filament partially submerged in liquid helium. The current through the sensor maintains the filament in the normal state in the helium gas and in the superconducting state in the liquid. The resulting voltage along the sensor is proportional to the length of the wire above the liquid helium and provides a continuous measure of the helium depth. Voltage measurements are made using a four wire technique to eliminate errors resulting from variations in the length of the leads. The small amount of heat generated in the probe is dissipated primarily in the helium gas rather than in the liquid helium. The sensor is rugged, reliable and insensitive to magnetic fields to at least 10 tesla.

FIGURE 5-23
Illustration of Liquid
Helium Level Sensor

Source: American Magnetics, Inc.

load factor   Ratio of the average load over a designated period of time to the peak load occurring in that period.

logic chips   Integrated circuits consisting of arrays of gates each of which implements a Boolean function such as AND, OR, NOR, NAND. Computer processors are built from logic chips, as are many specialized digital systems.

low-temperature superconductor (LTS)   Materials that become superconducting only when cooled to a few degrees above absolute zero. All superconductors discovered before 1986 were low-temperature materials, with 23°K (–418°F) the highest known transition temperature.

luminosity   A term used to specify the intensity of the beams in a super collider. The higher the luminosity, the greater the rate of collisions.

technology, fusion plasma heating, and other fields, superconducting radiofrequency devices will proliferate.

M

magnetically-levitated train (maglev)   Trains suspended and propelled by magnetic forces offer the prospects of much higher speeds than can be achieved by conventional wheel-on-rail technologies. A prototype superconducting maglev train in Japan (also called a linear motor car) has achieved speeds of over 300 miles per hour.

magnetic circuit   The completely closed path described by a given set of lines of magnetic flux.

magnetic difference of potential   The difference between the magnetic states of two points in a magnetic field. It equals the line-integral of the magnetizing force between the two points.

magnetic field dependence A slight magnetic field will reduce the critical current of the Josephson junction. We can use this fact to make ultrafast switches, memory cells, and logic elements.

magnetic field strength Formerly called magnetic intensity. The magnitude of a magnetic field, measured in A m-1.

magnetic flux A measure of the total size of a magnetic field, defined as the scalar product of the flux density and the area. It is measured in webers.

magnetic flux density The magnetic flux passing through unit area of a magnetic field in a direction at right angles to the magnetic force. The vector product of the magnetic flux density and the current in a conductor gives the force per unit length of conductor. It is measured in teslas.

magnetic hysteresis A phenomenon observed in ferromagnetic materials below the Curie point where the magnetization of material varies nonlinearly with the magnetic field strength and also lags behind it. The magnetic susceptibility of such materials is large and positive, a large value of magnetization (M) being produced for comparatively small fields. A characteristic plot of either magnetization, M, or magnetic flux density, B, against magnetic field strength, H, demonstrates the hysteresis effect and is termed a hysteresis loop.

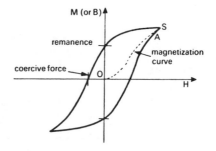

FIGURE 5-24
Illustration of Magnetic
Hysteresis Loop

magnetic levitation Magnetic levitation could be used by mounting a superconducting ring on a railroad car that runs on a magnetized rail. The current induced in the ring leads to a repulsive interaction with the rail, and levitation is possible. Magnetically levitated trains have been running on an experimental basis in Japan for several years and similar projects are under development in Europe.

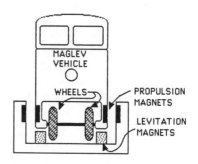

FIGURE 5-25
Magnetic Levitation Rail
System Concept

magnetic resonance imaging Magnetic resonance imaging (MRI) is a new diagnostic imaging technique that produces cross-sectional images. Unlike its predecessor, x-ray computed tomography (CT), it does not use ionizing radiation (x-rays) but rather utilizes interactions between a magnetic field, radio waves and hydrogen atoms (protons). Subtle differences in the magnetic properties of different tissues are magnified to produce images of striking detail and clarity. Magnetic resonance imaging is a superior diagnostic modality for imaging abnormalities of the brain, spine, pelvis, joints, and other organs and soft-tissue structures. With "gating" and surface coils, the clinical application of MRI is rapidly expanding to include organs such as the heart. Magnetic resonance imaging is a radically new technique in medical diagnosis and treatment, and its full impact is yet to be realized. Much more widespread availability of MRI systems can be anticipated, with concomitant reductions in cost and enhancement of features. The use of high-temperature superconducting materials would likely bring further small reductions in the costs of manufacture and operation. The redesign of MRI systems with liquid nitrogen cooling would also make them more user-friendly and reliable by reducing cooling system complexity.

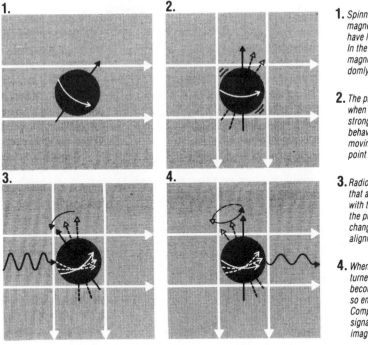

1. Spinning protons have small magnetic fields and thus behave like small bar magnets. In the absence of a strong magnetic field they are randomly aligned.

2. The protons become aligned when in the presence of a strong magnetic field. They behave like spinning tops, moving around a central point (precession).

3. Radio waves are introduced that are timed to coincide with the movements of the protons, temporarily changing their magnetic alignment.

4. When the radio waves are turned off, the protons again become aligned and in doing so emit a faint radio signal. Computer analysis of these signals produces the MRI image.

FIGURE 5-26
Magnetic Resonance
Imaging Conceptual
Illustration

magnetic separator   A device used for separating minerals or other particulate materials based on their density and magnetic properties.

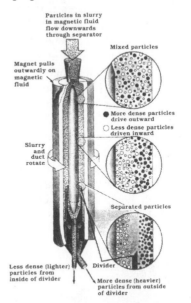

FIGURE 5-27
Illustration of Magnetic
Separator

Source: Intermagnetics General
Corporation

magnetic shielding   Both superconducting wires and superconducting sheets have been used for many years to create regions free from all magnetic fields or to shape magnetic fields. The advent of high-temperature superconductors may extend the range of this application. Like niobium-tin, high-temperature superconductors may be plasma-sprayed, permitting their use on surfaces of complex shape.

magnetization curve   A graphical curve showing the relation between magnetic induction B and magnetizing force H for a magnetic material. Also known as B-H curve.

magnetometer   Sensors which measure magnetic field strength. Because magnetic fields accompany so many physical phenomena, magnetometers, including ultrasensitive versions made from superconducting devices, have many uses.

majority carrier   In a semiconductor, the type of carrier constituting more than half of the total charge carrier concentration.

manufacturing   Techniques and equipment needed to achieve superconductor production to scale and quality needed for commercial applications.

mask   In the manufacture of semiconductor components and integrated circuits, a means of shielding selected areas of the semiconductor during the various processing steps. The circuit layout is described on a set of photographic masks which are used during photolithography processes to define the patterns of openings in the oxide layer through which the various diffusions are made, the windows through which the metal contacts are formed, and the pattern in which the desired metal interconnections are etched.

mass driver   A track running through a series of electromagnetic rings used to rapidly accelerate magnetic materials.

mechanical properties Present ceramic high-temperature superconducting materials can be strong, but they are always brittle. Hence, it may be that high-temperature superconductors wire will be wound into magnets prior to the final high-temperature oxidation step in its fabrication, after which it becomes very brittle. Other conductor fabrication techniques might be feasible, however—for example, those used for producing flexible tapes of Nb_3Sn. An elastic strain tolerance of 0.5 percent may be achieved in a multifilamentary conductor by a fine filament size and by induced compressive stresses. Currently available ceramic technology allows the fabrication of the kinds of complicated pieces that may be needed for such applications as radiofrequency cavities. There are some indications that the new materials may be deformable above 800 C and can then be shaped. The development of a mechanical forming process, however, is constrained by the parallel need for the process to optimize J_cs, both by aligning anisotropic crystal grains and by increasing the strength of the intergranular electrical coupling. Life testing will also be necessary to understand the performance of materials under realistic conditions such as temperature cycling and induced stresses due to transient fields. The adhesion of high-temperature superconductors to other materials is important in microelectronics, in which temperature cycling results in thermal expansion and contraction that cause stresses at the interface. More attention is needed to this problem.

Meissner Effect The expulsion of a magnetic field from a superconductor. Walther Meissner and R. Ochsenfeld learned that a superconductor is more than a perfect conductor of electricity—it also has surprising magnetic properties. A superconductor will not allow a magnetic field to penetrate its interior. If a superconductor is approached by a magnetic field, it sets up screening currents on its surface. These screening currents create an equal but opposite magnetic effect, thereby cancelling the magnetic field and leaving a net of zero inside the superconductor. Reversing the sequence gives you the same result: if you first place the material in a magnetic field and then cool it to the superconducting state, it sets up screening currents that expel the magnetic field. This phenomenon is known as the Meissner Effect. The Meissner Effect occurs only if the magnetic field is relatively small. If the magnetic field becomes too great, it penetrates the interior of the metal and the metal loses its superconductivity.

FIGURE 5-28
Photograph of the
Meissner Effect

mercury   Mercury is the only metallic element that is liquid not only at room temperature but down to –39°C (–38°F); it forms a monatomic vapor and can be compressed to a soft solid state. It is unique among liquids in its very high surface tension which causes it to fragment into globules rather than to flow; this property is responsible for the trivial name quick-silver. Elemental mercury is light gray in color, with a specific gravity of 13.6 and high electrical conductivity. It may have toxic effects as a result of long contact with the skin or inhalation of vapor; while oral intake should be avoided. Symbol - Hg, Atomic Wt. 200.59, State - Liquid, Valence 1,2, Group - IIB, Isotopes 7 stable, Atomic No. 80.

metal   All metals are elements; however, they rarely occur in elemental form in nature, but rather as compounds mixed with rocky material (ores), from which they are extracted by heat and chemical processing. Metals comprise about 75% of the elements; all but three (mercury, gallium, and cesium) are solid at room temperature. They display a broad range of physical properties with respect to conductivity, weight, strength, corrosion resistance, hardness, etc.; most have a crystalline structure and a characteristic sheen or luster. The atoms of metals are joined by bonds having free electrons, which account for their electrical and thermal conductivity. The most electrically conductive are silver, copper, gold, and aluminum. Most metals are quite reactive chemically, though a few are not; they form positive ions and exert electrochemical potential, lithium and potassium having the highest potential. Many metals oxidize (corrode) readily; a few, such as chromium, gold, tantalum, and vanadium, are quite resistant to corrosion. A few metals are toxic (beryllium, barium, cadmium) or combustible (magnesium, socium) in elemental form; many finely divided metals are flammable when dispersed in air.

metallic solid   A metallic solid has one or more electrons detached from each atom that can wander freely through the lattice.

micron   A unit of length in the metric system equivalent to one-millionth meter, or 10,000 angstrom units; a millimicron is one-thousandth micron, or 10 angstrom units. These units are used in physical chemistry and microscopy for measurement of macromolecular and colloidal sizes. The symbol conventionally used for micron is the Greek letter mu (μ), and for millimicron, mμ1 micron is generally accepted as the upper limit of the colloidal size range. A micron is also called a micrometer (μm), and a millimicron, a nanometer (nm).

mirror confinement   A magnetic fusion containment system.

MITI   Japan's Ministry of International Trade and Industry.

Mossbauer effect   Low temperatures are frequently used in experiments involving samples having short spin relaxation times. Where both low temperatures and magnetic fields are planned, it is natural to consider the use of a superconducting magnet. Many experiments of this type are performed using proportional counters that are relatively insensitive to magnetic fields and can be used in conjunction with type simple solenoids. Mossbauer effect measurements can be made without the application of a magnetic field. Additional information can be gained, however, with magnetic hyperfine splitting by placing the source or absorber in a magnetic field. Applied fields are also useful in Mossbauer effect studies of magnetic materials.

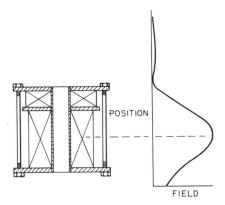

POSITION

FIELD

FIGURE 5-29
Illustration of Mossbauer
Effect

muon   A lepton apparently identical to the electron except for being 200 times heavier.

Multicore Project   Established by Japan's Science and Technology Agency to link nine laboratories and government organizations working on high-temperature superconductors with one another and with industry.

N

neutrino   Any of three uncharged, apparently massless particles associated with the electron, the muon, and the tau lepton. Neutrinos feel only the weak force, and are very hard to detect.

neutron   An uncharged particle, very similar in mass to the proton. It is a composite of three quarks, and is found in all atomic nuclei except that of common hydrogen (which consists of a single proton).

New Superconductivity Materials Research Association   Generally called the "superconductivity forum," this association was set up by Japan's Science and Technology Agency. It provides workshops, symposiums, and other opportunities for interaction among corporations, universities, and national laboratories.

niobium   The element niobium was originally named columbium by its first discoverer, but some years later it was "rediscovered" by another chemist, who named it after the goddess Niobe, the daughter of Tantalus, because of the strong similarity of tantalum and niobium and the close relationship between them. The earlier name is still used by metallurgists, but niobium is the approved name. It is a soft, malleable metal, having only one stable form, m.p. 2468°C (4475°F). It is quite stable at room temperature to attack by oxygen and strong acids; it oxidizes readily at high temperature. Symbol - Nb, Atomic No. 41, State - Solid, Atomic Wt. 92.9064, Group - VB, Valence 2,3,4,5.

nitrogen   Nitrogen is a comparatively unreactive element which comprises 78% by volume of air, though it is unavailable to plants and man from this source. It is as essential to vital processes as are oxygen or carbon, for it is a constituent of amino acids and proteins, the basis of all animal life. In elemental form, nitrogen is used as a protective blanket for highly reactive chemicals, and as a scavenging and carrier gas. It has no active toxicity and is noncombustible. Symbol - N, Atomic Wt. 14.0067, State - Gas, Valence 1,2,3,4,5, Group - VA, Isotopes 2 stable, Atomic No. 7.

TABLE 5-3 Comparison of Nitrogen, Neon, and Helium Cryogen

Cryogen	Boiling Temp	Relative Heat of Vaporization	Power
HELIUM	4.2K	1*	1,000 W/W
NEON	27.2K	41	140 W/W
NITROGEN	77.0K	64	30 W/W

* 28 milliwatts boils 1 liter liquid helium per day

nuclear magnetic resonance   An effect observed when radio-frequency radiation is absorbed by matter. A nucleus with a spin has a nuclear magnetic moment. In the presence of an external magnetic field this magnetic moment precesses about the field direction. Only certain orientations of the magnetic moment are allowed and each of these has a slightly different energy. This resonance produces a signal in the detector coil. A plot of the detected signal against the magnetic field gives nuclear magnetic resonance spectrum which can be used for determining nuclear magnetic moments. The energy levels of the nuclei depend to some extent on the surrounding orbital electrons.

nucleus   The tiny, dense central core of an atom. It is an aggregation of protons and neutrons, bound together by the strong force. In physical chemistry, the total number of protons and neutrons, which constitutes virtually the entire mass of an atom. These particles are held together by forces of exceedingly high magnitude known as the binding energy, a portion of which is released by fission. In crystallography, any microscopic particle that serves as a basis for crystal formation (nucleation).

O

ohm   The Standard International unit of electric resistance, defined as the resistance between two points on a conductor through which a current of one ampere flows as a result of a potential difference of one volt applied between the points, the conductor not being a source of electromagnetic force.

ohmic loss   Power dissipation in an electrical circuit arising from its resistance rather than from other causes such as back electromotive force.

Ohm's law   The electrical current in any conductor is proportional to the potential difference between its ends, other factors remaining constant. As the ratio of potential difference to current in a conductor is termed the resistance of the conductor, Ohm's law is often expressed as $I = E/R$, where I is the current, E is the potential difference, and R the resistance.

oxide An extensive class of compounds formed by chemical combination of oxygen with another element or with an organic compound; the most common examples of inorganic oxides are water, an oxide of hydrogen, and carbon dioxide, both of which are products of combustion. Virtually all the elements, except the noble gases, form oxides, especially when heated. They are relatively stable compounds, and in the case of metals, oxygen often acts preferentially, displacing other combined elements. The formation of oxides is called oxidation, though this term has a wider and more sophisticated significance.

P

particle accelerator A device used to accelerate electrically charged particles to high speeds using magnetic fields.

particle physics The branch of physics concerned with understanding the properties, behavior, and structure of elementary particles, especially through study of collisions or decays involving energies of hundreds of MeV or more.

Pauli exclusion principle An important generalization, discovered by Wolfgang Pauli, an Austrian-born physicist (1900–1958), Nobel Prize 1945, relating to atomic structure, which states that it is impossible for two electrons in the same atom to have the same value for all four quantum numbers; two electrons can occupy the same orbital only if they have opposite spin, or direction of rotation, i.e., +1/2 and –1/2. This principle explains the arrangement of elements in the Periodic Table and accounts for the maximum number of electrons that can occur in the seven shells, i.e., 2 in the first, 8 in the second, 18 in the third, and 32 in the fourth.

peak load Maximum power consumed or produced by a unit or group of units in a stated period.

Periodic Table A systematic classification of the chemical elements based on the Periodic Law; it was originally worked out by the Russian chemist, Dimitri Mendeleef (1834–1907), in 1869 by listing the then-known elements in the order of increasing atomic weight. It was later modified by the English physicist, Henry G.J. Moseley (1887–1915), and others, who showed that the periodicity actually depends on the atomic number. Thus, the location of an element in the table is an indication of its electronic structure and chemical properties; elements having a similar electronic pattern occur on a regularly repeating (periodic) basis. There are seven horizontal divisions, called periods, containing, respectively, 2, 8, 8, 18, 18, 32, and 19 elements. Each period begins with an element having one outermost electron and ends with one having eight outermost electrons (except helium). The vertical columns of the table comprise nine major groups; Groups I through VII are divided into two subgroups designated A and B. Thus, Group IA is a vertical series of univalent, reactive elements, while those in the last (noble gas) group are inert, or nearly so, the first three having zero valence.

1A	2A	3B	4B	5B	6B	7B	8			1B	2B	3A	4A	5A	6A	7A	0
1 H																	2 He
3 Li	4 Be											5 B	6 C	7 N	8 O	9 F	10 Ne
11 Na	12 Mg		←		transition elements				→			13 Al	14 Si	15 P	16 S	17 Cl	18 Ar
19 K	20 Ca	21 Sc	22 Ti	23 V	24 Cr	25 Mn	26 Fe	27 Co	28 Ni	29 Cu	30 Zn	31 Ga	32 Ge	33 As	34 Se	35 Br	36 Kr
37 Rb	38 Sr	39 Y	40 Zr	41 Nb	42 Mo	43 Tc	44 Ru	45 Rh	46 Pd	47 Ag	48 Cd	49 In	50 Sn	51 Sb	52 Te	53 I	54 Xe
55 Cs	56 Ba	57* La	72 Hf	73 Ta	74 W	75 Re	76 Os	77 Ir	78 Pt	79 Au	80 Hg	81 Tl	82 Pb	83 Bi	84 Po	85 At	86 Rn
87 Fr	88 Ra	89† Ac															

*lanthanides	57 La	58 Ce	59 Pr	60 Nd	61 Pm	62 Sm	63 Eu	64 Gd	65 Tb	66 Dy	67 Ho	68 Er	69 Tm	70 Yb	71 Lu
†actinides	89 Ac	90 Th	91 Pa	92 U	93 Np	94 Pu	95 Am	96 Cm	97 Bk	98 Cf	99 Es	100 Fm	101 Md	102 No	103 Lr

FIGURE 5-30
Periodic Table of the Elements

permanent magnet   A magnetized mass of steel or other ferromagnetic substance, of high retentivity and stable against reasonable handling. It requires a definite demagnetizing field to destroy the residual magnetism.

perovskite   A type of crystal. Refers to the crystal structure shared by the 1-2-3 and other high-temperature superconductors.

persistent mode operation   After it has been energized, a superconducting magnet can be operated in the persistent mode by short-circuiting the magnet with a superconductor. This is accomplished by connecting a section of superconducting wire across the terminals of the magnet. This section of superconductor can be heated to drive it into the resistive state so a voltage can be established across the terminals and the magnet can be charged or discharged. During the persistent mode of operation, the heater is turned off and the switch is permitted to cool into the superconducting state. In this condition, the power supply may be turned off and the magnet current will circulate through the magnet and the persistent switch. The decay of the magnet is given by $H = H_o e{\text -}t/t$ where t is the usual L/R time constant. The small residual resistance in the magnet occurs either from resistance in the joints or from the flux motion resistance discussed earlier. To get the best persistence the magnet must be operated at fields lower than those causing flux flow in the conductor. This requires that the magnet be operated at lower current densities and lower fields when extreme persistence is required. Consequently, such magnets are bulkier, more costly and require more liquid helium for cool-down than magnets having somewhat less persistence. Nevertheless, these magnets are quite desirable where great persistence is required, such as in nuclear magnetic resonance projects. In addition to the limitation on the current density, it is necessary that the resistance in joints between conductors be as low as possible. Such joints are most readily made in conductors containing only one filament of superconductor. These magnets of course, are susceptible to flux jumping and must be charged and discharged at a slower rate than magnets constructed with multi-filament conductors.

persistent switches   Persistent switches are provided on many magnets to increase their stability over long periods of time or to reduce the rate of helium boil-off associated with continually supplying current to the magnet. A persistent switch is comprised of a short section of superconducting wire connected across the input terminals of a magnet and an integral heater used to drive the wire into the resistive, normal state. When the heater is turned on and the wire is resistive, a voltage can be established across the terminals of the magnet and the magnet can be energized. Once energized, the heater is turned off, the wire becomes superconducting and further changes in the magnet current cannot be made. In this *persistent* mode of operation, the external power supply can be turned off to reduce the heat input to the helium bath and the current will continue to circulate through the magnet and the persistent switch. Persistent mode current switches are installed on and become an integral part of the magnet. This is necessary since special care must be taken in making the joints between the switch and the magnet leads. The electrical heater in the persistent switch has a nominal resistance of 60 Ohms and typically requires 35 mA of current to drive the superconductor into the resistive state. The superconductive wire typically has 9 Ohms of resistance in the normal state.

phonon   The atomic mechanism causing electron pairing on the BCS theory.

photon   The massless boson, the carrier of the electromagnetic interaction. All electromagnetic radiation consists of photons.

picosecond   One trillionth of a second.

plasma spraying   A technique used to coat objects with superconducting material.

polarity reversing switch   In superconducting magnet projects where the sign of the field will be changed, a polarity reversing switch should be incorporated to avoid the onerous, hazardous and potentially fatal task of disconnecting and reversing the leads manually.

pole   the place towards which lines of magnetic flux converge, or from which they diverge. It usually exists near a surface of magnetic discontinuity and in the material of higher permeability. A north pole of a magnet is that which, if free, tends to move towards the north (magnetic) pole of the earth.

pole face   The end surface of the core of a magnet through which surface the useful magnetic flux passes. In particular, in an electrical machine it is that surface of the core or pole piece of a field magnet which directly faces the armature.

positron   The positively charged antiparticle of the electron.

powder   Any finely divided solid material which has been precipitated by a chemical reaction or mechanically reduced by grinding; particle sizes range from less than 1 micron to several millimeters. Typical powders are carbon black, precipitated barium sulfate (blanc fixe), and fine-ground clays, whitings, and talcs. Many metals are comminuted to powders by various methods for use in making compressed products having microscopically small pores (powder metallurgy); and they are also used for catalysts (e.g., nickel).

prime mover   The mechanical energy that drives a generator.

processing   Preparation of new superconducting materials with desired properties on a laboratory or industrial scale.

proton   A positively charged particle, made of three quarks, found in all atomic nuclei. The Superconducting Super Collider (SSC) will collide protons.

Q

quanta   A number indicating a quantum condition and given to each energy level in an atom.

quantum mechanics   The modern theory of matter, of electromagnetic radiation, and of the interaction between matter and radiation; it differs from classical physics, which it generalizes and supersedes, mainly in the realm of atomic and subatomic phenomena.

quark   The six different quarks and their antiquarks are elementary particles that carry fractional amounts of electric charge. They are bound together by the strong force in twos or threes, to make protons, neutrons, kaons, pions, and many other composite particles.

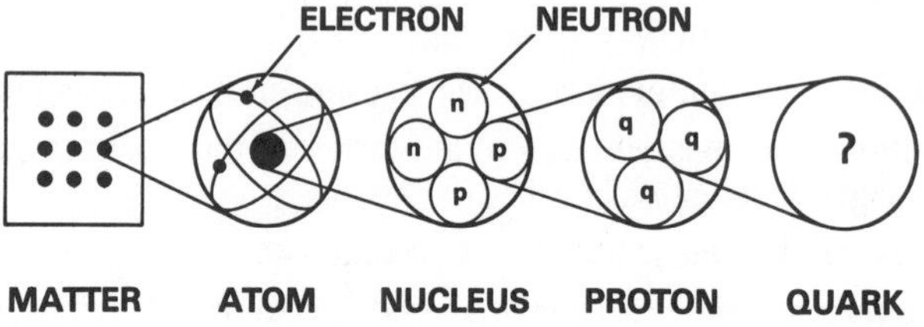

FIGURE 5-31
Fundamental Building Blocks—From Matter to Quark

Source: DOE

quenching   An unavoidable phenomenon in superconducting magnets is quenching. Any superconducting magnet can be quenched by increasing the current and field indiscriminately. A quench in a well-encapsulated magnet typically occurs at the location of the highest field in the magnet. Resistance is restored to the conductor at this point and heating occurs in the magnet. This heat spreads to adjacent areas and drives more conductor normal, and the normal zone continues to spread until the magnet is completely discharged. If the resistance across the terminals of the magnet due to the power supply is low, the power supply may be ignored to a first approximation in analyzing a quench. Thus, the quench may be viewed as the discharge of an indicator into a time varying resistance. The resistive voltage is counteracted by an inductive voltage. Unlike the few volts used in charging the magnet, the voltages encountered during a quench discharge can be measured in kilovolts. Initially, the voltage is confined to the layers of windings near the point where the quench initiated and internal arcing can occur between layers, if sufficient insulation has not been provided. The voltage during a quench can be decreased by increasing the ratio of copper to superconductor in the magnet. However, this has the undesirable effect of reducing the current density in the magnet and results in a larger magnet generating the same field. This approach also localizes the quench and causes greater thermally induced mechanical stresses that can result in cracking of the epoxy or breaking of the conductor.

R

radiation detector Superconducting microwave and far-infrared radiation detectors (quasiparticle mixers, superconducting bolometers) already exist using conventional superconductors. In spite of a loss of sensitivity due to increased electrical noise at higher temperatures, the increased energy gap of high-temperature superconductors would offer sensitive detection in a largely inaccessible frequency range, and the simplified refrigeration allows increased ease of use. Other microwave applications include high-Q waveguides, phase shifters, and antenna arrays.

radiation effects High-temperature superconductors appear to be somewhat more sensitive to radiation than conventional superconductors. High sensitivity to radiation damage could pose a difficult, although not insurmountable, problem for application to magnetic fusion machines. For electronic applications the substitution of either conventional or high-temperature superconducting devices for those employing semiconductors would result in an improvement of several orders of magnitude in resistance to radiation damage.

rail gun See **coil/rail gun.**

RAM chips Integrated circuits that provide random access memory for computers and other digital systems.

resistance That physical property of a network element that accounts for permanent energy loss in the circuit, such as heat generation. The ratio of the potential difference across a conductor to the current flowing through it. If the current is alternating, the resistance is the real part of the electrical impedance Z, i.e. $Z = R + iX$, where X is the reactance. Resistance is measured in ohms.

resistivity The resistance of unit length of the substance with uniform unit cross-section. The product of the resistivity and the density is sometimes called the mass resistivity.

TABLE 5-4 Resistivity of Conductors, Semiconductors, and Insulators

Material	Resistivity (ohm metres)
Conductors	10^{-8} to 10^{-6}
Semiconductors	10^{-6} to 10^{-7}
Insulators	10^{-7} to 10^{23}

rotor The rotating part of a generator.

S

saturation current The portion of the static characteristic of an electronic device in which further increases in the voltage do not lead to a corresponding increase in the current, until breakdown is reached. The actual value of the saturation current is a function of the device and the external circuit.

Science and Technology Agency (STA) Under the Prime Minister's Office in Japan.

semiconductor An element or compound whose electrical properties are midway between a conductor and an insulator. Semiconduction was first discovered in the metal germanium and subsequently in the non-metals selenium and silicon and in the compounds silicon carbide, lead sulfide, and a few others. It is caused by two factors: free electrons ejected from low-energy covalent bonds at the site of a lattice defect, leaving an energy deficit called a hole (positively charged); and electrons supplied by atoms of impurity elements. Electrons from both sources pass through the crystal from one energy deficit location to another, so that the deficit itself can be regarded as a flow of positive electricity. If enough electrons are present to satisfy the energy deficit by saturating the covalent bond, the electrons become the conducting medium, and the semiconductor is said to be n-type; but if too few electrons are available, the conductivity is by the positively charged holes, which move through the crystal as the electrons shift location from one bond to another. In this case the semiconductor is p-type.

shell A term applied by physical chemists to the various energy levels or groups of orbitals occupied by electrons as they revolve around atomic nuclei. There are seven possible shells, designated as K, L, M, N, etc.; the number varies with the element from 1 in the lightest (hydrogen and helium) to 7 in the heaviest (uranium). Each shell can contain only a certain number of electrons, for example, 2 for K, 8 for L, 18 for M, and 32 for N; the shells may or may not be completely filled. The outer shell contains the valence electrons, which are directly involved in chemical bonding.

short circuit Abnormal connection of relatively low resistance, whether made accidentally or intentionally, between two points of different potential in a circuit.

siemens The Standard International unit of electrical conductance, defined as the conductance of an element that possesses a resistance of one ohm. The unit used to be called the mho or reciprocal ohm.

signal-to-noise ratio An important parameter for sensors, the signal-to-noise ratio compares the signal the sensor is intended to measure with background noise (one source of which is thermal, rising with temperature).

silicon The element silicon, m.p. 1420°C (2588°F), is the most abundant solid element, being second only to oxygen in prevalency; it rarely occurs in elemental form, virtually all of it existing as compounds (silicon dioxide, silicates, etc.). Silicon is a nonmetal, like carbon, to which it is chemically similar; it has the same valence and is next below carbon in the Periodic Table. Silicon forms single bonds with itself and with carbon, oxygen, hydrogen, and halogens, but it does not form double or triple bonds nor chains of more than six silicon atoms. This similarity to carbon accounts for its ability to form silanes (with hydrogen), siloxanes (with oxygen), and the industrially important silicone compounds (with oxygen and organic groups). Silicon is one of the few elements that have semiconducting properties. Symbol - Si, Atomic Wt. 28.086, State - Solid, Valence 4, Group - IVA, Isotopes 3 stable, Atomic No. 14.

simple solenoid The most frequently currently requested superconducting magnets are in the form of simple solenoids. They are designed with uniform current densities throughout the windings. At higher fields, it is necessary to increase the conductor diameter or to incorporate a Nb_3Sn conductor in the high field region in order to achieve the necessary current. This can result in two or more sections of the coil with differing current densities. Typically, these magnets have bore diameters of a few centimeters and homogeneities in the range of 1.0-0.1% in a 1 cm diameter spherical volume (DSV), although much larger magnets have been constructed (up to 2 m I.D. and 3 m long). Fields up to 9 Tesla are generated with NbTi solenoids. Higher fields generally require Nb_3Sn conductors when operated at the normal boiling temperature of liquid helium (4.2 K). These magnets are quite versatile and can be used for a great variety of purposes, including: Superconductivity Applications; Hall Effect; Fermi Surface Studies; Magnetroresistivity; Magnetocaloric, Magnetostriction Studies; Faraday Rotation; Bubble Chambers; Nuclear Beam Spin Flipping; Magnetization Measurements; Energy Storage; Magnetic Separation; Homopolar Motors; Beam Focussing; and Mossbauer Effect.

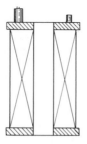

FIGURE 5-32
Illustration of Simple Solenoid

solid-state physics The branch of physics centering on the physical properties of solid materials; it is usually concerned with the properties of crystalline materials only, but it is sometimes extended to include the properties of glasses or polymers.

specific heat The ratio of the amount of heat required to raise a mass of material 1 degree in temperature to the amount of heat required to raise an equal mass of a reference substance, usually water, 1 degree in temperature; both measurements are made at a reference temperature, usually at constant pressure or constant volume. The quantity of heat required to raise a unit mass of homogeneous material one degree in temperature in a specified way; it is assumed that during the process no phase or chemical change occurs.

TABLE 5-5 Specific Heat for Liquid Helium

T(K)	Cp(J/gK)
3.0	3.05
3.5	3.54
3.75	4.30
4.0	4.50
4.5	6.23
5.0	13.50

Source: American Magnetics, Inc. *"Selection Guide."*

splice   Cable joint between two or more separate lengths of cable with the conductors in one length and with the protecting sheaths so connected as to extend protection over the joint.

stability   When used with reference to a power system, that attribute of the system, or part of the system, that enables it to develop restoring forces among the elements thereof equal to, or greater than, the disturbing forces so as to restore a state of equilibrium between the elements.

standard model   The combined theories of the strong, electromagnetic, and weak forces, incorporating all the known quarks, leptons, and force-carrying bosons. The electromagnetic and weak interactions have been mathematically unified into a single electroweak force, but Grand Unification, which includes the strong force too, has not yet been achieved. The standard model embodies everything we know about nuclear and particle physics.

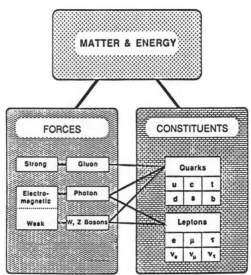

FIGURE 5-33
The Standard Model

Source: DOE

stator   The fixed part of a generator.

strong force   A short range force, the strongest of the fundamental interactions. Quarks and all the composite particles made of quarks feel its influence, but leptons do not.

superconducting alloy   An alloy capable of exhibiting superconductivity, such as an alloy of niobium and zirconium or an alloy of lead and bismuth.

superconducting cables   Superconducting cables are power transmission cables that can be buried underground and cooled to temperatures near absolute zero, which reduces heat dissipation to a very low level. This enables the conductors to carry about five times more current than conventional cables. The technical characteristics will permit very long cables to operate at high power levels. This technology provides an alternative to overhead transmission in situations where new overhead transmission is unacceptable, such as the penetration of suburban areas around a load center.

superconducting circuit   An electric circuit having elements which are in a superconducting state at least part of the time, such as a cryotron.

superconducting computer   A high-performance computer whose circuits employ superconductivity and the Josephson effect to reduce computer cycle time.

superconducting electromagnets   Superconducting electromagnets have been applied
in programs during the past several years. Once a current is established in the coil of
such a magnet, no additional power input is required because there is no resistive
energy loss. The coils can also be made more compact because there is no need to
provide channels for the circulation of cooling fluids. Therefore, superconducting
electromagnets can attain very large fields more easily and less expensively than
conventional magnets. The most outstanding feature of a superconducting magnet is
its ability to support a very high current density with a vanishingly small resistance.
This characteristic permits magnets to be constructed that generate intense magnetic
fields with little or no electrical power input. This feature also permits steep magnetic
field gradients to be generated at fields so intense that the use of ferromagnetic
materials for field shaping is limited in effectiveness. Since the current densities are
high, superconducting magnet systems are quite compact and occupy only a small
amount of laboratory space. Another outstanding feature of superconducting magnets
is the stability of the magnetic field in the persistent mode of operation. In the
persistent mode of operation, the time constant is extremely long and the magnet can
be operated for days or even months at a nearly constant field; a feature of great
significance where signal averaging must be performed over an extended period of
time. Small superconducting magnets are frequently used to attain field intensities,
stabilities or profiles that are not attainable with alternative magnets—or because their
cost is less than the cost of conventional magnets offering comparable or inferior
performance. In large magnets, the trade-off is frequently made in favor of superconducting magnets based on the relative costs of power for operating the magnets. The
cost trade-off becomes more favorable for superconducting magnets as the period of
operation increases. Magnetic field intensities of 1 Tesla or less, without demanding
stability requirements, are frequently better generated with water cooled copper coils
with or without iron.

superconducting energy gap   At low temperatures, so-called Cooper pairs condense
into an electrical superfluid, with energy levels a discrete amount below those of
normal electron states.

superconducting magnet   An electromagnet wound with superconducting wire. Essentially all the power consumed goes for refrigeration to keep the coil windings below
their superconducting transition temperatures. An electromagnet whose coils are made
of a Type II superconductor with a high transition temperature and extremely high
critical field, such as niobium tin, Nb_3Sn; it is capable of generating magnetic fields of
100,000 oersteds and more with no steady power dissipation.

superconducting magnetic energy storage system (SMES)   A coil or solenoid of
superconducting wire in which an electric current can circulate, storing energy until
needed for purposes such as feeding an electric utility grid or powering a free-electron
laser.

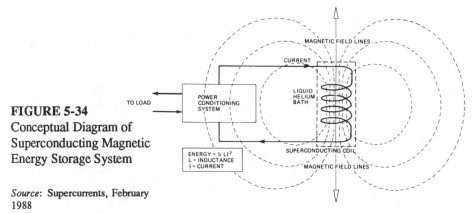

FIGURE 5-34
Conceptual Diagram of
Superconducting Magnetic
Energy Storage System

Source: Supercurrents, February
1988

superconducting magnetic field A magnetic field cannot exist inside a superconducting material. Attempts to establish such a magnetic field result in induced eddy currents that cancel the applied field everywhere inside the material.

superconducting memory A computer memory made up of a number of cryotrons, thin-film cryotrons, superconducting thin films, or other superconducting storage devices; these operate under cryogenic conditions and dissipate power only during the read or write operation, which permits construction of large, dense memories.

superconducting metal A metal capable of exhibiting superconductivity.

superconducting quantum interference device (SQUID) A very sensitive instrument, built with Josephson junctions, used to detect magnetic signals. Superconducting quantum interference devices (SQUIDs), operating at liquid helium temperatures as sensitive magnetic field detectors, are already of value in many disciplines including medical diagnostics, geophysical prospecting, undersea communications, and submarine detection. SQUIDs made with the new high-temperature materials have been operated at liquid nitrogen temperatures. Relatively inexpensive SQUID-based magnetometers operating at 77 K or higher would be deployed in large numbers if electrical noise can be held to acceptably low levels.

FIGURE 5-35
Photograph of a High-
Temperature Supercon-
ducting Thin Film
(SQUID)

Source: IBM

superconducting radiofrequency cavity If microwave alternating current loss characteristics are tolerable, the new superconductors may greatly improve the performance of superconducting radiofrequency cavities by allowing them to operate at higher fields. Indeed, the potential impacts embrace all of microwave power technology, especially in the promising millimeter-wave region. Accelerator technology might also be significantly advanced by the availability of liquid nitrogen-cooled superconducting cavities. The applicability of superconducting technology to recirculating linear accelerators, on the other hand, is an accepted fact. In addition to providing high-quality beams for nuclear physics research, these machines are natural candidates for continuous beam injectors used in free-electron lasers. As technology matures and industrial applications develop for high-power, high-efficiency tuneable lasers in biotechnology, fusion plasma heating, and other fields, superconducting radiofrequency devices will proliferate.

superconducting thin film A thin film of indium, tin, or other superconducting element, used as a cryogenic switching or storage device, as in a thin-film cryotron.

superconductivity Total loss of resistance to direct electrical currents. The ability of some materials to carry an electric current with no power loss, owing to the complete absence of electrical resistance. A property of many metals, alloys, and chemical compounds at temperatures near absolute zero by virtue of which their electrical resistivity vanishes and they become strongly diamagnetic. Superconductivity was discovered in 1911 by the Dutch physicist H. Kamerlingh-Onnes while experimenting on the temperature dependence of resistivity of materials.

superconductor Any material capable of exhibiting superconductivity; examples include iridium, lead, mercury, niobium, tin, tantalum, vanadium, and many alloys. Also known as cryogenic conductor; superconducting material. Superconductors are the materials that lose all their electrical resistivity below a certain temperature and become diamagnetic. High values of externally applied magnetic field are required to destroy the superconductivity. These electrical and magnetic properties of superconducting materials have found applications in lossless electrical transmission and generation of high magnetic fields. The superconducting magnets are used where normal iron magnets are inadequate. These magnets are used as exciter magnets for homopolar generators or rotors in large alternators, where significant gain in efficiency and power density is obtained. Future fusion reactors can use superconducting magnets for confining the deuterium and tritium plasma. Superelectron pairs in a superconductor can tunnel through a nonconducting thin layer. Based on this "Josephson effect," superconducting Josephson junctions are used as sensors, as high-energy electromagnetic radiation detectors, and in high-speed digital signal and data processing.

1-2-3 superconductor One of a new class of high-temperature superconductors, typified by yttrium-barium-copper-oxide and called 1-2-3 because of their generic chemical formula: $RBa_2CO_3O_{7-x}$, with R almost any one of the rare-earth elements. Much of the research on the new superconductors has focused on the 1-2-3 materials, which typically have transition temperatures above 90°K.

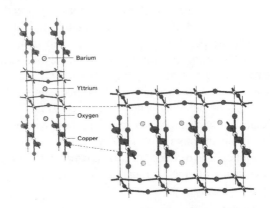

FIGURE 5-36
Diagram of the Yttrium-
Barium-Copper Oxide
High-Temperature 1-2-3
Superconductor

Source: Argonne National
Laboratory

superconductor, Type I   Material with perfect electrical conductivity for direct current
that also possesses perfect diamagnetism (i.e., magnetic flux is totally excluded from
the material): most metal element superconductors are Type I. When an external
magnetic field is applied on this superconductor, the transition from superconducting
to normal state is sharp.

superconductor, Type II   Material with perfect electrical conductivity for direct current (at least for low currents) that does not possess perfect diamagnetism, (i.e., flux
penetration of the material is possible); most metal alloys and compounds that are
superconductors are Type II. When an external magnetic field is applied, the transition from superconducting to normal state occurs after going through a broad "mixed
state" region.

supercooling   Cooling of a substance below the temperature at which a change of state
would ordinarily take place without such a change of state occurring, for example, the
cooling of a liquid below its freezing point without freezing taking place; this results
in a metastable state. The process by which liquids, by slow and continuous cooling,
are reduced to a temperature below the normal freezing point. A supercooled liquid is
a metastable state and the introduction of the smallest quantity of the solid at once
starts solidification. Small mechanical disturbance may also initiate solidification which,
once started, will continue with the evolution of heat until the normal freezing point is
reached.

superfluidity   Superfluidity is a term used to describe a property of condensed matter
in which a resistanceless flow of current occurs. The mass-four isotope of helium in
the liquid state plus some twenty-five metallic elements are presently known to exhibit this phenomenon.

surge   Transient variation in the current and/or potential at a point in the circuit.

synchrotron   An accelerator in which the energy of charged particles is increased as
they travel around a circular orbit of fixed radius. The Superconducting Super Collider (SSC) will be a proton synchrotron.

synchrotron radiation   Electromagnetic radiation emitted by any charged particle when
it is forced off a straight path, as in a circular accelerator.

T

tau lepton A fundamental particle apparently identical to the electron and the muon, except for its greater mass.

temperature The temperature of an object is a property that determines the direction of heat flow when the object is brought into thermal contact with other objects: heat flows from regions of higher to those of lower temperatures. A value representing a measurement of the degree of hotness or coldness of any matter; the thermal state or condition of a substance or mixture (air). There are several temperature scales, each of which is a measurement of a thermal condition compared with an arbitrary reference point, for example, the freezing and boiling points of water, absolute zero, etc. The best-known are the Fahrenheit, centigrade, and Kelvin scales, which are separately defined. The term *ambient temperature* means the temperature of any local environment in which an experiment is conducted or in which an operation is carried out.

The temperature at which the resistance vanishes in a superconductor is reduced when a magnetic field is applied. The maximum field that can be applied to a superconductor at a particular temperature and still maintain superconductivity is called the critical field.

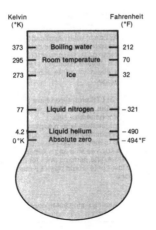

FIGURE 5-37
Comparison of Kelvin/
Fahrenheit Temperature
Scales

Source: Office of Technology
Assessment, 1988

terahertz A unit of frequency, equal to 10_{12} hertz, or 1,000,000 megahertz. Abbreviated THz.

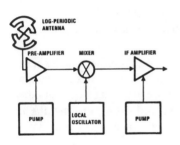

- Fully integrated
- Small size
- Low noise
- High frequency scalable
- Highly survivable

FIGURE 5-38
Superconductive Terahertz
Receiver Block Diagram

Source: TRW

tesla   A unit used to describe the strength of magnetic fields.

Tevatron   The collider recently constructed at Fermilab, smashing together protons and antiprotons of energy 1 TeV.

thallium   Thallium is a relatively heavy metal element of low melting point, 303°C (577°F), occurring at an average concentration of about 1 gram per ton in the earth's crust; its highest concentrations, as in potassium-containing minerals, are too few to be economically workable. Thallium compounds can be recovered from by-products of zinc smelting, and the metal obtained by various separation techniques. The metal oxidizes in air to form a protective oxide coating. Its uses in metallic form are chiefly in alloys with bismuth, cadmium, mercury, etc. It is quite poisonous. Symbol - Tl, Atomic Wt. 204.37, State - Solid, Valence 1,3, Group - IIIA, Isotopes 2 stable, Atomic No. 81.

thermal conductivity   The heat flow across a surface per unit area per unit time, divided by the negative of the rate of change of temperature with distance in a direction perpendicular to the surface. Also known as heat conductivity.

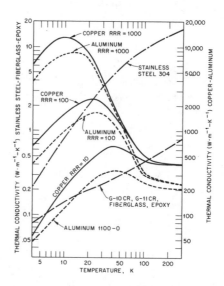

FIGURE 5-39
Illustration of Thermal
Conductivity vs
Temperature

Source: American Magnetics,
Inc. "Selection Guide."

thermal equilibrium   The condition of a system in which the net rate of exchange of heat between the components is zero.

thin film   A very thin layer of a material usually deposited on a substrate.

three-terminal electronic device   One which, like a transistor, can amplify a signal substantially.

tin   Tin is an element obtained by roasting or smelting the ore cassiterite, which occurs chiefly in Malaya, Bolivia, and Central Africa. It is relatively soft and has a comparatively low melting point 232°C (449°F). Its primary uses are in plating steel and other metals and as a component of such alloys as terne plate (led + tin), low-melting or fusible alloys (solders), bearing alloys (babbitt), and type metal. Tin also has a number of specialized engineering uses. The metal itself is not toxic, but its organic compounds are quite poisonous; they are used as polymerization catalysts. Symbol - Sn, Atomic Wt. 118.69, State - Solid, Valence 2,4, Group - IVA, Isotopes 10 stable, Atomic No. 50.

tokamak   The tokamak concept is to contain a stable high-temperature plasma in an evacuated torus (doughnut-shaped container). However, the wall of the torus is not used to confine the plasma. Confinement is accomplished by three different magnetic fields: the toroidal field, the poloidal field, and the equilibrium/stability field. The toroidal magnetic field, which is directed around the inside of the torus, is the basic confining field. It is produced by coils of wire wrapped around the body of the torus. The poloidal field produces magnetic pressure, which forces the plasma toward the middle of the toroidal ring. It has a circular shape in each radial cross-section perpendicular to the plane of the ring. The poloidal field is generated by a plasma current made to flow inside the toroidal ring by increasing the electrical current in the magnetic transformer coils. The toroidal and poloidal fiends combine to produce the total resulting magnetic field shown in the figure. A third set of magnetic fields, used to maintain plasma equilibrium and stability, is generated by smaller coils that run along the outside periphery of the torus (not shown in the figure). Three different techniques are used to heat the plasma. Initial heating is provided by the plasma current (axial current), which also supplies the poloidal magnetic field. Further heating occurs by either the injection of high energy deuterium atoms (called neutral beam injection) or by transferring energy to plasma electrons or ions through RF or microwave electromagnetic radiation. The objectives of tokamak development are to demonstrate the scientific feasibility of the tokamak concept and then to exploit that concept for commercial reactor development.

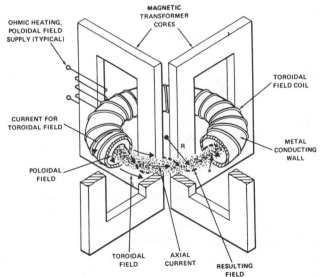

FIGURE 5-40
Schematic Diagram of a
Tokamak Device

Source: Handbook of Energy
Technology, V. Daniel Hunt

training effect The training effect in magnets is caused by wire motion. The heat capacities of the materials in a superconducting magnet at 4 K are several orders of magnitude lower than the heat capacities of the same materials at room temperature. Thus, only a small amount of heat dissipated inside the magnet can raise the temperature of the conductor above its critical temperature at the ambient field and current density. One source of heating is wire motion caused by the Lorentz force on the conductor in the magnet. Imperceptible motions of the wire can result in frictional heating sufficient to drive the conductor normal at fields well below the anticipated maximum field of the magnet. Upon reenergizing the magnet, it is frequency observed that it will "train" to successively higher fields before quenching, ultimately achieving the design field. In some cases, the wires will remain in their shifted positions and the magnet will perform well. In other cases, however, retraining is required after the magnet is warmed to room temperature. To avoid this training effect, it is necessary that the conductors be bonded securely in place to prevent the wire from moving.

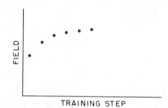

FIGURE 5-41
Illustration of the Training
Effect in Magnets

transient stability Condition that exists in a power system if, after an aperiodic disturbance has taken place, the system regains steady-state stability.

transition element Any of a large number of metallic elements occurring in the fourth, fifth, sixth, and seventh periods of the Periodic Table. Some authorities prefer the term transitional element. Specifically, they are as follows:

Period 4	Period 5	Period 6	Period 7
scandium	yttrium	lanthanum	actinium
titanium	zirconium	lanthanide series	actinide series
vanadium	niobium	hafnium	
chromium	molybdenum	tantalum	
manganese	technetium	tungsten rhenium	
iron	ruthenium	osmium	
cobalt	rhodium	iridium	
nickel	palladium	platinum	
copper	silver	gold	

The transition elements or metals are so named because they represent a gradual shift from the strongly electropositive elements of Groups IA and IIA to the electronegative elements of Groups VIA and VIIA. They are also more versatile in their chemical bonding properties than are other elements, using the two outer shell orbitals instead of only those of the outermost shell. Most are active complexing agents, particularly cobalt, iron, and chromium.

transition temperature   The highest temperature at which a material becomes a super-conductor, also known as the critical temperature. The transition temperature drops as the magnetic field anc current density increase.

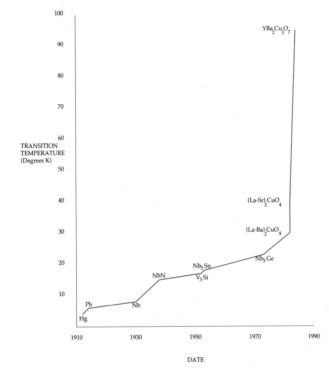

FIGURE 5-42

Transition Temperature vs Time

Source: Supercurrents, January 1988

two-terminal electronic device   One which, like a Josephson junction, can serve only as a weak amplifier.

U

Umklapp process   The interaction of three or more waves in a solid, such as lattice waves or electron waves, in which the sum of the wave vectors is not equal to zero but, rather, is equal to a vector in the reciprocal lattice. Also known as flip-over process.

unavailable energy   That part of the energy which, when an irreversible process takes place, is initially in a form completely available for work and is converted to a form completely unavailable for work.

unit magnetic pole   Two equal magnetic poles of the same sign have unit value when they repel each other with a force of 1 dyne if placed 1 centimeter apart in a vacuum.

unity coupling   Perfect magnetic coupling between two coils, so that all magnetic flux produced by the primary winding passes through the entire secondary winding.

unity power factor   Power factor of 1.0, obtained when current and voltage are in phase, as in a circuit containing only resistance or in a reactive circuit at resonance.

upper critical magnetic field $YBa_2Cu_3O_7$ samples generally exhibit extremely high upper critical fields (H_{c2}). Preliminary measurements indicate that for single crystals H_{c2} is anisotropic, that is, dependent upon field direction relative the a-, b-, or c-axes of the orthorhombic lattice. Values ranging from 30 T (c-axis) to 150 T (a- or b-axis) are reported at 4.2 K. The mechanical stresses associated with the confinement of such high magnetic fields in typical compact geometries are frequently beyond the yield or crushing strengths of known materials. Hence, improving these intrinsic H_{c2} values is less important that increasing T_c or J_c values. In fact, materials with higher Tcs should exhibit higher H_{c2} values if the performance of known materials is any guide. However, developing materials that can practically be fabricated into magnets and that retain useful J_cs at fields approaching H_{c2} even at 77 K is an important challenge.

V

vacancy   A defect in the form of an unoccupied lattice position in a crystal.

vacuum   Theoretically, a space in which there is no matter. Practically, a space in which the pressure is far below normal atmospheric pressure so that the remaining gases do not affect processes being carried on in the space.

vacuum bore split coil magnets   Many projects require that the radial and axial bore of the magnet be evacuated to avoid absorption of optical or nuclear beams. These magnets are made by assembling the coil sections on a helium leak-tight vacuum bobbin that is sized to fit a specific dewar design. Such magnets are supplied only as a part of an integrated system including the dewar since the vacuum bobbin becomes an integral part of the helium reservoir.

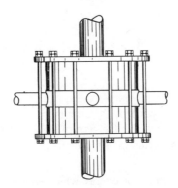

FIGURE 5-43
Illustration of Vacuum
Bore Split Coil Magnets

valence band   The highest electronic energy band in a semiconductor or insulator which can be filled with electrons.

valence electrons   Electrons in the outermost shell of an atom that are involved in chemical changes.

van der Waals bond   The van der Waals bond is a much weaker dipole—dipole interaction than the Pauli exclusion bond. The hydrogen bond uses the positive charge of the hydrogen nucleus to form an attractive interaction.

van der Waals forces Extremely weak forces of interaction between unexcited atoms or molecules of gases; they account for the fact that gases do not behave strictly in accordance with theory, i.e., their behavior varies slightly from that required by the ideal or perfect gas laws. They are weaker than hydrogen bonds and are not involved in chemical bonding. They were discovered and named for a Dutch scientist, Johannes Diderik van der Waals (1937–1923), Nobel Prize 1910, who presented his conclusions in 1873. Weak intermolecular and interatomic forces that are electrostatic in origin. If two molecules have permanent dipole moments and are in random thermal motion then some of their relative orientations cause repulsion and some attraction. On average there will be a net attraction. A molecule with a permanent dipole can also induce a dipole in a neighboring molecule and cause mutual attraction. These dipole-dipole and dipole-induced dipole interactions cannot occur between atoms. The van der Waals forces between single atoms arise because of small instantaneous dipole moments in the atoms themselves, resulting from their fluctuating electronic distribution. This instantaneous dipole can polarize a neighbouring atom, producing the weak attraction. Van der Waals forces are responsible for departures from ideal gas behavior in real gases.

varindor Inductor in which the inductance varies markedly with the current in the winding.

VHSIC program An R&D program begun by the Department of Defense in 1979 to develop advanced integrated circuits for military systems.

VLSI program Joint government-industry R&D effort in Japan for developing very large-scale integrated circuits (VLSI), in existence from 1976 to 1980.

volt The unit of potential difference or electromotive force in the meter-kilogram-second system, equal to the potential difference between two points for which 1 coulomb of electricity will do 1 joule of work in going from one point to the other. Symbolized V.

voltage The difference in electrical potential. Measured in volts.

voltage coefficient For a resistor whose resistance varies with voltage, the ratio of the fractional change in resistance to the change in voltage.

voltage drop The voltage between any two specified points of an electrical conductor is equal to the product of the current in amperes and the resistance in ohms between the two points (for direct current) or the product of the current in amperes and the impedance in ohms between the two points (for alternating current). In the case of alternating current, the product of the current and the resistance gives the resistance drop which is in phase with the current, whereas the product of the current and the reactance gives the reactance drop which is in quadrature with the current.

voltage standard Many countries now maintain a voltage standard in terms of the voltage generated across a low-temperature superconducting Josephson junction irradiated by microwaves at a precise frequency. This standard could be more cheaply maintained and more widely available with no significant loss of accuracy by operating at 77 K with the new materials.

volt-ampere The unit of apparent power; it is equal to the apparent power in a circuit when the product of the root-mean-square value of the voltage, expressed in volts, and the root-mean-square value of the current, expressed in amperes, equals 1. Abbreviated VA.

volt-ampere reactive The unit of reactive power; it is equal to the reactive power in a circuit carrying a sinusoidal current when the product of the root-mean-square value of the voltage, expressed in volts, by the root-mean-square value of the current, expressed in amperes, and by the sine of the phase angle between the voltage and the current, equals 1. Abbreviated var. Also known as reactive volt-ampere.

vortices Elementary matter in constant vertical turning or whirling motion, arranged in hypothetical rings.

W

warm superconductors Semiconducting materials that carry current with no loss of power and that do so at reasonably high temperatures. Research has demonstrated superconductors that in 1987 worked at 125 kelvin.

watt The Standard International unit of power (mechanical, thermal, and electrical), defined as the power resulting from the dissipation of one joule of energy in one second. In electrical circuits one watt is the product of one ampere and one volt.

weak force A feeble, short range force that affects all particles, both quarks and leptons. It is responsible for the decay of many particles.

Wiedemann-Franz law The ratio of the thermal to electrical conductivity of all pure metals at a given temperature is approximately constant. Except at very low temperatures the law is fairly well obeyed. The ratio is also proportional to thermodynamic temperature though this does not hold at very low temperatures.

winding Of an electrical machine, transformer, or other piece of apparatus. A complete group of insulated conductors designed to produce a magnetic field or to be acted upon by a magnetic field. A winding may consist of a number of separate conductors connected together electrically at their ends or may consist of a single conductor (wire or strip) which has been shaped or bent to form a number of loops or turns.

W^+, W^-, and Z^o The three massive bosons, discovered at CERN in 1983, that carry the weak force.

X

x-ray lithography Creation of patterns for fabricating integrated circuits using x-rays. Because x-rays have shorter wave lengths than visible light, they can produce finer patterns, hence denser circuits.

Y

yttrium The element yttrium is not one of the rare-earth metals, yttrium, m.p. 1509°C (2748°F), it is very similar to them and is always associated with them in such ores as monazite and gadolinite (Scandinavia, Idaho, Australia, Brazil). It forms compounds with oxygen and the halogen elements, and with carbon. The major use of the metal is as an additive or alloying ingredient with a number of other metals; additions of small percentages increase resistance to oxidation at high temperatures and have other beneficial effects upon the alloy systems to which it is added. Yttrium has only one stable form and is nontoxic. It is classed as a rare metal. Symbol Y, Atomic Wt. 88.9059, State - Solid, Valence 3, Group - IIIB, Atomic No. 39.

CHAPTER 6

ACRONYMS AND ABBREVIATIONS

AAAS	—	American Association for the Advancement of Science
AC	—	Alternating current
ACS	—	American Ceramics Society
ASM	—	American Society for Metals
BCS	—	Bardeen, Cooper, and Schrieffer Theory
CERN	—	The European Laboratory for Particle Physics, in Geneva, Switzerland
CVD	—	Chemical Vapor Deposition
DARPA	—	Defense Advanced Research Projects Agency
DC	—	Direct current
DESY	—	The German Electron Synchrotron Laboratory, in Hamburg, West Germany
DOE	—	Department of Energy
DSV	—	Diameter Spherical Volume
EPMF	—	European Powder Metallurgy Federation
ETL	—	Japan's Electrotechnical Laboratory
FCC	—	Japan's Fine Ceramics Center
Fermilab	—	The Fermi National Accelerator Laboratory, in Batavia, Illinois
GeV	—	One billion electron volts
H_c	—	Critical magnetic field

HFTF	—	High Field Test Facility
HTS	—	High-Temperature Superconductor
IEEE	—	Institute of Electrical and Electronics Engineers
ISR	—	International Society of Radiology
ISTEC	—	International Superconductivity Technology Center
J_c	—	Critical Current Density
JJ	—	Josephson Junction
JSAP	—	Japan Society of Applied Physics
KEK	—	The National High Energy Physics Laboratory of Japan, in Tsukuba, near Tokyo
LCTF	—	Large Coil Test Facility
LFPG	—	Laser-heated Pedestal Growth Method
linac	—	Linear Accelerator
LTS	—	Low-Temperature Superconductor
MagLev	—	Magnetic Levitation
MHD	—	Magnetohydrodynamics
MIT	—	Massachusetts Institute of Technology
MRI	—	Magnetic Resonance Imaging
MRS	—	Materials Research Society
NASA	—	National Aeronautics and Space Administration
NMR	—	Nuclear Magnetic Resonance
NSF	—	National Science Foundation
ONR	—	DOD Office of Naval Research
ORN	—	Oak Ridge National Laboratory
SAE	—	Society of Automotive Engineers
SAMPE	—	Society for the Advancement of Material and Process Engineering
SLAC	—	The Stanford Linear Accelerator Center, in Stanford, California.
SMES	—	Superconducting Magnetic Energy Storage
SQUID	—	Superconducting Quantum Interference Device
SSC	—	Superconducting Super Collider
T_c	—	Critical Temperature
TeV	—	One trillion electron volts, or one thousand GeV
THz	—	Terahertz
TMS	—	The Metallurgical Society

CHAPTER 7

PERIODICAL LITERATURE

A better guess at the savings from superconductors. Author, Emily T. Smith. *Business Week*, pg. 107, February 15, 1988.

Academies select panel to judge ideal SSC site. Author, Irwin Goodwin. *Physics Today*, Volume 40:52, August 1987.

Advances in superconductivity. *Electrical Construction and Maintenance*, Volume 86:16+, June 1987.

Advances in superconductivity challenge APS communications. *Physics Today*, Volume 40:82-3, May 1987.

Aerospace agencies foster research, application studies on superconductors. Author David Hughes. *Aviation Week & Space Technology*, Volume 127:89, November 23, 1987.

American Physical Society Meeting on Novel Materials and High-Temperature Superconductivity, New York City, March 18. *Science*, Volume 235:1571, March 27, 1987.

And the winner in the super atom smasher derby is . . . Author, Lois Therrien. *Business Week*, pg. 36, September 14, 1987.

A new compound joins the superconducting lineup. *Business Week*, pg. 77, February 29, 1988.

A new route to oxide superconductors. Author, Arthur L. Robinson. *Science*, Volume 236:1526, June 19, 1987.

Animals to supercomputers—what's new with patents. *Design News*, Volume 43:48, July 20, 1987.

Antiferromagnetism observed in La2Cu04 (research by David Johnston and others). Author, Arthur L. Robinson. *Science*, Volume 236:780, May 15, 1987.

A powder for testing conductivity. Author, William Stockton. *New York Times*, Volume 137:34, January 20, 1988.

Appearance of the superconducting phase transition of solders in magnetoresistance measurements. Authors, A.W. Buiji, H.M.A. Schleijpen, and F.A.P. Blom. *Journal of Applied Physics*, Volume 59:4108-12, June 15, 1986.

Applications of High-Temperature Superconductivity. Authors, A.P. Malozemoff, W.J. Gallagher, and R.F. Schwall. *Physics Today*, Special Issue: Superconductivity, March 1986.

Applied Superconductivity Conference and Exhibition: Baltimore, Maryland, September 29–October 3 (with program and list of exhibitors). *Cryogenics*, Volume 26:494-519, August/September 1986.

Applied superconductivity. Author, E.W. Collings. Reviewed by D. Dew-Hughes, *Cryogenics*, Volume 26:492-3, August/September 1986.

Assuring U.S. commercial leadership in superconductors. Author, Anthony Mavrinac. *Business America*, Volume 10:2, August 17, 1987.

A superconductivity dream come true. *Newsweek*, Volume 110:98, October 12, 1987.

Bednorz and Muller win Nobel Prize for new superconducting materials. Author, Anil Khurana. *Physics Today*, Volume 40:17, December 1987.

Berkeley lab marshals superconductor research in Bay Area. *Aviation Week & Space Technology*, Volume 127:92, November 23, 1987.

Big federal effort urged for superconductivity. *Chemical & Engineering News*, Volume 65:6, June 15, 1987.

Breakthroughs in superconductivity. Author, F.M. Mueller. *Journal of Metals*, Volume 39:6-8, May 1987.

Breakthroughs in superconductivity take a look into the future. Authors, Marla Walker and Peggy Davis. *Black Enterprise*, Volume 18:37, January 1988.

California is losing its high-tech edge. Author, Andrew Pollack. *New York Times*, Volume 137, Section 4, February 14, 1988.

The chemistry of superconductivity. Author, Janet Raloff. *Science News*, Volume 131:247, April 18, 1987.

This chipmaking technique may work for superconductors, too. *Business Week*, pg. 135, November 23, 1987.

Collision over the super collider. Author Gary Taubes. *Discover*, Volume 6:60, July 1985.

The coming of the SSC. Author, John G. Cramer. *Analog Science Fiction-Science Fact*, Volume 108:153, March 1988.

Computing superconductivity. Author, Sharon D. Stewart. *Simulation*, Volume 47:219-20, November 1986.

Condensed matter physics. Authors, Paul A. Fleury, Bertram Batlogg, Elihu Abrahams, Sebastian Doniach, Zachery Fisk, Myron B. Salamon, Daniel S. Fisher, R.A. Webb, and Stanislas Leibler. *Physics Today*, Volume 41:518, January 1988.

Covering superconductivity. Author, Irwin Goodwin. *Physics Today*, Volume 40:54, September 1987.

Critical Fields and Currents in Superconductors. Author, J. Bardenn. *Rev. Mod. Phys.*, Volume 34, p. 667, 1962.

Crystallography. Authors, Paula Fitzgerald and D.E. Cox. *Physics Today*, Volume 41:526, January 1988.

Cuomo and Moynihan feud over loss of project. Author, Jeffrey Schmaiz. New York Times, Volume 137:15, January 21, 1988.

Cuomo, under fire, drops bid for supercollider near Rochester. Author, Clifford D. May. *New York Times*, Volume 137:81, January 15, 1988.

Cuomo withdraws bid for supercollider. *New York Times*, Volume 137:11, January 15, 1988.

Current news about superconductors. Author, Dietrick E. Thomsen. *Science News*, Volume 131:308, May 16, 1987.

Dawn of the new stone age. Author Lowell Fonte. *Reader's Digest*, Volume 131:128, July 1987.

Detecting rust before it can be seen. Author, Otis Port. *Business Week*, pg. 107, July 20, 1987.

Developments in superconductivity—high T_c oxides. Author, Fred Fletcher. *Journal of Metals*, Volume 39:9-11, May 1987.

The discovery of a class of high-temperature superconductors. Authors, K. Alex Muller and J. Georg Bednorz. *Science*, Volume 237:1133, September 4, 1987.

DOE backs NAS choices for Super Collider site. Author, Mark Crawford. *Science*, Volume 239:458, January 29, 1988.

DOE submits 36 SSC site bids while House seeks to micro-manage project. Author, Irwin Goodwin. *Physics Today*, Volume 40:45, November 1987.

'DOE underreports accelerator costs.' Author, Ted Agres. *Research & Development*, Volume 28:39, July 1986.

Dreams into reality (high temperature superconductors that carry high currents). *Time*, Volume 129:61, May 25, 1987.

Effect of plastic deformation on the thermal conductivity of bismuth-thallium and bismuth-lead systems between 1.5 and 300K. Authors, C.K. Subramaniam and K.D. Chaudhuri. *Cryogenics*, Volume 26:628-33, November 1986.

Electrifying process. *U.S. News & World Report*, Volume 102:15, March 2, 1987.

Elementary particle physics. Authors, Eugene W. Beier, Robert N. Cahn, Roy F. Schwitters, D.R. Getz, R. Stiening, and A.I. Sanda. *Physics Today*, Volume 41:S33, January 1988.

Engine design held a breakthrough; motor is said to be the first applying superconductor. *New York Times*, Volume 137:7, January 2, 1988.

Enhancement of superconductivity in Pb-Sb system alloy filament produced by glass-coated metal spinning. Author, T. Goto. *Cryogenics*, Volume 26:346-51, June 1986.

Even lanthanum copper oxide is superconducting. Author, Anil Khurana. *Physics Today*, Volume 40:17, September 1987.

Excitement filled the air. Author, Maxim Karpinski. *Soviet Life*, pg. 23, October 1987.

Exit laughing (superconductors used in fishing tackle). Author Ed Zern. *Field & Stream*, Volume 92:138, June 1987.

Experimenting with 40 trillion electron-volts; it takes hundred physicists several years to design experimental detectors for the Superconducting Super Collider. Author, Dietrick E. Thomsen. *Science News*, Volume 132:314, November 14, 1987.

Fast-food physics. *Life*, Volume 10:38, September 1987.

The first superconducting motor has started to spin. *Business Week*, pg. 68, January 18, 1988.

Flexible A-15 superconducting tape via the amorphous state. Authors, Mireille Treuil Clapp and Donglu Shi. Journal of *Applied Physics*, Volume 57:4672-7, May 15, 1985.

Flying carpets could come true (levitation using superconductivity and magnetic fields; research by William Little). *USA Today* (Periodical), Volume 111:14-15, June 1983.

Frenzied hunt for the right stuff; Washington pushes for U.S. superconductor supremacy. Author, John Greenwald. *Time*, Volume 130:26, August 10, 1987.

Frictionless electricity (superconductive synthetic metals). Author, Alex Kozlov. *Science Digest*, Volume 94:28, April 1986.

Getting warmer .. . (high temperature superconductors). Author Michael Rogers. *Newsweek*, Volume 110:42-3, July 6, 1987.

Getting warmer; research in superconductivity posts more remarkable advances. Author, Tim M. Beardsley. Scientific American, Volume 257:32, October 1987.

Getting warmer . . . the race is on to find materials that superconduct at room temperature. Author, Michael Rogers. *Newsweek*, Volume 110:42, July 6, 1987.

Hardware hacker: superconductors for the hacker and more! Author, Don Lancaster. *Radio-Electronics*, Volume 59:73, February 1988.

Heavy-electron metals: new highly correlated states of matter. Authors, Z. Fisk, D.W. Hess, C.J. Pethick, D. Pines, J.L. Smith, J.D. Thompson, and J.O. Willis. *Science*, Volume 239:33, January 1, 1988.

Helium-3 researcher makes 1-mK measurements attainable. Author, T.A. Heppenheimer. *Research & Development*, Volume 29:62, December 1987.

A helping hand from Washington: how much is too much? Author, Evert Clark. *Business Week*, Volume 125, May 18, 1987.

Herrington backs Super Collider. Author, Mark Crawford. *Science*, Volume 234:1493, December 19, 1986.

High-field superconductivity. Authors, David Larbalestier, Gene Fisk, and Bruce Montgomery. *Physics Today*, Volume 39:24-33, March 1986.

High-powered discussions on high-temperature superconductivity. Author, Karen Hartley. *Science News*, Volume 132:359, December 5, 1987.

High Tc may not need phonons: supercurrents increase. Author, Anil Khurana. *Physics Today*, Volume 40:17, July 1987.

High-temperature superconductivity: what's here, what's near, and what's unclear. Author, Karen Hartley. *Science News*, Volume 132:106, August 15, 1987.

High-temperature superconductor hints. Author, Arthur L. Robinson. *Science*, Volume 236:1431, June 12, 1987.

High-temperature superconductivity in Y-Ba-Cu-O: identification of a copper-rich superconducting phase. Authors, Angelica M. Stacy, John V. Badding, and Margret J. Geselbracht. *Journal of the American Chemical Society*, Volume 109:2528-30, April 15, 1987.

High Temperature Superconductors with T$_c$ over 30 K (Symposium, Anaheim, California, April 21–25). *Journal of Metals*, Volume 39:11, June 1987.

Hot questions in superconductivity (research by Paul C.W. Chu and others). Author, Stefi Weisburd. *Science News*, Volume 131:164-5, March 14, 1987.

Hot superconductors force a change in theory. *New Scientist*, Volume 114:37, June 18, 1987.

How to make your own superconductors. Author, Bruce Schechter. *Omni*, Volume 10:72, November 1987.

IBM reveals more superconductivity advances. *Journal of Metals*, Volume 39:5-6, June 1987.

IBM sets off the stampede to superconducting. Author, Anthony Ramirez. *Fortune*, Volume 117:39, January 4, 1988.

IBM's Zurich lab is "flower" in Europe. Author, David Dickson. *Science*, Volume 237:125, July 10, 1987.

Institute of Physics Low Temperature Conference, London, England, May 13, 1986. *Cryogenics*, Volume 26:569, October 1986.

International Conference on Materials and Mechanisms of Superconductivity (Ames, Iowa, May 29–31). *Cryogenics*, Volume 25:598, October 1985.

In the trenches of science. Author, James Gleick. New York *Times Magazine*, Volume 136:28, August 16, 1987.

Ironing the wrinkles out of hot superconductors. *Business Week*, pg. 110, December 21, 1987.

Is Nb3Si a 25 K superconductor? Author, D. Dew-Hughes. *Cryogenics*, Volume 26:660-4, December 1986.

Japan edges up in the superconductor race. Author, Bob Johnstone. *New Scientist*, Volume 114:22, April 16, 1987.

The Journal talks with Congressman Don Ritter (Task Force on High Technology and Competitiveness). *Journal of Metals*, Volume 39:60-1, June 1987.

Local foes hamper bid by Albany for collider. Author, Clifford D. May. *New York Times*, Volume 137, Section I:13, January 10, 1988.

Low-temperature magnetic and superconducting behavior of some magnetic impurity substituted Y9Co7 compounds. Authors, B.V.B. Sarkissian and J.L. Tholence. *Journal of applied Physics*, Volume 55 ptIIA:2025-7, March 15, 1984.

Lucky seven in SSC site competition. Author, Dietrick E. Thomsen. *Science News*, Volume 133:68, January 30, 1988.

The lure of subatomic violence. Author, Stanley N. Wellborn. *U.S. News & World Report*, Volume 102:78, April 27, 1987.

Magic trick moves out of the lab and into our everyday lives. Author, James S. Trefil. *Smithsonian*, Volume 15:78-82+, July 1984.

Magnetic fields induce superconductivity. Author, Bruce Schechter. *Physics Today*, Volume 38:21-3, December 1985.

Magnetoencephalography and epilepsy research. Authors, D.F. Rose, P.D. Smith, and S. Sato. *Science*, Volume 238:329, October 16, 1987.

Materials science. Author, Philip H. Abelson. *Science*, Volume 239:125, January 8, 1988.

Medical physics. Authors, Samuel J. Williamson, Lloyd Kaufman, and R.D. Hichwa. *Physics Today*, Volume 41:S52, January 1988.

The microstructure of high-critical current superconducting films. Authors, P. Chaudhari, F.K. LeGouss, and Armin Segmuller. *Science*, Volume 238:342, October 16, 1987.

Ministers ponder superconductor research. *New Scientist*, Volume 114:20, April 23, 1987.

Moment formation in solids. NATO Advanced Study Institute on Moment Formation in Solids; North Atlantic Treaty Organization/Scientific Affairs Division. *Plenum Press*, p. 336, 1984. ISBN: 0-306-41834-7

More superconducting surprises. *New Scientist*, Volume 114:37, June 4, 1987.

More superconductivity questions than answers. Author, Arthur L. Robinson. *Science*, Volume 231:248, July 17, 1987.

Motor uses new ceramics. *New York Times*, Volume 137:16(N), January 2, 1988.

National Academy of Sciences Seminars. Washington, D.C. *Chemical & Engineering News*, Volume 65:4, March 30, 1987.

A new electrical revolution (superconducting ceramics). Author, William D. Marbach. *Newsweek*, Volume 109:74, May 25, 1987.

Neutrons clarify superconductors: neutron scattering experiments reveal a two-dimensional antiferromagnetic behavior that is consistent with an electron spin model of high-temperature superconductors. Author, Arthur L. Robinson. *Science*, Volume 237:1115, September 4, 1987.

New evidence at Wayne State for superconductivity at 240 K. Author, Arthur L. Robinson. *Science*, Volume 236:28, April 3, 1987.

New heights in superconductivity. Author, Ivars Peterson. *Science News*, Volume 131:23, January 10, 1987.

New hunt for ideal energy storage system: a coil powering 'Star Wars' weapons could also help electric utility companies. Author, Malcolm W. Browne. *New York Times*, Volume 137:28(N), January 6, 1988.

New organic superconductor (BEDT-TTF). Author, Thomas H. Maugh, II. *Science*, Volume 226:37, October 5, 1984.

A new route to oxide superconductors. Author, Arthur L. Robinson. *Science*, Volume 236:1526, June 19, 1987.

The new superconductivity. *Scientific American*, Volume 256:32-3, June 1987.

New York drops out of SSC sweepstakes. Author, Mark Crawford. *Science*, Volume 239:347, January 22, 1988.

Nobel committee is quick to recognize superconductivity work. *Research & Development*, Volume 29:45, December 1987.

Nobel Prize in physics. Author, Phillip F. Schewe. *Physics Today*, Volume 41:S72, January 1988.

No resistance to superconductivity. Author, Karen Hartley. *Science News*, Volume 132:84, August 8, 1987.

Novel types of superconductivity in f-electron systems. Author, M. Brian. *Physics Today*, Volume 39:72-80, March 1986.

Nyaah-nyaah. Author, Charles C. Mann. *Omni*, Volume 9:27, September 1987.

Ohmic, superconducting, shallow AuGe/Nb contacts to GaAs. Authors, M. Gurvitch, A. Kastalsky, and S. Schwarz. *Journal of Applied Physics*, Volume 60:3204-10, November 1, 1986.

Organic superconductivity secret revealed. *Cryogenics*, Volume 24:167, March 1984.

Our life has changed. *Business Week*, Volume 94-100, April 6, 1987.

Oxygen isotope effect in high-temperature oxide superconductors. Authors, Hans-Conrad zur Loye, Kevin J. Leary, Steven W. Keller, William K. Ham, Tanya A. Faltens, James N. Michaels, and Angelica M. Stacy. *Science*, Volume 238:1558, December 11, 1987.

Particle physics for everybody. Author, Paul Davies. *Sky and Telescope*, Volume 74:582, December 1987.

Pentagon boosts research spending to develop practical superconductors. Author, David Hughes. *Aviation Week & Space Technology*, Volume 127:57, November 16, 1987.

Percolation, localization, and superconductivity. Reviewed by J.C. Garland. NATO Advanced Study Institute on Percolation, Localization, and Superconductivity. *American Scientist*, Volume 73:384-5, July/August 1985.

Physics: IBM duo wins for seminal work in superconducting ceramics. Author, John Horgan. *Scientific American*, Volume 257:46, December 1987.

Physics in industry. Authors, A.P. Malozemoff, L.A. Feldkamp, R.B. Dorshow, Robert L. Swofford, R. Rosenberg, D. Grischkowsky, M.B. Ketchen, J.-M. Halbout, and C.-C. Chi. *Physics Today*, Volume 41:S47, January 1988.

Piece de no resistance. Author, Michelle Citron. *Discover*, Volume 9:24, January 1988.

The pivotal role of materials. *Design News*, Volume 44:52, January 18, 1988.

Plan for atom smasher riles a town. Author, Philip S. Gutis. *New York Times*, Volume 137:81, January 12, 1988.

Possible evidence for superconducting layers in single crystal YBa2Cu307-x by field ion microscopy. Authors, A.J. Melmed, R.D. Shull, C.K. Chiang, and H.A. Fowier. *Science*, Volume 239:176, January 8, 1988.

Predicting new solids and superconductors. Author, Marvin L. Cohen. *Science*, Volume 234:549-53, October 31, 1986.

The President's superconductivity initiative. *Business America*, Volume 10:5, August 17, 1987.

Pressure to construct SSC builds in House. Author, Mark Crawford. *Science*, Volume 237:840, August 21, 1987.

Production and superconducting properties of laboratory scale Nb-Ti and Nb-Ti-Ta multifilamentary cables. Authors, E. Olzi, G. Rondelli, and P. Bruzzone. *Cryogenics*, Volume 24:115-18, March 1984.

Properties of superconducting NbTi superfine filament composites with diameters 0.1 m. Authors, I. Hlasnik, S. Takacs, and V.P. Vurjak. *Cryogenics*, Volume 25:558-65, October 1985.

Putting superconductors to work—superfast. Author, Emily T. Smith. **Business Week**, Volume 124-6, May 18, 1987.

The race for a high-temperature superconductor hots up. Author, Christine Sutton. *New Scientist*, Volume 113:33, January 29, 1987.

Race for the ring: DOE reacts to Congress's anxieties on SSC. Author, Irwin Goodwin. *Physics Today*, Volume 40:47, August 1987.

Ralph E. Gomory: IBM's chief scientist has led on the world's most productive R&D organizations. *Research & Development*, Volume 29:96, October 1987.

Reactive diffusion and superconductivity of Nb3A1 multilayer films. Authors, J.M. Vandenberg, M. Hong, and R.A. Hamm. *Journal of Applied Physics*, Volume 58:618-19, July 1, 1985.

Reagan hails new age of superconductivity at 'pep rally.' Author, Irwin Goodwin. *Physics Today*, Volume 40:51, September 1987.

Reagan proposes a bold initiative. Author, Charles L. Cohen. *Electronics*, Volume 60:33, August 6, 1987.

Reagan unveils 11-point plant to commercialize superconductors. *Research & Development*, Volume 29:57, September 1987.

Record high-temperature superconductors claimed. Author, Arthur L. Robinson. *Science*, Volume 235:531-3, January 30, 1987.

Researchers face daunting problems in bringing superconductors to market. Author, Skip Derra. *Research & Development*, Volume 29:57, September 1987.

Research on high-critical-temperature superconductivity in Japan. Authors, Shoji Tanaka and Sadao Nakajima. *Physics Today*, Volume 40:53, December 1987.

The resonating valence bond state in La2CuO4 and superconductivity. Author, Philip Warren Anderson. *Science*, Volume 235:1196-8, March 6, 1987.

Revolutionary superconductor foreseen (lanthanum-barium-copper-oxide system). *Machine Design*, Volume 59:16, February 12, 1987.

The Ross Priory Meeting on Superconducting electronics. Strathclyde, United Kingdom, June 11-14. *Cryogenics*, Volume 25:286, May 1985.

Science committee okays supercollider. Author, Mark Crawford. *Science*, Volume 238:477, October 23, 1987.

Seeking the perfect wire. Author, William J. Cook. *U.S. News & World Report*, Volume 102:66-71, May 11, 1987.

Self-biased Josephson junctions. Authors, O.B. Hyun and D.K. Finnemore. *Journal of Applied Physics*, Volume 60:1214-16, August 1, 1986.

Seven states are finalists for atom smasher. *New York Times*, Volume 137:17(N), January 20, 1988.

Signs of a new high in ceramic superconductivity. Author, Ivars Peterson. *Science News*, Volume 132:356, December 5, 1987.

A simple instrument for determining superconducting transition temperatures. Author, M.L. Norton. *Journal of Physics. E, Scientific Instruments*, Volume 19:268-70, April 1986.

Simple method to realize the zero-field transition temperature of a superconductive fixed point. Author, D. Hechtfischer. *Cryogenics*, Volume 26:665-6, October 1986.

Special issue: superconductivity 75th anniversary. *Physics Today*, Volume 39:22-62, 72-80, March 1986.

SSC: an iffy proposition in Congress. Author, Dietrick E. Thomsen. *Science News*, Volume 132:374, December 12, 1987.

SSC, fusion machine hit a roadblock. Author, Mark Crawford. *Science*, Volume 237:1559, September 25, 1987.

SSC sites narrowed. *Science News*, Volume 132:247, October 17, 1987.

SSC sites: then there were eight. Author, Mark Crawford. *Science*, Volume 239:133, January 8, 1988.

The stability behaviour of a Cu-stabilized NbTi-multifilamentary conductor under different cooling conditions. Author P. Turowski. *Cryogenics*, Volume 24:629-35, November 1984.

Standard Theory of superconductivity called into question. *Chemical & Engineering News*, Volume 65:11, May 11, 1987.

States race SSC site-proposal deadline. Author, Dietrick E. Thomsen. *Science News*, Volume 132:167, September 12, 1987.

Stumbling on superconductors. Author, Marjorie Sun. *Science*, Volume 237:477, July 31, 1987.

The Supercollider: big bang for big bucks. Author, Alex Kozlov. *Discover*, Volume 9:27, January 1988.

Supercollider faces budget barrier. Author, Mark Crawford. *Science*, Volume 236:246, April 17, 1987.

The super collider: not in the other guys backyard. *Business Week*, pg. 88, February 1, 1988.

Super collider: steps to reality. Author, Dietrick E. Thomsen. *Science News*, Volume 132:103, August 15, 1987.

The supercollider sweepstakes. Author, Eliot Marshall. *Science*, Volume 237:1288, September 11, 1987.

Superconducting critical-current densities of commercial multifilamentary Nb3Sn(Ti) wires made by the bronze process. Authors, M. Suenaga, K. Tsuchiya, and N. Higuchi. *Cryogenics*, Volume 25:123-8, March 1985.

Superconducting Magnets. Author, A.L. Luiten. *Phillips Tech. Rev.*, Volume 29:309-322, 1968.

Superconducting Magnets. Author, P.F. Chester. *Repts. Progr. Phys.*, Volume 30, Part II, p. 561, 1967.

Superconducting materials. Authors, Malcolm R. Beasley and Theodore H. Geballe. *Physics Today*, Volume 37:60-8, October 1984.

Superconducting pottery. Author, Elizabeth A. Thomson. *Technology Review*, Volume 91:11, January 1988.

Superconducting spray formulated. *Design News*, Volume 43:56, July 6, 1987.

Superconduction possible at room temperatures? *Radio-Electronics*, Volume 58:5, July 1987.

Superconductive barriers surpassed. Author, Stefi Weisburd. *Science News*, Volume 131:116-17, February 21, 1987.

The superconductive computer in your future. Author, Stephen G. Davis. *Datamation*, Volume 33:74, August 15, 1987.

Superconductivity. Author, D.E. Anderson. M.W. Lads, Philadelphia, Pennsylvania. *Magnetic Materials Digest*, pp. 196-217, 1964.

Superconductivity. Authors, T.H. Gaballe and B.T. Matthias. *Ann. Rev. Phys. Chem.*, Volume 14:141-160, 1964.

Superconductivity. Author, J. Volger. *Philips Tech. Rev.*, Volume 29:1-16, 1968.

Superconductivity. Author, Michael Tinkham. *Physics Today*, Volume 39:22-3, March 1986.

Superconductivity above liquid nitrogen temperature: preparation and properties of a family of perovskite-based superconductors. Authors, E.M. Engler, V.Y. Lee, and A.I. Nazzal. *Journal of the American Chemical Society*, Volume 109:2848-9, April 29, 1987.

Superconductivity Above the Transition Temperature. Author, R.E. Glover, III. Progr. in *Low Temp. Phys.*, Volume 6:291-332, 1970.

Superconductivity: ACS session highlights the chemistry. *Chemical & Engineering News*, Volume 65:4-5, April 13, 1987.

Superconductivity: a generation away. *Electrical World*, Volume 201:39, July 1987.

Superconductivity: a hard frost. Author, Dietrick E. Thomsen. *Science News*, Volume 313:215, April 4, 1987.

Superconductivity and magnetism in organic metals. Authors, Paul M. Chaikin and Richard L. Greene. *Physics Today*, Volume 39:24-32, May 1986.

Superconductivity and the Periodic System. Author, B.T. Matthias. *American Scientist*, Volume 58:80, 1970.

Superconductivity: a physics rush. Author, Dietrick E. Thomsen. *Science News*, Volume 131:196-7, March 28, 1987.

Superconductivity: a revolution beckons microelectronics. Author, Eric Brus. *Microwaves & RF*, Volume 26:35-8+, July 1987.

Superconductivity: a revolution in electricity is taking shape. Author, Ron Dagani. *Chemical & Engineering News*, Volume 65:7-16, May 11, 1987.

Superconductivity at 52.5 K in the lanthanum-barium-copper-oxide system. Authors, C.W. Chu, P.H. Hor, and R.L. Meng. *Science*, Volume 235:567-9, January 30, 1987.

Superconductivity at 40 K in the oxygen-defect perovskites La2-xSrxCuO4-y. Authors, J.M. Tarascon, L.H. Greene, and W.R. McKinnon. *Science*, Volume 235:1373-6, March 13, 1987.

Superconductivity: betwixt and between. Author, Dietrick E. Thomsen. *Science News*, Volume 125:212, April 7, 1984.

Superconductivity breakthroughs. Author, Brian C. Fenton. *Radio-Electronics*, Volume 59:43, February 1988.

Superconductivity consortium pursues aerospace applications. *Aviation Week & Space Technology*, Volume 127:59, November 16, 1987.

Superconductivity: current-carrying capacity soars. *Chemical & Engineering News*, Volume 65:4-5, May 18, 1987.

Superconductivity drive gets hotter every day. Author, Tobias Naegele. *Electronics*, Volume 60:49+, April 2, 1987.

Superconductivity: do all applications have to wait? Author, John Teresko. *Industry Week*, Volume 234:47, September 21, 1987.

Superconductivity finds work in industry. *Machine Design*, Volume 58:18, October 23, 1986.

Superconductivity glimpsed near 300 K. Author, Dietrick E. Thomsen. *Science News*, Volume 132:4, July 4, 1987.

A superconductivity happening. Author, Arthur L. Robinson. *Science*, Volume 235:1571, March 27, 1987.

Superconductivity heats up. Author, Michael D. Lemonick. *Time*, Volume 129:62, March 2, 1987.

Superconductivity Heats Up. Author, Sharon Begley, *Newsweek*, Volume 77, March 21, 1988.

Superconductivity: hype vs. reality. Author, Gina Maranto. *Discover*, Volume 8:22, August 1987.

Superconductivity in alkaline earth-substituted La2CuO4y. Authors, J. Georg Bednorz, K. Alex Muller, and Masaaki Takashige. *Science*, Volume 236:73-5, April 3, 1987.

Superconductivity in inhomogeneous (Mo,Rh)-B glasses. Authors, Klaus Zoltzer and Herbert C. Freyhardt. *Journal of Applied Physics*, Volume 58:1910-15, September 1, 1985.

Superconductivity in Semiconductors and Semi-Metals. Authors, J.K. Hulm, D.W. Ashkin, and C.K. Jones. *Prog. in Low Temp. Phys.*, Volume 6:205-242, 1970.

Superconductivity moves from laboratory to the classroom: with inexpensive kits, teachers can give demonstrations. Author, Malcolm W. Browne. *New York Times*, Volume 137:18(N), January 12, 1988.

Superconductivity, Nobel Prize winners highlight AVS '87. Author, Gary L. Parr. *Research & Development*, Volume 29:107, October 1987.

Superconductivity of alloy films of Mo and simple metal elements. Author, Yuji Asada. *Journal of Applied Physics*, Volume 58:3162-5, October 15, 1985.

Superconductivity on the road. Author, William C. Shumay, Jr. *Advanced Materials & Processes Incorporating Metal Progress*, Volume 131:35-6, May 1987.

Superconducting organic solids. Authors, K. Bechgaard and D. Jerome. *Chemtech*, Volume 15, 682-5, November 1985.

Superconductivity seen above the boiling point of nitrogen. Author, Anil Khurana. *Physics Today*, Volume 40:17-23, April 1987.

Superconductivity seen at record high temperatures in metal oxides. Author, Ron Dagani. *Chemical & Engineering News*, Volume 65:29-30, February 2, 1987.

Superconductivity shows promise in undersea warfare. Author, Larry L. Booda. *Sea Technology*, Volume 26:10-11+, November 1985.

The superconductivity story continues. *Journal of Metals*, Volume 39:4, March 1987.

Superconductivity: temperature war heats up. Author, Gail M. Robinson. *Design News*, Volume 43:24-5, May 25, 1987.

Superconductivity—the state that came in from the cold. Authors, T.H. Geballe and J.K. Hulm. *Science*, Volume 239:367, January 22, 1988.

Superconductivity war hots up. *Electronics and Power*, Volume 33:296, May 1987.

Superconductivity without BCS (heavy-fermion superconductivity). *Science News*, Volume 128:361, December 7, 1985.

Superconductor advance (synthetic metal). *Electrical Construction and Maintenance*, Volume 84:16+, November 1985.

Superconductor claim raised to 94 K. Author, Arthur L. Robinson. *Science*, Volume 235:1137-8, March 6, 1987.

Superconductorelativisticexpiatethedosage. Author, Frederic B. Jueneman. *Research & Development*, Volume 29:15, August 1987.

Superconductor frenzy. Author, Arthur Fisher. *Popular Science*, Volume 231:54, July 1987.

Superconductor race attracts MCC. Authors, J. Robert Lineback and Tobias Naegele. *Electronics*, Volume 60:33-4, June 25, 1987.

The superconductor race heats up. Author, Tobias Naegele. *Electronics*, Volume 60:37, January 22, 1987.

Superconductor race heats up. Author, Arthur L. Robinson. *Science*, Volume 236:664, May 8, 1987.

Superconductor R&D moves ahead on several world fronts. Author, Wesley R. Iversen. *Electronics*, Volume 60:54-6, February 19, 1987.

Superconductor research warms up. Author, Jeff Hecht. *New Scientist*, Volume 114:27, June 4, 1987.

Superconductors. Author, Michael D. Lemonick. *Readers Digest*, Volume 131:13, November 1987.

Superconductors. Author, Michael D. Lemonick. *Time*, Volume 129:64, May 11, 1987.

Superconductor's critical current at a new high. Author, M. Mitchell Waldrop. *Science*, Volume 238:1655, December 18, 1987.

Superconductors: early visions. Author, Tim Cole. *Popular Mechanics*, Volume 164:16, August 1987.

Superconductors: facing reality. Author, Tim Cole. *Popular Mechanics*, Volume 164:32, November 1987.

Superconductors gain. Author, Arthur Fisher. *Popular Science*, Volume 231:10, August 1987.

Superconductors get into business. Author, Anthony Ramirez. *Fortune*, Volume 115:114-16+, June 22, 1987.

Superconductors heat up. Author, Malcolm Gray. *Maclean's*, Volume 100:44, April 6, 1987.

Superconductors heat up. *Scientific American*, Volume 256:64-5, March 1987.

Superconductors' promise. *World Press Review*, Volume 34:54, September 1987.

Superconductor stability, 1983: a review. Author, L. Dresner. *Cryogenics*, Volume 24:283-92, June 1984.

Superconductors: The startling breakthrough that could change our world. Author, Michael D. Lemonick. *Time*, Volume 129:64, May 11, 1987.

Superelectricity. Author, Mel Ray. *Electrical World*, Volume 201:8, July 1987.

Super excitement: supercollider race is on, but funding is iffy. Author, William H. Miller. *Industry Week*, Volume 234:22, September 21, 1987.

Super promise. Author, Richard Sharpe. *Management Today*, pg. XII, September 1987.

Superscramble for a supercollider. *U.S. News & World Report*, Volume 103:ii, September 14, 1987.

'Super' science squeezes 'small' science. Author, Philip W. Anderson. *New York Times*, Volume 137:17(N), February 8, 1988.

Surface properties of metal-nitride and metal-carbide films deposited on Nb for radio-frequency superconductivity. Authors, E.L. Garwin, F.K. King, and R.E. Kirby. *Journal of Applied Physics*, Volume 61:1145-54, February 1, 1987.

Synthesis, structure, and physical properties of a novel tetratellurafulvalene electron donor. Authors, Knud Lerstrup, Dwaine O. Cowan, and Thomas J. Kistenmacher. *Journal of the American Chemical Society*, Volume 106:8303-4, December 26, 1984.

Technique cools disc gradually, until it becomes superconducting: entrepreneur produces 90K superconductor for sale to engineers-scientists. Author, Frank Yeaple. *Design News*, Volume 44:122, January 18, 1988.

The good news about U.S. R&D. Author, Stuart Gannes. *Fortune*, Volume 117:48, February 1, 1988.

The U.S. has the advances, but Japan may have the advantage. Author, Evert Clark. *Business Week*, Volume 97, April 6, 1987.

The world's biggest machine. Author, Arthur Fisher. *Popular Science*, Volume 230:56, June 1987.

The 1987 Nobel Prize for physics. Author, M. Mitchell Waldrop. *Science*, Volume 238:481, October 23, 1987.

Tired of grinding your own superconductors? *Business Week*, pg. 88, February 1, 1988.

U.S. must be aggressive in superconductor research. *Research & Development*, Volume 29:45, November 1987.

Vacuum physics. Authors, R.H. Hammond, Gary M. McClelland, and R.M. Osgood. *Physics Today*, Volume 41:S69, January 1988.

Venture capital's new gold rush. Author, Jonathan B. Levine. *Business Week*, pg. 66, October 5, 1987.

Warm superconductivity: have we even grasped the potential? Author, John Teresko. Industry Week, Volume 236:55, January 18, 1988.

Warm superconductors. Author, John G. Cramer. *Analog Science Fiction-Science Fact*, Volume 107:125, October 1987.

Warm superconductors could have big impact on many industries. *Research & Development*, Volume 29:60, September 1987.

What's new in the superconductivity business. Authors, David E. Gumpert and Stanley R. Rich. *New York Times*, Volume 137, Section 3, January 24, 1988.

When protons — and politics — collide: is the SSC a $4.4 billion 'quark barrel'? Author, William D. Marbach. *Newsweek*, Volume 110:44, July 6, 1987.

White House spotlights new superconductors: Reagan preaches on the virtues of pursuing high-temperature materials, unveils program to assist American industry. Author, Mark Crawford. *Science*, Volume 237:593, August 7, 1987.

Who'd have picked them for a Nobel a year ago? Author, Tobias Naegele. *Electronics*, Volume 60:12, October 29, 1987.

Will high-Tc superconductivity affect the SSC's design? Author, Irwin Goodwin. *Physics Today*, Volume 40:50, August 1987.

Will 1988 see a 92k superconductor IC? Author, Larry Waller. *Electronics*, Volume 61:32, January 7, 1988.

Yb or not Yb? That is the question (work of Paul Chu and others). Author, Gina Kolata. *Science*, Volume 236:663-4, May 8, 1987.

CHAPTER 8

JOURNAL PAPERS

Jun Akimitsu, Toshikazu Ekino, Hiroshi Sawa, Koukichi Tomimoto, Toshimasa Nakamichi, Masanori Oshiro, Yuji Matsubara, Hideo Fujiki, and Norihisa Kitamura. **High-T$_c$ Superconductivity of Y-Ba-Cu-O System.** *Jpn. J. Appl. Phys.*, Pt. 2, 26, L449 (1987).

P.W. Anderson. **The Resonating Valence Bond State in La$_2$CuO$_4$ and Superconductivity.** *Science*, 235, 1196 (1987).

B. Batlogg, A.P. Ramirez, R.J. Cava, R.B. van Dover, and E.A. Rietman. **Electronic properties of La$_{2-x}$Sr$_x$CuO$_4$ high-T$_c$ superconductors.** *Phys. Rev. B*, 35, 5340 (1987).

J. Georg Bednorz and K.A. Muller. **Possible High T$_c$ Superconductivity in the Ba-La-Cu-O System.** *Z. Phys.*, 64, 189 (1986).

J. Georg Bednorz, K. Alex Muller, and Masaaki Takashige. **Superconductivity in Alkaline-Earth Substituted La$_2$CuO$_{4-\delta}$.** *Science*, 236, 73 (1987).

J. Georg Bednorz, M. Takashige, and K.A. Muller. **Susceptibility Measurements Support High T$_c$ Superconductivity in the Ba-La-Cu-O System.** *Europhys. Letters*, 3, 379 (1987).

Kenichi Bushida, Makio Akimoto, Norio Kobayashi, Naoki Toyota, and Yoshio Muto. **Upper Critical Field of Sr$_x$La$_{2-x}$CuO$_{4-y}$.** *Jpn. J. Appl. Phys.*, Pt. 2, 26, L458 (1987).

D.W. Capone II, D.G. Hinks, J.D. Jorgensen, and K. Zhang. **Upper critical fields and high superconducting transition temperatures of La$_{1.85}$Sr$_{0.15}$CuO$_4$ and La$_{1.85}$Ba$_{0.15}$CuO$_4$.** *Appl. Phys. Lett.* 50, 543 (1987).

R.J. Cava, B. Batlogg, R.B. van Dover, D.W. Murphy, S. Sunshine, T. Siegrist, J.P. Remeika, E.A. Rietman, S. Zahurak, and G.P. Espinosa. **Bulk superconductivity at 91K in Single-Phase Oxygen-Deficient Perovskite $Ba_2YCu_3O_9$-delta.** *Phys. Rev. Lett.*, 58, 1676 (1987).

R.J. Cava, R.B. van Dover, B. Batlogg, and E.A. Rietman. **Bulk Superconductivity at 36K in $La_{1.8}Sr_{0.2}CuO_4$.** *Phys. Rev. Lett.*, 58, 408 (1987).

C.W. Chu, P.H. Hor, R.L. Meng, L. Gao, and Z.J. Huang. **Superconductivity at 52.5K in the Lanthanum-Barium-Copper-Oxide System.** *Science*, 235, 567 (1987).

C.W. Chu, P.H. Hor, R.L. Meng, L. Gao, Z.J. Huang, and Y.Q. Wang. **Evidence for Superconductivity above 40 K in the La-Ba-Cu-O Compound System.** *Phys. Rev. Lett.*, 58, 405 (1987).

A.H. Davies and R.J.D. Tilley. **New layer structures in the La-Cu-O system.** *Nature*, 326, 859 (1987).

Toshikazu Ekino and Jun Akimitsu. **Superconducting tunneling in Y-Ba-Cu-O System.** *Jpn. J. Appl. Phys.*, Pt. 2, 26, L452 (1987).

D.U. Gubser, R.A. Hein, S.H. Lawrence, M.S. Osofsky, D.J. Schrodt, L.E. Toth, and S.A. Wolf. **Superconducting phase transitions in the La-M-Cu-O layered perovskite system, M=La,Ba,Sr, and Pb.** *Phys. Rev. B*, 35, 5350 (1987).

Tetsuya Hasegawa, Kohji Kishio, Makoto Aoki, Naoki Ooba, Koichi Kitazawa, Kazuo Fueki, Shinichi Uchida, and Shoji Tanaka. **High-T_c Superconductivity of $(La_{1-x}Sr_x)_2CuO_4$-Effect of Substitution of Foreign Ions for Cu and La on Super-conductivity.** *Jpn. J. Appl. Phys.*, Pt. 2, 26, L337 (1987).

Yasumasa Hasegawa and Hidetoshi Fukuyama. **Fermi Surface Instability of Quasi-Two-Dimensional Tight Binding Electrons: A Possible Phase Diagram of $(La_{1-x}M_x)_2CuO_4$.** *Jpn. J. Appl. Phys.*, Pt. 2, 26, L322 (1987).

D.K. Finnemore, R.N. Shelton, J.R. Clem, R.W. McCallum, H.C. Ku, R.E. McCarley, S.C. Chen, P. Klavins, and V.G. Kogan. **Magnetization of Superconducting Lanthanum Copper Oxides.** *Phys. Rev. B*, 35, 5319 (1987).

Toshizo Fujita, Yuji Aoki, Yoshiteru Maeno, Junji Sakurai, Hitoshi Fukuba, and Hiron-obu Fujii. **Crystallographic, Magnetic, and Superconductive Transitions in $(La_{1-x}B_{ax})_2CuO_{4-y}$.** *Jpn. J. Appl. Phys.*, Pt. 2, 26, L368 (1987).

Hidetoshi Fukuyama and Kei Yosida. **Critical Temperature of Superconductivity Caused by Strong Correlation.** *Jpn. J. Appl. Phys.*, Pt. 2, 26, L371 (1987).

C.E. Gough, M.S. Colclough, E.M. Forgan, R.G. Jordan, M. Keene, C.M. Muirhead, A.I.M. Rae, N. Thomas, J.S. Abell, and S. Sutton. **Flux quantization in a high-T_c superconductor.** *Nature*, 326, 855 (1987).

Takeshi Hatano, Akiyuki Matsushita, Keikichi Nakamura, Kinichi Honda, Takehiko Matsumoto, and Keiichi Ogawa. **Identification of Phases in High T, Oxide Super-conductor $Ba_{0.7}Y_{0.3}Cu_1O_x$.** *Jpn. J. Appl. Phys.*, Pt. 2, 26, L374 (1987).

Yoshikazu Hidaka, Youichi Enomoto, Minoru Suzuki, Migaku Oda, and Toshiaki Mu-rakami. **Anistropic Properties of Superconducting Single-Crystal $(La_{1-x}Sr_x)_2CuO_4$.** *Jpn. J. Appl. Phys.*, Pt. 2, 26, L377 (1987).

Masayuki Hirabayashi, Hideo Ihara, Norio Terada, Kiyoshi Senzaki, Kunihiko Hayashi, Shinya Waki, Keizo Murata, Madoka Tokumoto, and Youichi Kimura. **Structure and Superconductivity in a New Type of Oxygen Deficient Perovskites $Y_1Ba_2Cu_3O_7$.** *Jpn. J. Appl. Phys.*, Pt. 2, 26, L454 (1987).

Shinobu Hikami, Takashi Hirai and Seiichi Kagoshima. **High Transition Temperature Superconductor: Y Ba Cu Oxide.** *Jpn. J. Appl. Phys.*, Pt. 2, 26, L314 (1987).

Shinobu Hikami, Seiichi Kagoshima, Susumu Komiyama, Takashi Hirai, Hidetoshi Minami, and Taizo Masumi. **High T_c Magnetic Superconductor: Ho Ba Cu Oxide.** *Jpn. J. Appl. Phys.*, Pt. 2, 26, L347 (1987).

Yoshikiko Hirotsu, Sigemaro Nagakura, Yuzo Murata, Takaharu Nishihara, Masasuke Takata, and Tsutomu Yamashita. **Electron Diffraction and Microscopy of the Structure of La Ba(Sr) Cu Oxides at Room Temperature.** *Jpn. J. Appl. Phys.*, Pt. 2, 26, L380 (1987).

P.H. Hor, R. L. Meng, Y.Q. Wang. L. Gao, Z.J. Huang, J. Bechtold, K. Forster, and C.W. Chu. **Superconductivity above 90 K in the Square-Planar Compound System $ABa_2Cu_3O_{61x}$ with A Y,La,Nd,Sm,Eu,Gd,Ho,Er, and Lu.** *Phys. Rev. Lett.*, 58, 1891 (1987).

P.H. Hor, L. Gao, R.L. Meng, Z.J. Huang, Y.Q. Wang, K. Forster, J. Vassilious, C.W. Chu, M.K. Wu, J.R. Ashburn, and C.J. Torng. **High-Pressure Study of the New Y-Ba-Cu-O Superconducting compound System.** *Phys. Rev. Lett.*, 58, 911 (1987).

P.H. Hor, R.L. Meng, C.W. Chu, M.K. Wu, E. Zirngiebl, J.D. Thompson, and C.Y. Huang. **Switching phenomena in a new 90-K superconductor.** *Nature*, 326, 669 (1987).

Syoichi Hosoya, Shin-ichi Shamoto, Masashige Onoda, and Masatoshi Sato. **High-T_c Superconductivity in New Oxide Systems II.** *Jpn. J. Appl. Phys.*, Pt. 2, 26, L456 (1987).

Shoichi Hosoya, Shin-ichi Shamoto, Masashige Onoda, and Masatoshi Sato. **High-T_c Superconductivity in New Oxide Systems.** *Jpn. J. Appl. Phys.*, Pt. 2, 26, L325 (1987).

Ienari Iguchi, Hirohito Watanabe, Yuji Kasai, Takashi Mochiku, Akimitsu Sugishita, Shun-ichi Narumi, and Eiso Yamaka. **Superconductivity of Y-Ba-Cu-O Compound above Liquid Nitrogen Temperature.** *Jpn. J. Appl. Phys.*, Pt. 2, 26 L327 (1987).

Hideo Ihara, Masayuki Hirabayashi, Norio Terada, Yoichi Kimura, Kiyoshi Senzaki, Makio Akimoto, Kenichi Bushida, Fumiyuki Kawashima, and Ryuichi Uzuka. **Electronic Structures and Superconducting Mechanisms of $Ba_2Y_1Cu_3O_7$.** *Jpn. J. Appl. Phys.*, Pt. 2, 26, L460 (1987).

Hideo Ihara, Masayuki Hirabayashi, Norio Terada, Yoichi Kimura, Kiyoshi Senzaki, and Madoka Tokumoto. **Superconductivity and Electronic Structure of $Sr_xLa_{2-x}CuO_{4-y}$ Prepared under Reducing Condition.** *Jpn. J. Appl. Phys.*, Pt. 2, 26, L463 (1987).

Toshiaki Iwazumi, Ryozo Yoshizaki, Hideaki Sawada, Hiroaki Hayashi, Hiroshi Ikeda, and Etsuyuki Matsuura. **Weak Flux-Pinning Effect between 230 K and 40 K in $La_{1.8}Sr_{0.2}CuO_4$.** *Jpn. J. Appl. Phys.*, Pt. 2, 26, L383 (1987).

Toshiaki Iwazumi, Ryozo Yoshizaki, Hideaki Sawada, Hiromoto Uwe, Tunetaro Sakudo, and Etsuyuki Matsuura. **Preparation and Property of $La_{1.85}Sr_{0.15}CuO_4$ Single Crystal.** *Jpn. J. Appl. Phys.*, Pt. 2, 26, L386 (1987).

J.D. Jorgenson, H.-B. Schuttler, D.G. Hinks, D.W. Capone II, K. Zhang, and M.B. Brodsky. **Lattice Instability and High-T_c Superconductivity in $La_{2-x}Ba_xCuO_4$.** *Phys. Rev. Lett.*, 58, 1024 (1987).

Seiichi Kagoshima, Shinobu Hikami, Yoshio Nogami, Takashi Hirai, and Koichi Kubo. **Characterization of the High-T_c Superconductor Y-Ba-Cu Oxide-Critical Current and Crystal Structure.** *Jpn. J. Appl. Phys.*, Pt. 2, 26, L318 (1987).

Seiichi Kagoshima, Kei-ichi Koga, Hiroshi Yasuoka, Yoshio Nogami, Koichi Kubo, and Shinobu Hikami. **Magnetic and Structural Properties of the High-T_c Superconductor Ho-Ba-Cu Oxide.** *Jpn. J. Appl. Phys.*, Pt. 2, 26, L355 (1987).

Masashi Kawasaki, Makoto Funabashi, Shunroh Nagata, Kazuo Fueki, and Hideomi Koinuma. **Compositional and Structural Analyses for Optimizing the Preparation Conditions of Superconducting $(La_{1-x}Sr_x)_yCuO_{4-delta}$ Films by Sputtering.** *Jpn. J. Appl. Phys.*, Pt. 2, 26, L388 (1987).

Kohji Kishio, Koichi Kitazawa, Shinsaku Kanbe, Isamu Yasuda, Nobuyuki Sugii, Hidenori Takagi, Shin-ichi Uchida, Kazuo Fueki, and Shoji Tanaka. **New High Temperature Superconducting Oxides, $(La_{1-x}Sr_x)_yCuO_{4-delta}$.** *Chem. Lett.*, (Feb. 5), 429 (1987).

Kohji Kishio, Koichi Kitazawa, Tetsuya Hasegawa, Makoto Aoki, Kazuo Fueki, Shin-ichi Uchida, and Shoji Tanaka. **Effect of Lanthanide Ion Substitutions for Lanthanum Sites on Superconductivity of $(La_{1-x}Sr_x)_yCuO_{4-delta}$.** *Jpn. J. Appl. Phys.*, Pt. 2, 26, L391 (1987).

Kohji Kishio, Nobuyuki Sugii, Koichi Kitazawa, and Kazuo Fueki. **Effect of Residual Water on Superconductivity in $(La_{1-x}Sr_x)_yCuO_{4-delta}$.** *Jpn. J. Appl. Phys.*, Pt. 2, 26, L466 (1987).

Yasuyuki Kitano, Kouichi Kifune, Ichiro Mukouda, Hirohiko Kamimura, Junji Sakurai, Yukitomo Komura, Kenichi Hoshino, Morihisa Suzuki, Asao Minami, Yoshiteru Maeno, Masatune Kato, and Toshizou Fujita. **Analysis of Crystal Structure of an Y-Ba-Cu Oxide.** *Jpn. J. Appl. Phys.*, Pt. 2, 26, L394 (1987).

Yoshio Kitaoka, Shigeru Hiramatsu, Takao Kohara, Kunisuke Asayama, Katsuyoshi Oh-ishi, Masae Kikuchi, and Norio Kobayashi. **[139]La Pure Quadrupole Resonance of High T_c Superconducting Materials $(La_{1-x}M_x)_2CuO_4$ (M = Ba,Sr).** *Jpn. J. Appl. Phys.*, Pt. 2, 26, L397 (1987).

Koichi Kitazawa, Kohji Kishio, Hidenori Takagi, Tetsuya Hasegawa, Shinsaku Kanbe, Shin-ichi Uchida, Shoji Tanaka, and Kazuo Fueki. **Superconductivity at 95 K in the New Yb-Ba-Cu Oxide System.** *Jpn. J. Appl. Phys.*, Pt. 2, 26, L339 (1987).

Koichi Kitazawa, Masayuki Sakai, Shin-ichi Uchida, Hidenori Takagi, Kohji Kishio, Shinsaku Kanbe, Shoji Tanaka, and Kazuo Fueki. **Specific Heat and Superconductivity in $(La_{1-x}Ca_x)_2CuO_4$.** *Jpn. J. Appl. Phys.*, Pt. 2, 26, L342 (1987).

Norio Kobayashi, Takaaki Sasaoka, Katsuyoshi Ohishi, Takako Sasaki, Masae Kikuchi, Akihiko endo, Kunio Matsuzaki, Akihisa Inoue, Koshichi Noto, Yasuhiko Syono, Yoshitami Saito, Tsuyoshi Masumoto, and Yoshio Muto. **Upper Critical Fields on High Temperature Superconductivity in La-Sr-Cu-O System.** *Jpn. J. Appl. Phys.*, Pt. 2, 26, L258 (1987).

Hideomi Koinuma, Takuya Hashimoto, Masashi Kawasaki, and Kazuo Fueki. **Preparation of $(La_{1-x}Sr_x)_2CuO_{4-delta}$ Superconducting Films by Screen Printing Method.** *Jpn. J. Appl. Phys.*, Pt. 2, 26, L399 (1987).

Makio Kurisu, Hideoki Kadomatsu, Hiroshi Fujiwara, Yoshiteru Maeno, and Toshizo Fujita. **Effect of Hydrostatic Pressure on the Superconducting Transition Temperature of $(La_{1-x}B_{ax})_2CuO_{4-y}$ with x = 0.075.** *Jpn. J. Appl. Phys.*, Pt. 2, 26, L361 (1987).

W. Kwestroo, H.A.M. van Hal, and C. Langereis. **Compounds in the System $BaO-Y_2O_3$.** *Mat. Res. Bull.*, 9, 1631 (1974).

W.K. Kwok, G.W. Crabtree, D.G. Hinks, D.W. Capone, J.D. Jorgensen, and K. Zhang. **Normal- and superconducting-state properties of $La_{1.85}Sr_{0.15}CuO_4$.** *Phys. Rev. B*, 35, 5343 (1987).

Lin Li, Bairu Zhao, Yong Lu, Huisheng Wang, Yuying Zhao, and Yinhuan Shi. **Superconductivity of Sr-La-Cu-O Thin Films.** *Chin. Phys. Lett.*, 4, 233, (1987).

Sadamichi Maekawa, Hiromichi Ebisawa, and Yoshimasa Isawa. **Far-Infrared Absorption to Test Anisotropy of Energy Gap in High T_c Superconducting Oxides.** *Jpn. J. Appl. Phys.*, Pt. 2, 26, L468 (1987).

Yoshiteru Maeno, Masatsune Kato, and Toshizo Fujita. **Preparation of Y-Ba-Cu Oxides with Superconducting Transition above Liquid-Nitrogen Temperature.** *Jpn. J. Appl. Phys.*, Pt. 2, 26, L329 (1987).

Yoshiteru Maeno, Yuji Aoki, Hirohiko Kamimura, Junji Sakurai, and Toshizo Fujita. **Transport Properties and Specific Heat of $(La_{1-x}Ba_x)_2CuO_{4-y}$.** *Jpn. J. Appl. Phys.*, Pt. 2, 26, L402 (1987).

A.P. Malozemoff, W.J. Gallagher, and R.E. Schwall. **Applications of High-Temperature Superconductivity.** *Physics Today*, Special Issue: Superconductivity, (March 1986).

Atsushi Masaki, Hisashi Sato, Shin-ichi Uchida, Koichi Kitazawa, Shoji Tanaka, and Kazuhiko Inoue. **Phonons in $(La_{1-x}Sr_x)_2CuO_4$ and $Bapb_{1-x}Bi_xO_2$.** *Jpn. J. Appl. Phys.*, Pt. 2, 26, L405 (1987).

Akiyuki Matsushita, Takeshi Hatano, Takehiko Matsumoto, Haruyoshi Aoki, Yuji Asada, Keikichi Nakamura, Kinichi Honda, Tamio Oguchi, and Keiichi Ogawa. **High-T_c Superconductor $Ba_{0.5}Y_{0.5}Cu_1O_x$.** *Jpn. J. Appl. Phys.*, Pt. 2, 26, L332 (1987).

Tamifusa Matsuura and Kazumasa Miyake. **Local-Phonon Model of High T_c Superconductor Based on Layered Perovskite Structure.** *Jpn. J. Appl. Phys.*, Pt. 2, 26, L407 (1987).

Kunio Matsuzaki, Akihisa Inoue, Hisamichi Kimura, Keiji Moroishi, and Tsuyoshi Masumoto. **Preparation of a High-T_c Superconductor by Oxidization of an Amorphous $La_{1.8}Sr_{0.2}Cu$ Alloy ribbon in Air.** *Jpn. J. Appl. Phys.*, Pt. 2, 26, L334 (1987).

L.F. Mattheiss. **Electronic Band Properties and Superconductivity in $La_{2-y}X_yCuO_4$.** *Phys. Rev. Lett.*, 58, 1028 (1987).

Hidetoshi Minami, Taizo Masumi, and Shinobu Hikami. **Direct Evidence of Superconductivity of Y-Ba-Cu Oxides—The Meissner Effect—.** *Jpn. J. Appl. Phys.*, Pt. 2, 26, L345 (1987).

Kazuyuki Moriwaki, Youichi Enomoto, and Toshiaki Murakami. **Josephson Junctions Observed in $La_{1.8}Sr_{0.2}CuO_4$ Superconducting Polycrystalline-films.** *Jpn. J. Appl. Phys.*, Pt. 2, 26, L521 (1987).

K.A. Muller, M. Takashige, and J.G. Bednorz. **Flux Trapping and Superconductive Glass State in $La_2CuO_{4-y}Ba$.** *Phys. Rev. Lett.*, 58, 1143 (1987).

Keizo Murata, Hideo Ihara, Madoka Tokumoto, Masayuki Hirabayashi, Norio Terada, Kiyoshi Senzaki, and Yoichi Kimura. **Pressure Dependence of Superconductivity of the 90 K-Superconductors, Ba-Y-Cu-O.** *Jpn. J. Appl. Phys.*, Pt. 2, 26, L471 (1987).

Keizo Murata, Hideo Ihara, Madoka Tokumoto, Masayuki Hirabayashi, Norio Terada, Kiyoshi Senzaki, and Yoichi Kimura. **Upper Critical Fields of the 40K- and 90 K-Superconductors Sr-La-Cu-O and Ba-Y-Cu-O.** *Jpn. J. Appl. Phys.*, Pt. 2, 26, L473 (1987).

D.W. Murphy, S. Sunshine, R.B. van Dover, R.J. Cava, B. Batlog, S.M. Zahurak, and L.F. Schneemeyer. **New Superconducting Cuprate Perovskites.** *Phys. Rev. Lett.*, 58, 1888 (1987).

Keigo Nagasaka, Masayuki Sato, Hideo Ihara, Madoka Tokumoto, Masayuki Hirabayashi, Norio Terada, Kiyoshi Senzaki, and Yoichi Kimura. **Far Infrared Measurements in High-T_c Superconductor $Sr_xLa_{2-x}CuO_{4-y}$.** *Jpn. J. Appl. Phys.*, Pt. 2, 26, L479 (1987).

Shunroh Nagata, Masashi Kawasaki, Makoto Funabashi, Kazuo Fueki, and Hideomi Koinuma. **High T_c Thin Films of $(La_{1-x}M_x)_yCuO_{4-delta}$ (M = Sr,Ba,Ca) Prepared by Sputtering.** *Jpn. J. Appl. Phys.*, Pt. 2, 26, L410 (1987).

Koichi Nakao, Noboru Miura, Shin-ichi Uchida, Hidenori Takagi, Shoji Tanaka, Kohji Kishio, Junichi Shimoyama, Koichi Kitazawa, and Kazuo Fueki. **High Field Measurement of the Critical Field in $(La_{1-x}Sr_x)_2CuO_4$ up to 40 T.** *Jpn. J. Appl. Phys.*, Pt. 2, 26, L413 (1987).

Yasukage Oda, Ichiroh Nakada, Takao Kohara, Hiroshi Fujita, Tetuyuki Kaneko, Haruhisa Toyoda, Eiji Sakagami, and Kunisuke Asayama. **AC Susceptibility of Superconducting La-Sr-Cu-O System.** *Jpn. J. Appl. Phys.*, Pt. 2, 26, L481 (1987).

Tamio Oguchi. **Electronic Property of La_2CuO_4 with Two Different Layered Structures.** *Jpn. J. Appl. Phys.*, Pt. 2, 26, L417 (1987).

Kohji Ohbayashi, Norio Ogita, Masayuki Udagawa, Yuji Aoki, Yoshiteru Maeno, and Toshizo Fujita. **Temperature Dependence of Infrared Spectra of $(La_{1-x}Ba_x)_2CuO_{4-y}$.** *Jpn. J. Appl. Phys.*, Pt. 2, 26, L423 (1987).

Kohji Ohbayashi, Norio Ogita, Masayuki Udagawa, Yuji Aoki, Yoshiteru Maeno, and Toshizo Fujita. **Infrared and Raman Spectroscopy of High T_c Superconducting System $(La_{1-x}Ba_x)_2CuO_4$.** *Jpn. J. Appl. Phys.*, Pt. 2, 26, L420 (1987).

Katsuyoshi Oh-ishi, Masae Kikuchi, Yasuhiko Syono, Kenji Hiraga, and Yoshiyuki Morioka. **Crystallochemical Studies on the High Temperature Superconductors** $La_{2-x}M_xCuO_{4-y}$ **(M = Ba and Sr).** *Jpn. J. Appl. Phys.*, Pt. 2, 26, L484 (1987).

Masashige Onoda, Shinichi Shamoto, Masatoshi Sato, and Syoichi Hosoya. **Crystal Structures of** $(La_{1-x}M_x)_2CuO_{4-delta}$ **(M = Sr and Ba).** *Jpn. J. Appl. Phys.*, Pt. 2, 26, L363 (1987).

T.P. Orlando, K.A. Delin, S. Foner, E.J. McNiff, Jr., J.M. Tarascon, L.H. Green, W.R. McKinnon, and G.W. Hull. **Upper critical fields of high-T_c superconducting** $Y_{2-x}Ba_xCuO_{4-y}$**: Possibility of 140 tesla.** *Phys. Rev. B*, 35 5347 (1987).

Hiroyuki Oyanagi, Hideo Ihara, Tadashi Matsushita, Madoka Tokumoto, Masayuki Hirabayashi, Norio Terada, Kiyoshi Senzaki, Yoichi Kimura, and Takafumi Yao. **X-Ray Absorption Spectra of High T_c Superconducting** $Sr_xLa_{2-x}CuO_{4-y}$. *Jpn. J. Appl. Phys.*, Pt. 2, 26, L488 (1987).

C. Politis, J. Geerk, M. Dietrich, and B. Obst. **Superconductivity at 40K in** $La_{1.8}Sr_{0.2}CuO_4$. *Z. Phys. B - Condensed Matter*, 66, 141 (1987).

C. Politis, J. Geerk, M. Dietrich, B. Obst, and H.L. Luo. **Superconductivity above 100 K in Multi-Phase Y-Ba-Cu-O.** *Z. Phys. B - Condensed Matter*, 66, 279 (1987).

C.N.R. Rao, P. Ganguly, A.K. Raychaudhuri, R.A. Mohan Ram, and K. Sreedhar. **Identification of the phase responsible for high-temperature superconductivity in Y-Ba-Cu oxides.** *Nature*, 326, 856 (1987).

F.S. Razavi, F.P. Koffyberg, and B. Mitrovic. **Magnetic properties of a superconducting Ba-La-Cu-oxide.** *Phys. Rev. B*, 35, 5323 (1987).

Yoshitami Saito, Takashi Noji, Akihiko Endo, Nozomu Matsuzaki, Masao Katsumata, and Naoaki Higuchi. **Superconducting Properties in Sr-La-Cu-O System.** *Jpn. J. Appl. Phys.*, Pt. 2, 26, L491 (1987).

Yoshitami Saito, Takashi Noji, Akihiko Endo, Nozomu Matsuzaki, Masao Katsumata, and Naoaki Higuchi. **Properties of the Superconductor (Sr,Ca)-La-Cu-O System in Various Sintering Atmosphere.** *Jpn. J. Appl. Phys.*, Pt. 2, 26, L366 (1987).

Hideaki Sawada, Yoshuke Saito, Toshiaki Iwazumi, Ryozo Yoshizaki, Yoshihito Abe, and Etsuyuki Matsuura. **Study of the Infrared Properties of** $(La_{1-x}Sr_x)_2CuO_4$. *Jpn. J. Appl. Phys.*, Pt. 2, 26, L426 (1987).

H.-B. Schuttler, M. Jarrell, and D.J. Scalapino. **Superconducting T_c Enhancement Due to Excitonic Negative-U Centers: A Monte Carlo Study.** *Phys. Rev. Lett.*, 58, 1147 (1987).

Z. Schlesinger, R.L. Greene, J. Bednorz, and K.A. Muller. **Far-infrared measurement of the energy gap of** $La_{1.8}Sr_{0.2}CuO_4$. *Phys. Rev. B*, 35, 5334 (1987).

Kouichi Semba, Shigeyuki Tsurumi, Makoto Hikita, Tsunekazu Iwata, Juichi Noda, and Susumu Kurihara. **Novel High-T_c Superconducting Phase of the Y-Ba-Cu-O Compound.** *Jpn. J. Appl. Phys.*, Pt. 2, 26, L429 (1987).

Shin-ichi Shamoto, Syoichi Hosoya, Masashige Onoda, and Masatoshi Sato. **Effect of Vacuum Annealing on the Superconducting Transition Temperature of La-Sr-Cu-O System.** *Jpn. J. Appl. Phys.*, Pt. 2, 26, L493 (1987).

Shelton, H.C. Ku, R.W. McCallum, and P. Klavins. **Superconductivity near 90 K in the Lu-Ba-Cu-O System.** *Phys. Rev. Lett.*, 58, 1885 (1987).

R. Somekh, M.G. Blamire, Z.H. Barber, K. Butler, J.H. James, G.W. Morris, E.J. Tomlinson, A.P. Schwarzenberger, W.M. Stobbs, and J.E. Evetts. **High superconducting transition temperatures in sputter-deposited YBaCuO thin films.** *Nature*, 326, 857 (1987).

Michael Stavola, R.J. Cava, and E.A. Rietman. **Evidence for a Peierls distortion in La_2CuO_4 from Vibrational Spectroscopy.** *Phys. Rev. Lett.*, 58, 1571 (1987).

Shunji Sugai, Masatoshi Sato, and Shoichi Hosoya. **Raman Study of High T_c Superconductors $(La_{1-x}Sr_x)_2CuO_4$.** *Jpn. J. Appl. Phys.*, Pt. 2, 26, L495 (1987).

P.E. Sulewski, A.J. Sievers, S.E. Russek, H.D. Hallen, D.K. Lathrop, and R.A. Buhrman. **Measurement of the superconducting energy gap in La-Ba-Cu oxide and La-Sr-Cu oxide.** *Phys. Rev. B*, 35, 5330 (1987).

J.Z. Sun, D.J. Webb, M. Naito, K. Char, M.R. Hahn, J.W.P. Hsu, A.D. Kent, D.B. Mitzi, B. Oh, M.R. Beasley, T.H. Geballe, R.H. Hammond, and A. Kapitulnik. **Superconductivity and Magnetism in the High-T_c Superconductor Y-Ba-Cu-O.** *Phys. Rev. Lett.*, 58, 1574 (1987).

Minoru Suzuki and Toshiaki Murakami. **Hall Effect in Superconducting $La_{1-x}Sr_x)_2CuO_4$ Single Crystal Thin Films.** *Jpn. J. Appl. Phys.*, Pt. 2, 26, L524 (1987).

Yasuhiko Syono, Masae Kikuchi, Katsuyoshi Ohishi, Kenji Hiraga, Hide Arai, Yoshito Matsui, Norio Kobayashi, Takaaki Sasaoka, and Yoshio Muto. **X-Ray and Electron Microscopic Study of a High Temperature Superconductor $Y_{0.4}Ba_{0.6}CuO_{222}$.** *Jpn. J. Appl. Phys.*, Pt. 2, 26, L498 (1987).

Setsuko Tajima, Shin-ichi Uchida, Shoji Tanaka, Shinsaku Kanbe, Koichi Kitazawa, and Kazuo Fueki. **Plasma Spectra of the New Superconductors $(La_{1-x}A_x)_2CuO_4$.** *Jpn. J. Appl. Phys.*, Pt. 2, 26, L432 (1987).

Toshiro Takabatake, Hiroyuki Takeya, Yasuhiro Nakazawa, and Masayasu Ishikawa. **A Study on the Superconductivity in $Ba_{1-x}Y_xCuO_{3-z}$.** *Jpn. J. Appl. Phys.*, Pt. 2, 26, L502 (1987).

Hidenori Takagi, Shin-ichi Uchida, Kohji Kishio, Koichi Kitazawa, Kazuo Fueki, and Shoji Tanaka. **High-T_c Superconductivity and Diamagnetism of Y-Ba-Cu Oxides.** *Jpn. J. Appl. Phys.*, Pt. 2, 26, L320 (1987).

Hidenori Takagi, Shin-ichi Uchida, Haruhiko Obara, Kohji Kishio, Koichi Kitazawa, Kazuo Fueki, and Shoji Tanaka. **Magnetic Susceptibility of High-T_c Superconducting Oxides $(La.A)_2CuO_4$ (A = Ba,Sr).** *Jpn. J. Appl. Phys.*, Pt. 2, 26, L434 (1987).

Hiroki Takahashi, Chizuko Murayama, Shusuke Yomo, Nobuo Mori, Kohji Kishio, Koichi Kitazawa, and Kazuo Fueki. **Pressure Effect on the Lattice Constant and Compressibility of a Superconductor $(La_{0.9}Sr_{0.1})2CuO_{4-y}$.** *Jpn. J. Appl. Phys.*, Pt. 2, 26, L504 (1987).

Takashi Takahashi, Fumihiko Maeda, Shoichi Hosoya, and Masatoshi Sato. **Ultraviolet Photoemission Study of High-T_c Superconductor $(La_{1-x}Sr_x)CuO_{4-delta}$.** *Jpn. J. Appl. Phys.*, Pt. 2, 26, L349 (1987).

Eiji Takayama-Muromachi, Yoshishige Uchida, Yoshio Matsui, and Katsuo Kato. **Identification of the High T_c Superconductor in the System Y-Ba-Cu-O.** *Jpn. J. Appl. Phys.*, Pt. 2, 26, L476 (1987).

Katsuhiko Takegahara, Hisatomo Harima, and Akira Yanase. **Electronic Structure of La_2CuO_4 by APW Method.** *Jpn. J. Appl. Phys.*, Pt. 2, 26, L352 (1987).

Koki Takita, Takashi Ipposhi, Takashi Uchino, Tetsuo Gochou, and Kohzoh Masuda. **Superconductivity Transition in High-T_c Y-Ba-Cu-O Mixed-Phase System.** *Jpn. J. Appl. Phys.*, Pt. 2, 26, L506 (1987).

J.M. Tarascon, L.H. Greene, W.R. McKinnon, G.W. Hull, and T.H. Geballe. **Superconductivity at 40L ion the Oxygen-Defect Perovskites $La_{2-x}Sr_xCuO_{4-y}$.** *Science*, 235, 1373 (1987).

Norio Terada, Hideo Ihara, Masayuki Hirabayashi, Kiyoshi Senzaki, Youichi Kimura, Keizo Murata, and Madoka Tokumoto. **Deposition of $Sr_xLa_{2-x}CuO_{4-y}$ Thin Films by Sputtering.** *Jpn. J. Appl. Phys.*, Pt. 2, 26, L508 (1987).

Norio Terada, Hideo Ihara, Masayuki Hirabayashi, Kiyoshi Senzaki, Youichi Kimura, Keizo Murata, Madoka Tokumoto, Osamu Shimomura, and Takumi Kikegawa. **Lattice Parameter and Its Pressure Dependence of High T_c $Sr_xLa_{2-x}CuO_{4-y}$ (x approx. 0.6) Prepared under Reducing Condition.** *Jpn. J. Appl. Phys.*, Pt. 2, 26, L510 (1987).

Kiyoyuki Terakura, Hiroshi Ishida, Key Taeck Park, Akira Yanase, and Noriaki Hamada. **Electronic Origin of Distortion of Oxygen Octahedron in $(La_{1-x}M_x)_2CuO_4$ with M = Ca,Sr and Ba.** *Jpn. J. Appl. Phys.*, Pt. 2, 26, L512 (1987).

Madoka Tokumoto, Masayuki Hirabayashi, Hideo Ihara, Keizo Murata, Norio Terada, Kiyoshi Senzaki, and Yoichi Kimura. **Magnetization and Meissner Effect in the High T_c Superconductor $Ba_xY_{1-x}CuO_{3-y}$.** *Jpn. J. Appl. Phys.*, Pt. 2, 26, L517 (1987).

Madoka Tokumoto, Hideo Ihara, Keizo Murata, Masayuki Hirabayashi, Norio Terada, Kiyoshi Senzaki, and Yoichi Kimura. **Diamagnetic Shielding and Meissner Effect in the High-T_c Superconductor $Sr_xLa_{2-x}CuO_{4-y}$.** *Jpn. J. Appl. Phys.*, Pt. 2, 26, L515 (1987).

Masayoshi Tonouchi, Yasufumi Fujiwara, Sadamu Kita, Takeshi Kobayashi, Masasuke Takata, and Tsutomu Yamashita. **Hall Coefficient of La-Sr-Cu Oxide Superconducting Compound.** *Jpn. J. Appl. Phys.*, Pt. 2, 26, L519 (1987).

Shin-ichi Uchida, Hidenori Takagi, Hideo Ishii, Hiroshi Eisaki, Tomoaki Yabe, Setsuko Tajima, and Shoji Tanaka. **Transport Properties of $(La_{1-x}A_x)_2CuO_4$.** *Jpn. J. Appl. Phys.*, Pt. 2, 26 (1987).

Shin-ichi Uchida, Hidenori Takagi, Kohji Kishio, Koichi Kitazawa, Kazuo Fueki, and Shoji Tanaka. **Superconducting Properties of $(La_{1-x}A_x)_2CuO_4$.** *Jpn. J. Appl. Phys.*, Pt. 2, 26, L443 (1987).

Shin-ichi Uchida, Hidenori Takagi, Hironori Yanagisawa, Kohji Kishio, Koichi Kitazawa, Kazuo Fueki, and Shoji Tanaka. **Electric and Magnetic Properties of La_2CuO_4.** *Jpn. J. Appl. Phys.*, Pt. 2, 26, L445 (1987).

R.B. van Dover, R.J. Cava, B. Batlogg, and E.A. Rietman. **Composition-dependent superconductivity in $La_{2-x}Sr_xCuO_{4-delta}$.** *Phys. Rev. B*, 35, 5337 (1987).

U. Walter, M.S. Sherwin, A. Stacy, P.L. Richards, and A. Zettl. **Energy gap in the high-T_c superconductor $La_{1.85}Sr_0.15CuO_4$.** *Phys. Rev. B*, 35, 5327 (1987).

Hau H. Wang, Urs Geiser, R.J. Thorn, K. Douglas Carlson, Mark A. Beno, Marilyn R. Monaghan, Thomas J. Allen, Roger B. Proksch, Dan L. Stupka, W.K. Kwok, G.W. Crabtree, and Jack M. Williams. **Synthesis, Structure, and Superconductivity of Single Crystals of High-T_c $La_{1.85}Sr_{0.15}CuO_4$.** *Inorganic Chemistry*, 26, 1190 (1987).

Werner Weber. **Electron-Phonon Interaction in the New Superconductors $La_{2-x}(Ba,Sr)_xCuO_4$.** *Phys. Rev. Lett.*, 58, 1371 (1987.

M.K. Wu, J.R. Ashburn, C.J. Torng, P.H. Hor, R.L. Meng, L. Gao, Z.J. Huang, Y.Q. Wang, and C.W. Chu. **Superconductivity at 93 K in a New Mixed-Phase Y-Ba-Cu-O Compound System at Ambient Pressure.** *Phys. Rev. Lett.*, 58, 908 (1987).

Yasuo Yamaguchi, Hiroshi Yamauchi, Masayoshi Ohashi, Hisao Yamamoto, Nobuyuki Shimoda, Masae Kikuchi, and Yasuhiko Syono. **Observation of the Antiferromagnetic Ordering in La_2CuO_4.** *Jpn. J. Appl. Phys.*, Pt. 2, 26, L447 (1987).

Ryozo Yoshizaki, Toshiaki Iwazumi, Hideaki Sawada, and Etsuyuki Matsuura. **Magnetization Property of Annealed $La_{1.85}Sr_{0.15}CuO_4$ from 300 K to 5 K.** *Jpn. J. Appl. Phys.*, Pt. 2, 26, L316 (1987).

Ryozo Yoshizaki, Toshiaki Iwazumi, Hideaki Sawada, Hiroshi Ikeda, and Etsuyuki Matsuura. **Superconducting Properties of $La_{1.85}Sr_{0.15}CuO_4$ Made by Hot-Press and Sinter Methods.** *Jpn. J. Appl. Phys.*, Pt. 2, 26, L311 (1987).

Jaejun Yu, A.J. Freeman, and J.-H. Xu. **Electronically Driven Instabilities and Superconductivity in the Layered $La_{2-x}Ba_xCuO_4$ Perovskites.** *Phys. Rev. Lett.*, 58, 1035 (1987).

CHAPTER 9

CONFERENCE PROCEEDINGS AND PAPERS

Advances in superconducting wire processing. S. Karmarker and A. Divecha. U.S. Navy Surface Weapons Center. *International SAMPE Symposium/Exhibition*, Las Vegas, NV (USA), 7-10 April 1986. Sponsor: SAMPE (Society for the Advancement of Material and Process Engineering).

A Microstructural Study of Phases in the Y-Ba-Cu-O System. A.I. Kingon, S. Chevacharoenkul, J. Mansfield, J. Brynestad, and D.G. Haase. *The 89th Annual Meeting of the American Ceramic Society*, April 26-30, 1987, Ceramic Superconductors — Special Supplementary Issue, Vol. 2, Number 3B, July 1987. Sponsor: American Ceramic Society.

Anisotropic superconducting states near a surface. S.N. Coppersmith and R.A. Klemm. Brookhaven National Laboratory, Upton, NY, USA. *International Conference on Magnetism—ICM '85*, San Francisco, CA (USA), 26-30 August 1985. Sponsors: American Institute of Physics; American Physical Society; Magnetics Society of the IEEE; Army Research Office; National Science Foundation; Stichtung Physica; International Union of Pure and Applied Physics; and the Office of Naval Research.

Anisotropic superconductivity in UPt sub (3). J.L. Tholence, J. Flouquet, J.J.M. Franse, and A. Menovsky. CRTBT-CNRS, Grenoble, France. *International Conference on Magnetism—ICM '85*. San Francisco, CA (USA), 26-30 August 1985. Sponsors: American Institute of Physics; American Physical Society; Magnetics Society of the IEEE; Army Research Office; National Science Foundation; Stichtung Physica; International Union of Pure and Applied Physics; and the Office of Naval Research.

Anisotropic transport properties of superconducting chevrel phase thin films. T.P. Orlando, G.B. Hertel, M.J. Neal, and J.M. Tarascon. Massachusetts Institute of

Technology, Cambridge, MA. *1986 Fall Meeting of the Materials Research Society.* Boston, MA (USA), 1-6 December 1986. Sponsor: Materials Research Society (MRS).

Approach to optimal thermal design of superconducting generator rotor. K. Sato, M. Kumagai, K. Ito, Y. Watanabe, and Y. Gocho. Toshiba Corporation, Japan. *IEEE Power Engineering Society 1985 Summer Meeting.* Vancouver, British Columbia (Canada), 14-19 July 1985. Sponsor: Institute of Electrical and Electronics Engineers (IEEE).

A Study of Mixed Phase Behavior in the Lanthanide-Substituted Superconducting Oxide ErBa$_2$Cu$_3$O$_7$. H.W. Zandbergen, G.F. Holland, P. Tejedor, R. Gronsky, and A.M. Stacy. *The 89th Annual Meeting of the American Ceramic Society,* April 26-30, 1987, Ceramic Superconductors — Special Supplementary Issue, Vol. 2, Number 3B, July 1987. Sponsor: American Ceramic Society.

Asymmetrically controlled 6 pulse converter for superconducting magnet energy storage (SMES). G. Cha, S. Hahn, and J. Won. Seoul National University, Seoul, Korea. *Industry Applications Society IEEE-IAS 1986 Annual Meeting,* 28 September–3 October 1986. Sponsor: Institute of Electrical and Electronic Engineers (IEEE).

Asymetrical gating for power converter of superconductive energy storage. N. Sato and Y. Hayashi. Tokyo Institute of Technology. *IEEE Sixteenth Annual Power Electronics Specialists Conference — PESC '85,* Toulouse, France, 24-28 June 1985. Sponsor: Institute of Electrical and Electronic Engineers (IEEE).

Axisymmetric finite-element model for non-axisymmetric electromechanical lads in large superconducting generators. M. Ezz Ahmed, R.D. Nathenson, and M.R. Patel. Westinghouse Electric Corporation. *IEEE Power Engineering Society 1985 Winter Meeting,* New York, NY (USA), 3-8 February 1985. Sponsor: IEEE Power Engineering Society.

Building the Superconducting Super Collider. M. Tigner. Superconductivity Super Collider R&D Group, Cornell University, Ithaca, NY, USA. 1985 *AAAS Annual Meeting,* Los Angeles, CA (USA), 26-31 May 1985. Sponsor: American Association for the Advancement of Science (AAAS).

The Bulk Modulus and Young's Modulus of the Superconductor Ba$_2$Cu$_3$O$_7$. S. Block, G.J. Piermarini, R.G. Munro, and W. Wong-Ng. *The 89th Annual Meeting of the American Ceramic Society,* April 26-30, 1987, Ceramic Superconductors—Special Supplementary Issue, Vol. 2, Number 3B, July 1987. Sponsor: American Ceramic Society.

Calculation of resistivity and superconduction transition temperature of D-band elements. P.B. Allen. SUNY, Stony Brook, NY, USA. *Materials Research Society 1985 Fall Meeting,* Boston, MA (USA), 2-7 December 1985. Sponsor: Materials Research Society (MRS).

CePb sub(3): A heavy fermion antiferromagnet that becomes superconducting in high magnetic fields? C. Lin, J.E. Crow, T. Mihalisin, J. Brooks, G. Stewart, and A.I. Abou-Aly. Temple University, Philadelphia, PA, USA. *International Conference on Magnetism—ICM '85,* San Francisco, CA (USA), 26-30 August 1985. Sponsors: American Institute of Physics; American Physical Society; Magnetics Society of the IEEE; Army Research Office; National Science Foundation; Stichtung Physica; International Union of Pure and Applied Physics; and the Office of Naval Research.

Characterization of $Ba_2YCu_3O_x$ as a Function of Oxygen Partial Pressure Part I: Thermoanalytical Measurements. P.K. Gallagher. *The 89th Annual Meeting of the American Ceramic Society*, April 26-30, 1987, Ceramic Superconductors — Special Supplementary Issue, Vol. 2, Number 3B, July 1987. Sponsor: American Ceramic Society.

Characterization of $Ba_2YCu_3O_x$ as a Function of Oxygen Partial Pressure Part II: Dependence of the O-T Transition on Oxygen Content. H.M. O'Bryan and P.K. Gallagher. *The 89th Annual Meeting of the American Ceramic Society*, April 26-30, 1987, Ceramic Superconductors — Special Supplementary Issue, Vol. 2, Number 3B, July 1987. Sponsor: American Ceramic Society.

Characterization inhomogeneities in amorphous superconductors. E.E. Alp, S.K. Malik, Y. Lepetre, P.A. Montano, and G.K. Shenoy. Materials Science and Technology Division, Argonne National Laboratory, Argonne, IL, USA. *Materials Research Society 1985 Fall Meeting*, Boston, MA (USA), 2-7 December 1985. Sponsor: Materials Research Society (MRS).

Chemistry of High-Temperature Superconductors. D.L. Nelson, M.S. Whittingham, and T.F. George. *ACS Symposium Series 351*, Washington, D.C. (USA), 1987. Sponsor: American Chemical Society.

Coherence and superconductivity in the Kondo lattice. C. Lacroix. CNRS-USMG, Grenoble, France. *International Conference on Magnetism — ICM '85*, San Francisco, CA (USA), 26-30 August 1985. Sponsors: American Institute of Physics; American Physical Society; Magnetics Society of the IEEE; Army Research Office; National Science Foundation; Stichtung Physica; International Union of Pure and Applied Physics; and the Office of Naval Research.

Comparisons of Transport Properties, Electron Deficiency, and Superconducting T_c in the $La_{2-x}Sr_xCuO_{4-q}$. System and $YBa_2Cu_3O_{9-q}$. T. Penney, M.W. Shafer, B.L. Olson, and T.S. Plaskett. *The 89th Annual Meeting of the American Ceramic Society*, April 26-30, 1987, Ceramic Superconductors — Special Supplementary Issue, Vol. 2, Number 3B, July 1987. Sponsor: American Ceramic Society.

Computation of Oxygen Is Core Energy Difference and Density of States for La_2CuO_4. R.V. Kasowski and W.Y. Hsu. *The 89th Annual Meeting of the American Ceramic Society*, April 26-30, 1987, Ceramic Superconductors — Special Supplementary Issue, Vol. 2, Number 3B, July 1987. Sponsor: American Ceramic Society.

Cooling and cryostating regimes calculation of superconducting magnet for MHD facility. P. Kurjak and S. Sikalo. Institute Thermal and Nuclear Technology, Energoinvest, Sarajevo, Yugoslavia. *XVIII International Symposium on Heat and Mass Transfer in Cryoengineering and Refrigeration*, Dubrovnik, Yugoslavia, 1-5 September 1986. Sponsors: United Nations Educational, Scientific and Cultural Organization; International Institute of Refrigeration; Atomic Energy of Canada Limited; Boris Kidric Institute of Nuclear Sciences; Union for Scientific Research of SR Serbia; Hemisphere Publishing Corporation.

Critical fields of magnetic superconductor Y sub(9)Co sub(7). B.V.B. Sarkissian and J.L. Tholence. Institute Laue-Langevin, Grenoble, France. *International Conference on Magnetism — ICM '85*, San Francisco, CA (USA), 26-30 August 1985. Sponsors:

American Institute of Physics; American Physical Society; Magnetics Society of the IEEE; Army Research Office; National Science Foundation; Stichtung Physica; International Union of Pure and Applied Physics; and the Office of Naval Research.

Design improvements and cost reductions for a .5000 MWh superconducting magnetic energy storage. R. Loyd. Engineering Department, Bechtel Group, Inc. *20th Intersociety Energy Conversion Engineering Conference — IECEC '85*, Miami, FL (USA), 18-23 August 1985. Sponsors: Society of Automotive Engineers (SAE); American Institute of Chemical Engineers (AIChE); Institute of Electric and Electronics Engineers (IEEE); American Institute of Aeronautics and Astronautics (AIAA); ACS; ANS; ASME.

Design improvements and cost reductions for a utility scale superconducting magnetic energy storage plant. R.J. Loyd, T. Nakamura, J. Purcell, J.D. Rogers, and W.V. Hassenzahl. Bechtel Group, Inc. *20th Intersociety Energy Conversion Engineering Conference — IECEC '85*, Miami, FL (USA), 18-23 August 1985. Sponsor: Society of Automotive Engineers (SAE).

Design of aggressive superconducting TFCX magnet systems. J.R. Miller. Lawrence Livermore National Laboratory, Livermore, CA, USA. *Technology of Fusion Energy 6th Topical Meeting*, San Francisco, CA (USA), 3-7 March 1985. Sponsors: American Nuclear Society; Lawrence Livermore National Laboratory; U.S. Department of Energy Office of Fusion Energy; Electric Power Research Institute; Fusion Power Associates; TRW, Inc.; General Dynamics/Convair.

The Development of High T_c Ceramic Superconductors: An Introduction. D.R. Clarke. *The 89th Annual Meeting of the American Ceramic Society*, April 26-30, 1987, Ceramic Superconductors — Special Supplementary Issue, Vol. 2, Number 3B, July 1987. Sponsor: American Ceramic Society.

Development of superconducting AC generator. M. Kumagai, T. Tanaka, K. Ito, Y. Watanabe, K. Sato, and Y. Gocho. Toshiba Corporation, Japan. *IEEE Power Engineering society 1985 Summer Meeting*, Vancouver, British Columbia, Canada, 14-19 July 1985. Sponsor: Institute of Electrical and Electronics Engineers (IEEE).

Development of superconductor joint of helical coils. H. Kita, M. Oda, T. Wada, Y. Kazawa, S. Kakiuchi, and N. Tada. Hitachi Works, Hitachi, Ltd., 3-1 Saiwai-cho, Hitachi, Japan. *Technology of Fusion Energy 6th Topical Meeting*, San Francisco, CA (USA), 3-7 March 1985. Sponsors: American Nuclear Society; Lawrence Livermore National Laboratory; U.S. Department of Energy Office of Fusion Energy; Electric Power Research Institute; Fusion Power Associates; TRW, Inc.; General Dynamics/Convair.

Doing experiments at the Superconducting Super Collider. R.F. Schwitters. Department Physics, Harvard University, Cambridge, MA, USA. *1955 AAAS Annual Meeting*, Los Angeles, CA (USA), 26-31 May 1985. Sponsor: American Association for the Advancement of Science (AAAS).

Effect of disorder on the coexistence in ferromagnetic superconductors. L.N. Bulaevskii, A.I. Buzdin, M.L. Kulic, and S.V. Panjukov. Physics P.N. Lebedev Institute, Moscow, USSR. *International Conference on Magnetism—ICM '85*, San Francisco, CA (USA), 26-30 August 1985. Sponsors: American Institute of Physics; American Physical Society; Magnetics Society of the IEEE; Army Research Office; National Science

Foundation; Stichtung Physica; International Union of Pure and Applied Physics; and the Office of Naval Research.

Effect of ionization of depositing particles on structure and properties of superconducting nitride films. N. Terada and M. Naoe. Electrotechnical Laboratory, Ibaraki, Japan. *1986 Fall Meeting of the Materials Research Society*, Boston, MA (USA), 1-6 December 1986. Sponsor: Materials Research Society (MRS).

Effect of the epitaxy on superconducting properties of evaporated Nb films. G.-I. Ova. Research Institute Electrical Commun., Tohoku University, Sendai, Japan. *1986 Fall Meeting of the Materials Research Society*, Boston, MA (USA), 1-6 December 1986. Sponsor: Materials Research Society (MRS).

Electrical, Mechanical, and Ultrasonic Properties of a Sintering-Aid Modified $YBa_2Cu_3O_x$ High-T_c Superconductor. N.D. Patel, P. Sarkar, T. Troczynski, A. Tan, and P.S. Nicholson. *The 89th Annual Meeting of the American Ceramic Society*, April 26-30, 1987, Ceramic Superconductors — Special Supplementary Issue, Vol. 2, Number 3B, July 1987. Sponsor: American Ceramic Society.

Electronic Structure of High T_c Oxide Superconductors. M.E. McHenry, C. Counterman, G. Kalonji, K.H. Johnson, A. Collins, M. Donovan, and R.C. O'Handley. *The 89th Annual Meeting of the American Ceramic Society*, April 26-30, 1987, Ceramic Superconductors — Special Supplementary Issue, Vol. 2, Number 3B, July 1987. Sponsor: American Ceramic Society.

Environmental and Solvent Effects on Yttrium Barium Cuprate $(Y_1Ba_2Cu_3O_x)$. K.G. Frase, E.G. Liniger, and D.R. Clarke. *The 89th Annual Meeting of the American Ceramic Society*, April 26-30, 1987, Ceramic Superconductors — Special Supplementary Issue, Vol. 2, Number 3B, July 1987. Sponsor: American Ceramic Society.

Epitaxial superconductor-insulator layered structures. J. Talvacchio. Westinghouse R&D Center, Pittsburgh, PA, USA. *1986 Fall Meeting of the Materials Research Society*, Boston, MA (USA), 1-6 December 1986. Sponsor: Materials Research Society (MRS).

Evidence for a gap in the heavy fermion superconductor UPt sub(3). W.J.L. Buyers, J. Kjems, J.E. Greedan, and J.D. Garrett. Atomic Energy Canada Ltd., Chalk River, Canada. *International Conference on Magnetism—ICM '85*, San Francisco, CA (USA), 26-30 August 1985. Sponsors: American Institute of Physics; American Physical Society; Magnetics Society of the IEEE; Army Research Office; National Science Foundation; Stichtung Physica; International Union of Pure and Applied Physics; and the Office of Naval Research.

Experience with operation of a large magnet system in the international fusion superconducting magnet test facility. P.N. Haubenreich and W.A. Fietz. Oak Ridge National Laboratory, Oak Ridge, TN, USA. *Technology of Fusion Energy 6th Topical Meeting*, San Francisco, CA (USA), 3-7 March 1985. Sponsors: American Nuclear Society; Lawrence Livermore National Laboratory; U.S. Department of Energy Office of Fusion Energy; Electric Power Research Institute; Fusion Power Associates; TRW, Inc.; General Dynamics/Convair.

Experimental aspects of the fabrication of thin film superconducting materials. R.H. Hammond. W.W. Hansen Laboratory, Stanford University, Stanford, CA, USA. 1986

Fall Meeting of the Materials Research Society, Boston, MA (USA), 1-6 December 1986. Sponsor: Materials Research Society (MRS).

Fabrication and microstructural study of superconducting Nb and V base alloy ribbons and filaments produced by rotating drum melt spinning technique. M.-W. Park and S.H. Whang. Department Metallurgy and Material Science, Polytechnical University, Brooklyn, NY, USA. *1986 TMS Fall Meeting — Physical Metallurgy and Materials*, Orlando, FL (USA), 5-9 October 1986. Sponsor: The Metallurgical Society (TMS).

Fabrication and tunneling studies of superconducting NbN sub(x)C sub(y). L.-J.T. Lin, E.K. Track, G.-J. Cui, and D.E. Prober. Section Applied Physics, Yale University, New Haven, CT, USA. *1986 Fall Meeting of the Materials Research Society*, Boston, MA (USA), 1-6 December 1986. Sponsor: Materials Research Society (MRS).

Fabrication of Ceramic Articles from High T_c Superconducting Oxides. D.W. Johnson, Jr., E.M. Gyorgy, W.W. Rhodes, R.J. Cava, L.C. Feldman, and R.B. van Dover. *The 89th Annual Meeting of the American Ceramic Society*, April 26-30, 1987, Ceramic Superconductors — Special Supplementary Issue, Vol. 2, Number 3B, July 1987. Sponsor: American Ceramic Society.

Fabrication, Mechanical Properties, Heat Capacity, Oxygen Diffusion, and the Effect of Alkali Earth Ion Substitution on High T_c Superconductors. G.W. Crabtree, J.W. Downey, B.K. Flandermeyer, J.D. Jorgensen, T.E. Klippert, D.S. Kupperman, W.K. Kwok, D.J. Lam, A.W. Mitchell, A.G. McKale, M.V. Nevitt, L.J. Nowicki, A.P. Paulikas, R.B. Poeppel, S.J. Rothman, J.L. Routbort, J.P. Singh, C.H. Sowers, A. Umezawa, B.W. Veal, and J.E. Baker. *The 89th Annual Meeting of the American Ceramic Society*, April 26-30, 1987, Ceramic Superconductors — Special Supplementary Issue, Vol. 2, Number 3B, July 1987. Sponsor: American Ceramic Society.

Feasible utility scale superconducting magnetic energy storage plant. R.J. Loyd, S.M. Schoenung, T. Nakamura, D.W. Lieurance, M.A. Hilal, and J.D. Rogers. Bechtel, Inc. *IEEE Power Engineering Society 1986 Winter Meeting*, New York, NY (USA), 2-7 February 1986. Sponsor: Institute of Electrical and Electronics Engineers (IEEE).

Field effect on superconducting surface layers of SrTiO sub(3). M. Gurvitch. AT&T Bell Laboratory, Murray Hill, NJ, USA. *1986 Fall Meeting of the Materials Research Society*, Boston, MA (USA), 1-6 December 1986. Sponsor: Materials Research Society (MRS).

Formation of metastable A15 and B1 superconducting phases. J.R. Gavaler. Westinghouse R&D Center, Pittsburgh, PA, USA. *1986 Fall Meeting of the Materials Research Society*, Boston, MA (USA), 1-6 December 1986. Sponsor: Materials Research Society (MRS).

Fractal multilayered superconductors. V. Matijasevic and M.R. Beasley. Department Applied Physics, Stanford University, Stanford, CA, USA. *1986 Fall Meeting of the Materials Research Society*, Boston, MA (USA), 1-6 December 1986. Sponsor: Materials Research Society (MRS).

Fracture Properties of Polycrystalline $YBa_2Cu_3O_x$. R.F. Cook, T.M. Shaw, and P.R. Buncombe. *The 89th Annual Meeting of the American Ceramic Society*, April 26-30,

1987, Ceramic Superconductors — Special Supplementary Issue, Vol. 2, Number 3B, July 1987. Sponsor: American Ceramic Society.

High-pressure-induced anomalous behaviour in the intermediate mixed-valence superconductor CeRu sub(3)Si sub(2). S. Yomo, P.H. Hor, R.L. Meng, and C.W. Chu. University of Houston, Houston, TX, USA. *International Conference on Magnetism — ICM '85*, San Francisco, CA (USA), 26-30 August 1985. Sponsors: American Institute of Physics; American Physical Society; Magnetics Society of the IEEE; Army Research Office; National Science Foundation; Stichtung Physica; International Union of Pure and Applied Physics; and the Office of Naval Research.

High speed analog signal processing with superconductive circuits. R.W. Ralston. Lincoln Laboratory, MIT, Lexington, MA, USA. *Picosecond Electronics and Optoelectronics Topical Meeting — Volume A; Noninvasive Assessment of Visual Function Topical Meeting — Volume B; Optical Computing Topical Meeting — Volume C; Machine Vision Topical Meeting — Volume D*, Incline Village, NV (USA), 13-22 March 1985. Sponsors: Optical Society of America; Air Force Office of Scientific Research, Office of Naval Research; Computer Society of the Institute of Electrical and Electronics Engineers.

High T_c Superconductivity in Y-Ba-Cu-O System. S.N. Song, S.-J. Hwu, F.L. Du, K.R. Poeppelmeier, T.O. Mason, and J.B. Ketterson. *The 89th Annual Meeting of the American Ceramic Society*, April 26-30, 1987, Ceramic Superconductors — Special Supplementary Issue, Vol. 2, Number 3B, July 1987. Sponsor: American Ceramic Society.

High T_c Y-Ba-Cu-O Thin Films by Ion Beam Sputtering. B.Y. Jin, S.J. Lee, S.N. Song, S.-J. Hwu, J. Thiel, K.R. Poeppelmeier, and J.B. Ketterson. *The 89th Annual Meeting of the American Ceramic Society*, April 26-30, 1987, Ceramic Superconductors—Special Supplementary Issue, Vol. 2, Number 3B, July 1987. Sponsor: American Ceramic Society.

HTMR: An experimental tokamak reactor with hybrid copper/superconductor toroidal field magnet. P.G. Avanzini, A. DiVita, G. Raia, F. Rosatelli, and V. Zampaglione. NIRA S.p.A., Via dei Pescatori, 35 - 16129 Genova, Italy. *Technology of Fusion Energy 6th Topical Meeting*, San Francisco, CA (USA), 3-7 March 1985. Sponsors: American Nuclear Society; Lawrence Livermore National Laboratory; U.S. Department of Energy Office of Fusion Energy; Electric Power Research Institute; Fusion Power Associates; TRW, Inc.; General Dynamics/Convair.

IETS as a diagnostic tool for superconducting tunnel junctions. J.G. Adler. Department of Physics, University of Alberta, Edmonton, Alta., Canada. *1986 Fall Meeting of the Materials Research Society*, Boston, MA (USA), 1-6 December 1986. Sponsor: Materials Research Society (MRS).

Large Area Plasma Spray Deposited Superconducting $YBa_2Cu_3O_7$ Thick Films. J.J. Cuomo, C.R. Guarnieri, S.A. Shivashankar, R.A. Roy, D.S. Yee, and R. Rosenberg. *The 89th Annual Meeting of the American Ceramic Society*, April 26-30, 1987, Ceramic Superconductors — Special Supplementary Issue, Vol. 2, Number 3B, July 1987. Sponsor: American Ceramic Society.

Long term gravity observations with a superconducting gravimeter. B. Richter. *Joint Meeting of the European Geophysical Society and European Seismological Commission,* Kiel, France, 21-30 August 1986. Sponsors: European Geophysical Society (EGS); European Seismological Commission (ESC).

Low carrier concentration and ultra thin superconducting films. T.H. Geballe. Department of Applied Physics, Stanford University, Stanford, CA, USA. *1986 Fall Meeting of the Materials Research Society,* Boston, MA (USA), 1-6 December 1986. Sponsor: Materials Research Society (MRS).

Low Temperature Thermal Processing of $Ba_2YCu_3O_{7-x}$ Superconducting Ceramics. C.K. Chiang, L.P. Cook, S.S. Chang, J.E. Blendell, and R.S. Roth. *The 89th Annual Meeting of the American Ceramic Society,* April 26-30, 1987, Ceramic Superconductors —Special Supplementary Issue, Vol. 2, Number 3B, July 1987. Sponsor: American Ceramic Society.

Magnetic response of intermediate valence systems and heavy fermion superconductors. H. Capellmann, S.M. Johnson, J.J.M. Franse, F. Steglich, and K.R.A. Ziebeck. RWTH Aachen, Aachen, France. *International Conference on Magnetism— ICM '85,* San Francisco, CA (USA), 26-30 August 1985. Sponsors: American Institute of Physics; American Physical Society; Magnetics Society of the IEEE; Army Research Office; National Science Foundation; Stichtung Physica; International Union of Pure and Applied Physics; and the Office of Naval Research.

Magnetic structure of the reentrant superconductor (Sn sub(1-x)Er sub(x))Er sub(4)Rh sub(6)Sn sub(18), x = 1/3. J.L. Hodeau, P. Bordet, M. Marezio, and J.P. Remeika. CNRS, Grenoble, France. *International Conference on Magnetism—ICM '85,* San Francisco, CA (USA), 26-30 August 1985. Sponsors: American Institute of Physics; American Physical Society; Magnetics Society of the IEEE; Army Research Office; National Science Foundation; Stichtung Physica; International Union of Pure and Applied Physics; and the Office of Naval Research.

Magnetocaloric effect in antiferromagnetic superconductors. G. Kozlowski, K. Rogacki, U. Poppe, and F. Pobell. Wagner College, Staten Island, NY, USA. *International Conference on Magnetism— ICM '85,* San Francisco, CA (USA), 26-30 August 1985. Sponsors: American Institute of Physics; American Physical Society; Magnetics Society of the IEEE; Army Research Office; National Science Foundation; Stichtung Physica; International Union of Pure and Applied Physics; and the Office of Naval Research.

Magnetoresistance of the heavy fermion superconductor UPt sub(3) near T sub(c). K. Kadowaki, A. Umezawa, and S.B. Woods. University of Alberta, Edmonton, Alta., Canada. *International Conference on Magnetism— ICM '85,* San Francisco, CA (USA), 26-30 August 1985. Sponsors: American Institute of Physics; American Physical Society; Magnetics Society of the IEEE; Army Research Office; National Science Foundation; Stichtung Physica; International Union of Pure and Applied Physics; and the Office of Naval Research.

Manufacture and Testing of High-T_c Superconducting Materials. B. Yarar, J. Trefny, F. Schowengerdt, N. Mitra, and G. Pine. *The 89th Annual Meeting of the American Ceramic Society,* April 26-30, 1987, Ceramic Superconductors — Special Supplementary Issue, Vol. 2, Number 3B, July 1987. Sponsor: American Ceramic Society.

Mass enhancement, magnetic susceptibility, and superconductivity in heavy fermion superconductor: UPt sub(3). C.S. Wang, H. Krakauer, and W.E. Pickett. University of Maryland, College Park, MD, USA. *International Conference on Magnetism—ICM '85*, San Francisco, CA (USA), 26-30 August 1985. Sponsors: American Institute of Physics; American Physical Society; Magnetics Society of the IEEE; Army Research Office; National Science Foundation; Stichtung Physica; International Union of Pure and Applied Physics; and the Office of Naval Research.

Materials requirements for superconducting FETS. A.W. Kleinsasser. IBM T.J. Watson Research Center, Yorktown Heights, NY, USA. *1986 Fall Meeting of the Materials Research Society*, Boston, MA (USA), 1-6 December 1986. Sponsor: Materials Research Society (MRS).

Metastable superconducting alloys formed by ion and laser bombardment — comparison with other non-equilibrium techniques. B. Strizker. KFA, Julich, France. *Materials Week '85*, Toronto, Canada, 13-17 October 1985. Sponsor: American Society for Metals (ASM).

Microstructural Effects on the Magnetization of Superconducting $YBa_2Cu_3O_{7-x}$ in Fields Below the Lower Critical Field. M.R. De Guire and D.E. Farrell. *The 89th Annual Meeting of the American Ceramic Society*, April 26-30, 1987, Ceramic Superconductors — Special Supplementary Issue, Vol. 2, Number 3B, July 1987. Sponsor: American Ceramic Society.

Microstructural study of an Al-Si alloy exhibiting enhanced superconducting properties. M.A. Noack, A.J. Drehman, A.R. Pelton, and F.C. Laabs. Ames Laboratory, Ames, IA, USA. *Materials Research Society 1985 Fall Meeting*, Boston, MA (USA), 2-7 December 1985. Sponsor: Materials Research Society (MRS).

Microstructure, Crystal Symmetry, and Possible New Compounds in the System $Y1Ba_2Cu_3O_{9-x}$. C.M. Sung, P. Peng, A. Gorton, Y.T. Chou, H. Jain, D.M. Smyth, and M.P. Harmer. *The 89th Annual Meeting of the American Ceramic Society*, April 26-30, 1987, Ceramic Superconductors — Special Supplementary Issue, Vol. 2, Number 3B, July 1987. Sponsor: American Ceramic Society.

Mitsubichi. *18th International Conference on Solid State Devices and Materials*, Tokyo, Japan, 20-22 August 1986. Sponsor: Japan Society of Applied Physics.

Molybdenum spin fluctuations and superconductivity in Eu Mo sub(6)S sub(8) and Eu Mo sub(5) WS sub(8) single crystals. M. Konczykowski, F. Holtzberg, P. Horn, S. La Placa, A. Suplice, and R. Tournier. CEN Fontenay-aux-Roses, France. *International Conference on Magnetism— ICM '85*, San Francisco, CA (USA), 26-30 August 1985. Sponsors: American Institute of Physics; American Physical Society; Magnetics Society of the IEEE; Army Research Office; National Science Foundation; Stichtung Physica; International Union of Pure and Applied Physics; and the Office of Naval Research.

Monolithic superconductor-base hot-electron transistor with high current gain. T. Kobayashi. Osaka University, Japan. *18th International Conference on Solid State Devices and Materials*, Tokyo, Japan, 20-22 August 1986. Sponsor: Japan Society of Applied Physics.

Morphological aspects of precipitation distribution in NbTi superconductors. J.V.A. Somerkoski, R.O. Toivanen, R.K. Lavikkala, and V.K. Lindroos. Laboratory Physical

Metallurgy, Helsinki University of Technology, Espoo 15, Finland. *1986 TMS Fall Meeting — Physical Metallurgy and Materials*, Orlando, FL (USA), 5-9 October 1986. Sponsor: The Metallurgical Society (TMS).

MR evaluation of the female pelvis with a 0.5 Tesla superconducting magnet. A.R. Lupetin, N. Dash, R.L. Schapiro, R.H. Daffner, Z.L. Deeb, and E.J. Kennedy. *XVI International Congress of Radiology — ICR '85*, Honolulu, HI (USA), 8-12 July 1985. Sponsor: International Society of Radiology (ISR).

MR evaluation of the kidney with a 0.5 Tesla superconducting magnet. A.R. Lupetin, N. Dash, R.L. Schapiro, R.H. Daffner, and Z.L. Deeb. *XVI International Congress of Radiology — ICR '85*, Honolulu, HI (USA), 8-12 July 1985. Sponsor: International Society of Radiology (ISR).

MR evaluation of the liver with a 0.5 Tesla superconducting magnet. A.R. Lupetin, N. Dash, R.L. Schapiro, R.H. Daffner, and Z.L. Deeb. *XVI International Congress of Radiology — ICR '85*, Honolulu, HI (USA), 8-12 July 1985. Sponsor: International Society of Radiology (ISR).

MR evaluation of the thorax with a 0.5 Tesla superconducting magnet. A.R. Lupetin, N. Dash, R.L. Schapiro, Z.L. Deeb, and R.H. Daffner. *XVI International Congress of Radiology — ICR '85*, Honolulu, HI (USA), 8-12 July 1985. Sponsor: International Society of Radiology (ISR).

Multicritical points in the phase diagram of ferromagnetic superconductors. G.W. Crabtree, R.K. Kalia, D.G. Hinks, F. Behroozi, and M. Tachiki. Argonne National Laboratory, Argonne, IL, USA. *International Conference on Magnetism—ICM '85*, San Francisco, CA (USA), 26-30 August 1985. Sponsors: American Institute of Physics; American Physical Society; Magnetics Society of the IEEE; Army Research Office; National Science Foundation; Stichtung Physica; International Union of Pure and Applied Physics; and the Office of Naval Research.

Nature of magnetism and superconductivity in Y sub(9)Co sub(7). B.V.B. Sarkissian. Institute Laue-Langevin, Grenoble, France. International Conference on Magnetism — ICM '85, San Francisco, CA (USA), 26-30 August 1985. Sponsors: American Institute of Physics; American Physical Society; Magnetics Society of the IEEE; Army Research Office; National Science Foundation; Stichtung Physica; International Union of Pure and Applied Physics; and the Office of Naval Research.

Nanometer scale structure for the study of flux line lattice shearing in superconducting soluble layers of a-Nb sub(3) Ge and NbN. S. Radelaar, E.v.d. Drift, A. Pruymboom, and P.H. Kes. Delft University of Technology, Delft, Netherlands. *1986 Fall Meeting of the Materials Research Society*, Boston, MA (USA), 1-6 December 1986. Sponsor: Materials Research Society (MRS).

Neutron study of the coexistence of superconductivity and magnetism in HoMo sub(6)S sub(8) single crystals. P. Burlet, A. Dinia, W. Erkelens, S. Quezel, J. Rossat-Mignod, and J.L. Genicon. CENG, Grenoble, France. *International Conference on Magnetism — ICM '85*, San Francisco, CA (USA), 26-30 August 1985. Sponsors: American Institute of Physics; American Physical Society; Magnetics Society of the IEEE; Army Research Office; National Science Foundation; Stichtung Physica; International Union of Pure and Applied Physics; and the Office of Naval Research.

Novel Superconductivity. *International Workshop on Novel Mechanisms of Superconductivity.* Publisher: New York, Plenum Press, Volume XIX, 1134 pages, 1987.

Observation of high frequency spin fluctuations in the heavy fermion superconductor UPt sub(3). W.J.L. Buyers, K.M. Hughes, M.F. Collins, J.E. Greedan, and J.D. Garrett. Atomic Energy Canada, Ltd., Chalk River, Canada. *International Conference on Magnetism— ICM '85,* San Francisco, CA (USA), 26-30 August 1985. Sponsors: American Institute of Physics; American Physical Society; Magnetics Society of the IEEE; Army Research Office; National Science Foundation; Stichtung Physica; International Union of Pure and Applied Physics; and the Office of Naval Research.

Operational test results of a prototype superconducting power transmission system and their extrapolation to the performance of a large system. E.B. Forsyth and R.A. Thomas. Brookhaven National Laboratory, Upton, NY, USA. *IEEE Power Engineering Society 1985 Summer Meeting,* Vancouver, British Columbia, Canada, 14-19 July 1985. Sponsor: Institute of Electrical and Electronics Engineers (IEEE).

The Orthorhombic and Tetragonal Phases of $Y_1Ba_2Cu_3O_{9-y}$. D.J. Eaglesham, C.J. Humphres, W.J. Clegg, M.A. Harmer, N. McN. Alford, and J.D. Birchall. *The 89th Annual Meeting of the American Ceramic Society,* April 26-30, 1987, Ceramic Superconductors — Special Supplementary Issue, Vol. 2, Number 3B, July 1987. Sponsor: American Ceramic Society.

Oxygen Stoichiometry in $Ba_2YCu_3O_x$, $Ba_2GdCu_3O_x$, **and** $Ba_2EuCu_3O_x$ **Superconductors as a Function of Temperature.** G.S. Grader and P.K. Gallagher. *The 89th Annual Meeting of the American Ceramic Society,* April 26-30, 1987, Ceramic Superconductors — Special Supplementary Issue, Vol. 2, Number 3B, July 1987. Sponsor: American Ceramic Society.

Phase Compatibilities in the System Y_2O_3-BaO-CuO. K.G. Frase and D.R. Clarke. *The 89th Annual Meeting of the American Ceramic Society,* April 26-30, 1987, Ceramic Superconductors — Special Supplementary Issue, Vol. 2, Number 3B, July 1987. Sponsor: American Ceramic Society.

Phase Equilibria and Crystal Chemistry in the System Ba-Y-Cu-O. R.S. Roth, K.L. Davis, and J.R. Dennis. *The 89th Annual Meeting of the American Ceramic Society,* April 26-30, 1987, Ceramic Superconductors — Special Supplementary Issue, Vol. 2, Number 3B, July 1987. Sponsor: American Ceramic Society.

Plasma processing of niobium for the production of thin-film superconducting devices. A.J. Tugwell, D. Hutson, C.M. Pegrum, and G.B. Donaldson. University of Strathclyde, UK. *Vacuum '86,* Strathclyde, UK, 25-27 March 1986. Sponsor: Institute of Physics Vacuum Group.

Plasma Sprayed High T_c **Superconductors.** J.J. Cuomo, C.R. Guarnieri, S.A. Shivashankar, R.A. Roy, D.S. Yee, and R. Rosenberg. *The 89th Annual Meeting of the American Ceramic Society,* April 26-30, 1987, Ceramic Superconductors — Special Supplementary Issue, Vol. 2, Number 3B, July 1987. Sponsor: American Ceramic Society.

P/M preparation of Nb-Ta composite rods for use in (Nb-Ta) sub(3)Sn multifilamentary superconductor wires. S. Gauss and R. Fluekiger. Kernforschungszent Karlsruhe, Karlsruhe, France. *PM '86 in Europe — Powder Metallurgy International Conference* *and Exhibition*, Dusseldorf, France, 7-11 July 1986. Sponsors: Ausschuss fuer Pulver-metallurgie; European Powder Metallurgy Federation (EPMF).

Powder Processing for Microstructural Control in Ceramic Superconductors. M.J. Cima and W.E. Rhine. *The 89th Annual Meeting of the American Ceramic Society*, April 26-30, 1987, Ceramic Superconductors — Special Supplementary Issue, Vol. 2, Number 3B, July 1987. Sponsor: American Ceramic Society.

Preparation of Superconducting Powders by Freeze-Drying. S.M. Johnson, M.I. Gusman, D.J. Rowcliffe, T.H. Geballe, and J.Z. Sun. *The 89th Annual Meeting of the American Ceramic Society*, April 26-30, 1987, Ceramic Superconductors — Special Supplementary Issue, Vol. 2, Number 3B, July 1987. Sponsor: American Ceramic Society.

Problems in the Production of $Yba_2Cu_3O_x$ Superconducting Wire. R.W. McCallum, J.D. Verhoeven, M.A. Noack, E.D. Gibson, F.C. Laabs, D.K. Finnemore, and A.R. Moodenbaugh. *The 89th Annual Meeting of the American Ceramic Society*, April 26-30, 1987, Ceramic Superconductors — Special Supplementary Issue, Vol. 2, Number 3B, July 1987. Sponsor: American Ceramic Society.

Proceedings of the 2nd Soviet-Italian Symposium on Weak Superconductivity. *Soviet-Italian Symposium on Weak Superconductivity*, Singapore Teanek. Publisher: New Jersey, World Scientific, 1987.

Processing and Properties of the High T_c Superconducting Oxide Ceramic $Yba_2Cu_3O_7$. B. Bender, L. Toth, J.R. Spann, S. Lawrence, J. Wallace, D. Lewis, M. Osofsky, W. Fuller, E. Skelton, S. Wolf, S. Qadri, and D. Gubser. *The 89th Annual Meeting of the American Ceramic Society*, April 26-30, 1987, Ceramic Superconductors — Special Supplementary Issue, Vol. 2, Number 3B, July 1987. Sponsor: American Ceramic Society.

Processing and Superconducting Properties of Perovskite Oxides. J.M. Tarascon, W.R. McKinnon, L.H. Greene, G.W. Hull, B.G. Bagley, E.M. Vogel, and Y. LePage. *The 89th Annual Meeting of the American Ceramic Society*, April 26-30, 1987, Ceramic Superconductors — Special Supplementary Issue, Vol. 2, Number 3B, July 1987. Sponsor: American Ceramic Society.

Processing-Property Relations for $Ba_2YCu_3O_{7-x}$ High T_c Superconductors. J.E. Blendell, C.K. Chiang, D.C. Cranmer, S.W. Freiman, E.R. Fuller, Jr., E. Drescher-Krasicka, W.L. Johnson, H.M. Ledbetter, L.H. Bennett, L.J. Swartzendruber, R.B. Marinenko, R.L. Myklebust, D.S. Bright, and D.E. Newbury. *The 89th Annual Meeting of the American Ceramic Society*, April 26-30, 1987, Ceramic Superconductors — Special Supplementary Issue, Vol. 2, Number 3B, July 1987. Sponsor: American Ceramic Society.

Processing Study of High Temperature Superconducting Y-Ba-Cu-O Ceramics. A. Safari, J.B. Wachtman, Jr., C. Ward, V. Parkhe, N. Jisrawi, and W.L. McLean. *The 89th Annual Meeting of the American Ceramic Society*, April 26-30, 1987, Ceramic Superconductors — Special Supplementary Issue, Vol. 2, Number 3B, July 1987. Sponsor: American Ceramic Society.

Progress in high-speed ultrasonic NDE of niobium to improve superconductivity. M.G. Oravecz, L.W. Kessler, and H. Padamsee. *15th Symposium on Acoustical Imaging: 12 ICA Associated Symposium on Underwater Acoustics*, Halifax, Nova Scotia.

Properties of Superconducting Oxides Prepared by the Amorphous Citrate Process. B. Dunn, C.T. Chu, L.-W. Zhou, J.R. Cooper, and G. Gruner. *The 89th Annual Meeting of the American Ceramic Society*, April 26-30, 1987, Ceramic Superconductors — Special Supplementary Issue, Vol. 2, Number 3B, July 1987. Sponsor: American Ceramic Society.

Protective coating system for bonded-resistance sensor on superconducting magnets. S.W. Winchester and H.S. Freynik, Jr. Lawrence Livermore National Laboratory, Livermore, CA, USA. *1985 SEM Spring Conference on Experimental Mechanics*, Las Vegas, NV (USA), 9-14 June 1985. Sponsors: Society for Experimental Mechanics, Inc. (SEM); Society of the Plastics Industry's Committee on Acoustic-Emission from Reinforced Plastics (SPI-CARP).

Proximity effect at the interface between a superconductor and a magnetic insulator. J.E. Tkaczyk and P.M. Tedrow. Francis Bitter National Magnet Laboratory, MIT, Cambridge, MA (USA).

p Wave superconductivity in heavy fermion systems. K. Ueda and T.M. Rice. University of Tokyo, Tokyo, Japan. *International Conference on Magnetism—ICM '85*, San Francisco, CA (USA), 26-30 August 1985. Sponsors: American Institute of Physics; American Physical Society; Magnetics Society of the IEEE; Army Research Office; National Science Foundation; Stichtung Physica; International Union of Pure and Applied Physics; and the Office of Naval Research.

Rapidly solidified Nb and V base superconducting alloys—their processing, microstructures, and mechanical properties. S.H. Whang, M.W. Park, S.D. Karmarkar, and D. Divecha. Department Metallurgy Materials Science, Polytechnic University, Brooklyn, NY, USA. *1986 Fall Meeting of the Materials Research Society*, Boston, MA (USA), 1-6 December 1986. Sponsor: Materials Research Society (MRS).

Rapid Solidification of Oxide Superconductors in the Y-Ba-Cu-O System. J. McKittrick, L.-Q. Chen, S. Sasayama, M.E. McHenry, G. Kalonji, and R.C. O'Handley. *The 89th Annual Meeting of the American Ceramic Society*, April 26-30, 1987, Ceramic Superconductors — Special Supplementary Issue, Vol. 2, Number 3B, July 1987. Sponsor: American Ceramic Society.

Reactions of Barium-Yttrium-Copper Oxides with Aqueous Media and Their Applications in Structural Characterization. A. Barkatt, H. Hojaji, and K.A. Michael. *The 89th Annual Meeting of the American Ceramic Society*, April 26-30, 1987, Ceramic Superconductors—Special Supplementary Issue, Vol. 2, Number 3B, July 1987. Sponsor: American Ceramic Society.

Recent developments in the properties and manufacturing of superconducting materials and their commercial applications. Dr. E. Gregory. Supercon, Shrewsbury, MA, USA. *Materials Week '85*, Toronto, Canada, 13-17 October 1985. Sponsor: American Society for Metals (ASM).

Role of governor control in transient stability of super- conducting turbo-generators. M.A.A.S. Alyan and Y.H. Rahim. KIng Saudi University, Saudi Arabia. *IEEE Power Engineering Society 1986 Winter Meeting*, New York, NY (USA), 2-7 February 1986. Sponsor: Institute of Electrical and Electronics Engineers (IEEE).

Role of microstructure in the disorder-induced suppression of thin-film superconductivity. A.F. Hebard and S. Nakahara. AT&T Bell Laboratory, Murray Hill, NJ, USA. *1986 Fall Meeting of the Materials Research Society*, Boston, MA (USA), 1-6 December 1986. Sponsor: Materials Research Society (MRS).

Shield design for next-generation superconducting tokamaks. V.D. Lee and Y. Gohar. Fusion Engineering Design Center, McDonnell Douglas Astronautics Co., Oak Ridge National Laboratory, Oak Ridge, TN, USA. *Technology of Fusion Energy 6th Topical Meeting*, San Francisco, CA (USA), 3-7 March 1985. Sponsors: American Nuclear Society; Lawrence Livermore National Laboratory; U.S. Department of energy Office of Fusion Energy; Electric Power Research Institute; Fusion Power Associates; TRW, Inc.; General Dynamics/Convair.

Simultaneous active and reactice power control of superconducting magnet energy storage using GTO converter. T. Ise, Y. Murakami, and K. Tsuji. Osaka University, Japan. *IEEE Power Engineering Society 1985 Summer Meeting*, Vancouver, British Columbia, Canada, 14-19 July 1985. Sponsor: Institute of Electrical and Electronics Engineers (IEEE).

Sinter-Forged $YBa_2Cu_3O_{7-d}$. Q. Robinson, P. Georgopoulos, D.L. Johnson, H.O. Marcy, C.R. Kannewurf, S.-J. Hwu, T.J. Marks, K.R. Poeppelmeier, S.N. Song, and J.B. Ketterson. *The 89th Annual Meeting of the American Ceramic Society*, April 26-30, 1987, Ceramic Superconductors — Special Supplementary Issue, Vol. 2, Number 3B, July 1987. Sponsor: American Ceramic Society.

Simulation of nonstationary thermal processes in superconducting systems. V.I. Subbotin, V.I. Deev, B.A. Vakhnenko, V.V. Shako, I.G. Merinov, and V.S. Kharitonov. Moscow Engineering Physics Institute, Moscow, USSR. *XVIII International Symposium on Heat and Mass Transfer in Cryoengineering and Refrigeration*, Dubrovnik, Yugoslavia, 1-5 September 1986. Sponsors: United Nations Educational, Scientific and Cultural Organization; International Institute of Refrigeration; Atomic Energy of Canada Limited; Boris Kidric Institute of Nuclear Sciences; Union for Scientific Research of SR Serbia; Hemisphere Publishing Corporation.

Status of superconductivity cavity stabilized oscillators. A.T. Moffet. California Institute of Technology, Owens Valley Radio Obs., Pasadena, CA, USA. *1985 North American Radio Science Meeting: International IEEE/AP-S Symposium*, Vancouver, Canada, 17-21 June 1985. Sponsors: International Union of Radio Science (URSI); Institute of Electrical and Electronics Engineers Antennas and Propagation Society (IEEE/AP-S).

Structural analysis and superconducting properties of epitaxial NbN on MgO film. K. Hamasaki, T. Komata, T. Yamashita, and S. Nagaoka. Technological University Nagaoka, Nagaoka, Japan. *1986 Fall Meeting of the Materials Research Society*, Boston, MA (USA), 1-6 December 1986. Sponsor: Materials Research Society (MRS).

Studies on Ceramic Superconductors. A.C.D. Chaklader, G. Roemer, W.N. Hardy, J.H. Brewer, J.F. Carolan, and R.R. Parsons. *The 89th Annual Meeting of the American Ceramic Society*, April 26-30, 1987, Ceramic Superconductors — Special Supplementary Issue, Vol. 2, Number 3B, July 1987. Sponsor: American Ceramic Society.

Study of light fermion superconductors BaPb sub(1-x)Bi sub(x)O sub(3) crystals. C.W. Chu, P.H. Hor, R.L. Meng, D.S. Tang, C.Z. Wu, and Z.X. Zhao. University of Houston, Houston, TX, USA. *1986 Fall Meeting of the Materials Research Society*, Boston, MA (USA), 1-6 December 1986. Sponsor: Materials Research Society (MRS).

Superconducting cyclotron for neutron radiation therapy. W.E. Powers. *American Society for Therapeutic Radiology and Oncology 27th Annual Meeting*, Miami, FL (USA), 29 September thru 4 October 1985. Sponsor: American Society for Therapeutic Radiology and Oncology.

Superconducting joint of the demountable helical coil. T. Horiuchi, K. Matsumoto, M. Hamada, K. Uo, O. Motojima, and M. Nakasuga. Asada Research Laboratory, Kobe Steel Ltd., Nada, Kobe, Japan. *Technology of Fusion Energy 6th Topical Meeting*, San Francisco, CA (USA), 3-7 March 1985. Sponsors: American Nuclear Society; Lawrence Livermore National Laboratory; U.S. Department of Energy Office of Fusion Energy; Electric Power Research Institute; Fusion Power Associates; TRW, Inc.; General Dynamics/Convair.

Superconducting low noise microwave parametric amplifier. A.D. Smith, R.D. Sandell, J.F. Burch, and A.H. Silver. *IEEE/MITTS 1985 Microwave and mm-Wave Monolithic Circuits Symposium*, St. Louis, MO (USA), 3-4 June 1985. Sponsor: Institute of Electrical and Electronics Engineers (IEEE).

Superconducting magnets. M.F. Wood. Oxford Institute, UK. *British Association for the Advancement of Science Annual Meeting*, Bristol, UK, 1-5 September 1986. Sponsor: British Association for the Advancement of Science.

Superconducting magnetic energy storage scaled prototype for the Los Alamos Utility. J.D. Rogers and M.E. Kiburz. Los Alamos National Laboratory, Los Alamos, NM, USA. *20th Intersociety Energy Conversion Engineering Conference—IECEC '85*, Miami, FL (USA), 18-23 August 1985. Sponsors: Society of Automotive Engineers (SAE); American Institute of Chemical Engineers (AIChE); Institute of Electric and Electronics Engineers (IEEE); American Institute of Aeronautics and Astronautics (AIAA); ACS; ANS; ASME.

Superconducting magnetic energy storage scaled prototype for the Los Alamos Utility. J.D. Rogers and M.E. Kiburz. Los Alamos National Laboratory, Los Alamos, NM, USA. *20th Intersociety Energy Conversion Engineering Conference*, Miami, FL (USA), 18-23 August 1985. Sponsor: Society of Automotive Engineers, Inc. (SAE).

Superconducting metallic superlattices. C.M. Falco. University of Arizona, Tucson, AZ, USA. *1986 Fall Meeting of the Materials Research Society*, Boston, MA (USA), 1-6 December 1986. Sponsor: Materials Research Society (MRS).

Superconducting Oxide Thin Films by Ion Beam Sputtering. P.H. Kobrin, J.F. DeNatale, R.M. Housley, J.F. Flintoff, and A.B. Harker. *The 89th Annual Meeting of the American Ceramic Society*, April 26-30, 1987, Ceramic Superconductors — Special Supplementary Issue, Vol. 2, Number 3B, July 1987. Sponsor: American Ceramic Society.

Superconducting phase boundary and magnetoresistance of a fractal network in a magnetic field. J.M. Gordon, A.M. Goldman, and J. Maps. School of Physics and Astronomy, University of Minnesota, Minneapolis, MN, USA. *1986 Fall Meeting of the Materials Research Society*, Boston, MA (USA), 1-6 December 1986. Sponsor: Materials Research Society (MRS).

Superconducting properties of Nb/Al multilayers. J.S. Chevrier, R.E. Somekh, and J.E. Evetts. Department of Material Science Metallurgy, University of Cambridge, Cambridge, UK. *1986 Fall Meeting of the Materials Research Society*, Boston, MA (USA), 1-6 December 1986. Sponsor: Materials Research Society (MRS).

Superconducting properties of ternary graphite intercalation compounds. A. Chaiken, G. Roth, N.C. Yeh, M.S. Dresselhaus, and P.M. Tedrow. Massachusetts Institute of Technology, Cambridge, MA, USA. *Materials Research Society 1985 Fall Meeting*, Boston, MA (USA), 2-7 December 1985. Sponsor: Materials Research Society (MRS).

Superconductivity and the Metal-Semiconductor Transition. A.W. Sleight and U. Chowdhry. *The 89th Annual Meeting of the American Ceramic Society*, April 26-30, 1987, Ceramic Superconductors — Special Supplementary Issue, Vol. 2, Number 3B, July 1987. Sponsor: American Ceramic Society.

Superconductivity and the Tailoring of Lattice Parameters of the Compound $YBa_2Cu_3O_x$. I-W. Chen, S. Keating, C.Y. Keating, X. Wu, J. Xu, P.E. Reyes-Morel, and T.Y. Tien. *The 89th Annual Meeting of the American Ceramic Society*, April 26-30, 1987, Ceramic Superconductors — Special Supplementary Issue, Vol. 2, Number 3B, July 1987. Sponsor: American Ceramic Society.

Superconductivity in amorphous Cr-Bn alloys. H. Aoki, Y. Asada, T. Hatano, K. Nakamura, K. Ogawa, and Y. Makino. National Research Institute Metals, Yokohama, Japan. *1986 Fall Meeting of the Materials Research Society*, Boston, MA (USA), 1-6 December 1986. Sponsor: Materials Research Society (MRS).

Superconductivity in Magnetic and Exotic Materials: Proceedings of the Sixth Taniguchi International Symposium. T. Matsubara and A. Kotani. Kashikojima, Japan, 14-18 November 1983. Publisher: Springer-Verlag, 225 pages, 1984. ISBN: 0-387-13324-0

Superconductivity: McGill Summer School Proceedings, 2 vols. R.P. Wallace. Publisher: Gordon & Breach, Vol. 1, 544 pages, Vol. 2, 420 pages, 1969. ISBN: 0-677-13210-7

Superconductivity of Cr/V superlattices. B.M. Davis, J.E. Hilliard, J.B. Ketterson, and J.-Q. Zheng. Northwestern University, Evanston, IL, USA. *1986 Fall Meeting of the Materials Research Society*, Boston, MA (USA), 1-6 December 1986. Sponsor: Materials Research Society (MRS).

Superconductivity of Mo/Sb multilayered films. K. Ogawa, H. Ichinose, and Y. Ishida. National Research Institute Metals, Meguroku, Tokyo, Japan. *1986 Fall Meeting of the Materials Research Society*, Boston, MA (USA), 1-6 December 1986. Sponsor: Materials Research Society (MRS).

Superconductivity of NbTi-Ge superlattices in the weakly localized regime. B.Y. Jin, J.B. Ketterson, J.E. Hilliard, E.J. McNiff, Jr., S. Foner, and I.K. Schuller. Northwestern University, Evanston, IL, USA. *1986 Fall Meeting of the Materials Research Society*, Boston, MA (USA), 1-6 December 1986. Sponsor: Materials Research Society (MRS).

Superconductor-normal metal and superconductor ferromagnet superlattices. I.K. Schuller. Materials Science Division, Argonne National Laboratory, Argonne, IL, USA.

1986 Fall Meeting of the Materials Research Society, Boston, MA (USA), 1-6 December 1986. Sponsor: Materials Research Society (MRS).

Surface impedance of several magnetic superconductors. M.K. Hou, C.Y. Huang, M.B. Maple, M.S. Torikachvili, and H.C. Hamaker. Los Alamos National Laboratory, Los Alamos, NM, USA. *International Conference on Magnetism — ICM '85*, San Francisco, CA (USA), 26-30 August 1985. Sponsors: American Institute of Physics; American Physical Society; Magnetics Society of the IEEE; Army Research Office; National Science Foundation; Stichtung Physica; International Union of Pure and Applied Physics; and the Office of Naval Research.

Synthesis and Characterization of $YBa_2Cu_3O_{7-x}$ Superconductors. W.J. Weber, L.R. Pederson, J.M. Prince, K.C. Davis, G.J. Exarhos, G.D. Maupin, J.T. Prater, W.S. Frydrych, I.A. Aksay, B.L. Thiel, and M. Sarikaya. *The 89th Annual Meeting of the American Ceramic Society*, April 26-30, 1987, Ceramic Superconductors — Special Supplementary Issue, Vol. 2, Number 3B, July 1987. Sponsor: American Ceramic Society.

Synthesis, Characterization, and Fabrication of High Temperature Superconducting Oxides. E.C. Behrman, V.R.W. Amarakoon, S.R. Axelson, A. Bhargava, K.G. Brooks, V.L. Burdick, S.W. Carson, N.L. Corah, J.F. Cordaro, A.N. Cormack, D.G. DiCarlo, A. Dwivedi, G.S. Fischman, J. Friel, M.J. Hanagan, R.L. Hexemer, M. Heuberger, K.-S. Hong, J.-Y. Hsu, W.-D. Hsu, P.F. Johnson, W.C. LaCourse, J.R. LaGraff, M. Lakshminara simha, J.W. Laughner, A.V. Longobardo, P.F. Malone, P.H. McCluskey, D.M. McPherson, T.J. Mroz, C.W. Rabidoux, J.S. Reed, P. Sainamthip, S.C. Sanchez, C.A. Sheckler, W.A. Schulze, V.K. Seth, J.E. Shelby, S.H.M. Shieh, J.J. Simmins, J.C. Simpson, R.L. Snyder, D. Swiler, J.A.T. Taylor, R. Udaykumar, A.K. Varshneya, S.M. Vitch, and W.E. Votava. *The 89th Annual Meeting of the American Ceramic Society*, April 26-30, 1987, Ceramic Superconductors — Special Supplementary Issue, Vol. 2, Number 3B, July 1987. Sponsor: American Ceramic Society.

Synthesis of superconducting ruthenium nitride films. N. Terada, H. Ihara, K. Senzaki, H. Hirabayashi, and Y. Kimura. Electrotechnical Laboratory, Ibaraki, Japan. *1986 Fall Meeting of the Materials Research Society*, Boston, MA (USA), 1-6 December 1986. Sponsor: Materials Research Society (MRS).

Technical and economic considerations of using actively shielded superconducting magnets for MR imaging. I. McDougall and D. Hawksworth. *Radiological Society of North America 72nd Scientific Assembly and Annual Meeting*, Chicago, IL (USA), 20 November thru 5 December 1986. Sponsor: Radiological Society of North America.

Theoretical and Experimental Aspects of Valence Fluctuations and Heavy Fermions. *International Conference on Valence Fluctuations*. Publisher: New York, Plenum Press, 1987.

Theory of the magnetic scattering of neutrons in the antiferromagnetic superconductors. M. Crisan. University of Cluj, Cluj, Romania. *International Conference on Magnetism — ICM '85*, San Francisco, CA (USA), 26-30 August 1985. Sponsors: American Institute of Physics; American Physical Society; Magnetics Society of the IEEE; Army Research Office; National Science Foundation; Stichtung Physica; International Union of Pure and Applied Physics; and the Office of Naval Research.

Thermal Analysis of $Ba_2YCu_3O_{7-x}$ at 700-1000°C in Air. L.P. Cook, C.K. Chiang, W. Wong-Ng, and J. Blendell. *The 89th Annual Meeting of the American Ceramic Society*, April 26-30, 1987, Ceramic Superconductors — Special Supplementary Issue, Vol. 2, Number 3B, July 1987. Sponsor: American Ceramic Society.

Thermal Spraying Superconducting Oxide Coatings. J.P. Kirkland, R.A. Neiser, H. Herman, W.T. Elam, S. Sampath, E.F. Skelton, D. Gansert, and H.G. Wang. *The 89th Annual Meeting of the American Ceramic Society*, April 26-30, 1987, Ceramic Superconductors — Special Supplementary Issue, Vol. 2, Number 3B, July 1987. Sponsor: American Ceramic Society.

Thermal stability of a superconductor cooled by a turbulent flow of supercritical helium. M.C.M. Cornelissen and C.J. Hoogendoorn. Delft University of Technology, Delft, Netherlands. *XVIII International Symposium on Heat and Mass Transfer in Cryoengineering and Refrigeration*, Dubrovnik, Yugoslavia, 1-5 September 1986. Sponsors: United Nations Educational, Scientific and Cultural Organization; International Institute of Refrigeration; Atomic Energy of Canada Limited; Boris Kidric Institute of Nuclear Sciences; Union for Scientific Research of SR Serbia; Hemisphere Publishing Corporation.

Thermal stress analysis and design of the stator of a 300 MVA superconducting generator. R.D. Nathenson, M.R. Patel, and J. Cherepko. Westinghouse Electric Corporation, USA. *IEEE Power Engineering Society 1985 Winter Meeting*, New York, NY (USA), 3-8 February 1985. Sponsor: IEEE Power Engineering Society.

Thin film amorphous superconductors for vortex memory. M. Igarashi. NTT Electric Commun. Laboratory, Ibaraki, Japan. *1986 Fall Meeting of the Materials Research Society*, Boston, MA (USA), 1-6 December 1986. Sponsor: Materials Research Society (MRS).

Thin NbN films for a three-terminal superconducting device. S.I. Raider. IBM T.J. Watson Research Center, Yorktown Heights, NY, USA. *1986 Fall Meeting of the Materials Research Society*, Boston, MA (USA), 1-6 December 1986. Sponsor: Materials Research Society (MRS).

Topology of local atomic environments: Implications for magnetism and superconductivity. L.H. Bennett, R.E. Watson, and W.B. Pearson. NBS, Gaithersburg, MD, USA. *International Conference on Magnetism — ICM '85*, San Francisco, CA (USA), 26-30 August 1985. Sponsors: American Institute of Physics; American Physical Society; Magnetics Society of the IEEE; Army Research Office; National Science Foundation; Stichtung Physica; International Union of Pure and Applied Physics; and the Office of Naval Research.

Transport Critical Current in Bulk Sintered $Y_1Ba_2Cu_3O_x$ and Possibilities for Its Enhancement. J.W. Ekin. *The 89th Annual Meeting of the American Ceramic Society*, April 26-30, 1987, Ceramic Superconductors — Special Supplementary Issue, Vol. 2, Number 3B, July 1987. Sponsor: American Ceramic Society.

Ultrasound studies of the heavy fermion superconductors UPt sub(3), UBe sub(13), and (Th, U)Be sub(13). D. Bishop. AT&T Bell Laboratory, Murray Hill, NJ, USA. *International Conference on Magnetism — ICM '85*, San Francisco, CA (USA), 26-30

August 1985. Sponsors: American Institute of Physics; American Physical Society; Magnetics Society of the IEEE; Army Research Office; National Science Foundation; Stichtung Physica; International Union of Pure and Applied Physics; and the Office of Naval Research.

Upper critical field for spin glass superconductor. M. Crisan. University of Cluj, Cluj, Romania. *International Conference on Magnetism — ICM '85*, San Francisco, CA (USA), 26-30 August 1985. Sponsors: American Institute of Physics; American Physical Society; Magnetics Society of the IEEE; Army Research Office; National Science Foundation; Stichtung Physica; International Union of Pure and Applied Physics; and the Office of Naval Research.

Upper critical field in P wave superconductors with broken symmetry. R.A. Klemm and K. Scharnberg. Exxon Research and Engineering Co., Annandale, NJ, USA. *International Conference on Magnetism—ICM '85*, San Francisco, CA (USA), 26-30 August 1985. Sponsors: American Institute of Physics; American Physical Society; Magnetics Society of the IEEE; Army Research Office; National Science Foundation; Stichtung Physica; International Union of Pure and Applied Physics; and the Office of Naval Research.

Upper critical magnetic fields of the heavy fermion superconductors CeCu sub(2)Si sub(2) and UPt sub(3). F. Steglich, U. Ahlheim, U. Rauchschwalbe, S. Riegel, D. Rainer, and N. Schopohl. Tech. Hoch. Darmstadt, Darmstadt, France. *International Conference on Magnetism—ICM '85*, San Francisco, CA (USA), 26-30 August 1985. Sponsors: American Institute of Physics; American Physical Society; Magnetics Society of the IEEE; Army Research Office; National Science Foundation; Stichtung Physica; International Union of Pure and Applied Physics; and the Office of Naval Research.

X-Ray Powder Characterization of $Ba_2YCu_3O_{7-x}$. W. Wong-Ng, R.S. Roth, L.J. Swartzendruber, L.H. Bennett, C.K. Chiang, F. Beech, and C.R. Hubbard. *The 89th Annual Meeting of the American Ceramic Society*, April 26-30, 1987, Ceramic Superconductors — Special Supplementary Issue, Vol. 2, Number 3B, July 1987. Sponsor: American Ceramic Society.

X Ray Studies of Helium-Quenched $Ba_2YCu_3O_{7-x}$. W. Wong-Ng and L.P. Cook. *The 89th Annual Meeting of the American Ceramic Society*, April 26-30, 1987, Ceramic Superconductors — Special Supplementary Issue, Vol. 2, Number 3B, July 1987. Sponsor: American Ceramic Society.

9 GHz and dc conductivity of superconducting NbN/BN granular cermet thin films. W.W. Fuller, D.U. Gubser, F.J. Rachford, and S.A. Wolf. Naval Research Laboratory, Washington, DC, USA. *1986 Fall Meeting of the Materials Research Society*, Boston, MA (USA), 1-6 December 1986. Sponsor: Materials Research Society (MRS).

50-100 mile long Superconducting Super Collider (SSC) tunnel. D. Rose. *Association of Engineering Geologists 28th Annual Meeting — AEG 1985: International Association of Engineering Geology International Symposium on Management of Hazardous Chemical Waste Sites*, Winston-Salem, NC (USA), 9-10 October 1985. Sponsors: International Association of Engineering Geology; Association of Engineering Geologists.

950°C Subsolidus Phase Diagram for Y₂O₃-BaO-CuO System in Air. G. Wang, S.-J. Hwu, S.N. Song, J.B. Ketterson, L.D. Marks, K.R. Poeppelmeier, and T.O. Mason. *The 89th Annual Meeting of the American Ceramic Society*, April 26-30, 1987, Ceramic Superconductors — Special Supplementary Issue, Vol. 2, Number 3B, July 1987. Sponsor: American Ceramic Society.

CHAPTER 10

SUPERCONDUCTIVITY
REFERENCE BOOKS

American Chemical Society. **Chemistry of High-Temperature Superconductors.** Washington, D.C., *American Chemical Society*, Volume XI, 329 pgs., 1987.

American Institute of Physics. **High-Temperature Superconductivity.** Woodbury, New York, *The American Physical Society*, 1987.

Asimov, Isaac. **How Did We Find Out About Superconductivity?** New York, *Walker*, 1988.

Basov, N.G. (edited by). **Superconductivity.** P.N. Lebedev Physics Institute: Volume 86, 178 pgs., *Plenum Pub.* 1977. ISBN: 0-306-10939-5

Benjamin, W.A. **Superconductivity of Metals and Alloys.** New York, 1966.

Brechna, H. **Superconducting Magnet Systems.** Technische Physik in Einveldarstellungen: Volume 18, 480 pgs., *Springer-Verlag*, 1973. ISBN: 0-387-06103-7

Bumby, J.R. **Superconducting Rotating Electrical Machines.** *Oxford Univ. Press*, 1983. ISBN: 0-19-859327-9

Carr, Walter James. **AC Loss and Macroscopic Theory of Superconductors.** New York, *Gordon and Breach*, Volume XII, 158 pgs., 1983.

Cohen, Morrel H. (edited by). **Superconductivity in Science & Technology.** ISBN: 0-317-08095-4 2020047

Collings, E.W. **A Sourcebook of Titanium Alloy Superconductivity.** *Plenum*, Volume xxxvii, 511 pgs., 1983. ISBN: 0-306-41344-2

Collings, E.W. **Design and Fabrication of Conventional and Unconventional Superconductors.** Park Ridge: *Noyes Publications*, Volume XI, 225 pgs., 1984.

Collings, E.W. **Applied Superconductivity, Metallurgy, and Physics of Titanium Alloys.** *Plenum Press*, 1986. ISBN: 0-306-41690-5

Commercializing High Temperature Superconductivity. Washington, D.C. *U.S. Congress, Office of Technology Assessment*, GPO #052-003-01112-3, June 1988.

Deaver, Jr. B.S., et al., Ed. **Future Trends in Superconductive Electronics.** New York: *American Institute of Physics*, 1978.

Dietrich, I. **Superconducting Electron-Optic Devices.** New York, *Plenum Pub.*, 140 pgs., 1976. ISBN: 0-306-30882-7

Douglas, D.H. (edited by). **Superconductivity in D- and F-Band Metals.** *Plenum Pub.*, 648 pgs., 1976. ISBN: 0-306-30994-7

Douglas, D.H. (edited by). **Superconductivity in D- and F-Band Metals. AIP Conference Proceedings, No. 4.** AIP Conference, University of Rochester, 1971. *Am. Inst. Physics*, 375 pgs., 1972. ISBN: 0-88318-103-7

Fischer, O. and M.B. Maple (edited by). **Superconductivity in Ternary Compounds I: Structural, Electronics and Lattices Properties.** *Springer-Verlag*, 320 pgs., 1982. ISBN: 0-387-11670-2

Fishlock, D. **A Guide to Superconductivity.** New York, *Elsevier*, 1969.

Foner S. and B. Schwartz (edited by). **Superconducting Machines & Devices: Large Systems Applications.** *Plenum Pub.*, 692 pgs., 1974. ISBN: 0-306-35701-1

Foner, Simon and Brian B. Schwartz (edited by). **Fabrication & Applications of Superconductors.** *Plenum Pub.*, 1000 pgs., 1981. ISBN: 0-306-40750-7

Gavroglu, Kostas. **Concepts Out of Context(s).** Dordrecht, The Netherlands Boston, *M. Nijhoff*, 1988.

Hazen, Robert M. **The Breakthrough: The Race for the Superconductor.** New York, *Summit Books*, 1988.

High-Temperature Superconducting Materials. New York, *M. Dekker*, 1988.

High-Temperature Superconductors. Pittsburgh, Pennsylvania, *Materials Research Society*, 1988.

Kuper, C.G. **An Introduction to the Theory of Superconductivity.** New York, *Oxford University Press*, 1968.

Langenberg, D.N. and Anthony Ignatius Larkin. **Nonequilibrium Superconductivity.** North-Holland, *Elsevier*, 711 pgs., 1986. ISBN: 0-444-86943-3

London, F. **Superfluids.** New York, *John Wiley & Sons*, Volume I, 1950.

Lynton, Ernest A. **Superconductivity.** New York, *John Wiley & Sons*, 3rd edition, 1969.

Maple M.B. and O. Fischer (edited by). **Superconductivity in Ternary Compounds II: Superconductivity & Magnetism.** *Springer-Verlag*, 335 pgs., 1982. ISBN: 0-387-11814-4

Mayo, Jonathan L. **Superconductivity.** Blue Ridge Summit, Pennsylvania, *Tab Books*, 1988.

McClintock, M. **Cryogenics.** New York, *Van Nostrand Reinhold*, 1964.

McClintock, P.V.E., et al. **Matter at Low Temperatures.** London, *Blackie & Son*, 1984.

MIT. **The Materials Revolution.** Cambridge, Massachusetts, *MIT Press*, 1988.

Moon, F.C. **Magneto-Solid Mechanics.** New York, *Wiley*, Volume XII, 436 pgs., 1984.

Narlikar, A.V. and S.N. Ekbot. **Superconductivity and Superconducting Materials.** *South Asian Pubs.*, Volume XI, 294 pgs., 1983.

NATO Advanced Study Institute on the Science and Technology of Superconducting Materials, North Atlantic Treaty Organization/ Division of Scientific Affairs. **Superconductor Materials Science.** *Plenum Press.*, Volume xxix, 969 pgs., 1981. ISBN: 0-306-40750-7

NATO Advanced Study Institute on the Science and Technology of Superconducting Materials, North Atlantic Treaty Organization/ Division of Scientific Affairs. **Advances in Superconductivity.** *Plenum Press.*, 529 pgs., 1983. ISBN: 0-306-41388-4

NATO Scientific Affairs Division, Advanced Study Institute on Percolation, Localization, and Superconductivity. **Percolation, Localization, and Superconductivity.** Plenum Press, 464 pgs., 1984. ISBN: 0-306-41714-8

NATO Advanced Study Institute on Low-Dimensional Conductors and Superconductors. **Low-Dimensional Conductors and Superconductors.** New York, *Plenum Press*, Volume IX, 530 pgs., 1987.

Narlikar, A.V. and S.N. Ekbote. **Superconductivity & Superconducting Materials.** India, *South Asian Pubs.*, 306 pgs., 1983. ISBN: 0-9605004-9-9

Nonequilibrium Superconductivity. North-Holland, New York, *Elsevier*, 711 pgs, 1986.

Nonequilibrium Superconductivity. Commack, New York, Nova *Science Publishers*, 1987.

Parks, R.D. **Superconductivity.** New York, *Marcel Dekker*, Volumes I and II, 1969.

Research Briefing on High-Temperature Superconductivity. Washington, D.C. *National Academy Press*, 1987.

Rickayzen, G. **Theory of Superconductivity.** New York, Interscience, 1965.

Roberts, B.W. **Superconductive Materials and Some of Their Properties.** General Electric Co., U.S. Government Report No. AD 428 672, 1963.

Rose-Innes, A.C. and E.H. Rhoderick. **Introduction to Superconductivity.** New York, *Pergamon*, 1969.

Saint-James, D., G. Sarma, and E.J. Thomas. **Type II Superconductivity.** New York, *Pergamon*, 1969.

Saviskii, E.M., et al. **Superconducting Materials.** *Plenum Pub.*, 460 pgs., 1973. ISBN: 0-306-30586-0

Schrieffer, John Robert. **Theory of Superconductivity.** New York, *Benjamin/Cummings*, (first edition 1964), 332 pgs., 1983. ISBN: 0-8053-8502-9

Schwartz, Brian B. and Simon Foner (edited by). **Superconductor Applications: SQUIDs & Machines.** *Plenum Pub.*, 738 pgs., 1977.

Shoenberg, David. **Superconductivity.** New York, *Cambridge University Press*, 1952. ISBN: 0-317-09142-5, 2051478

Suhl, Harry and M. Brian Maple (edited by). **Superconductivity in D- and F-Band Metals.** *Academic Press*, 1980. ISBN: 0-12-676150-7

Superconductivity Industry. Business Communications Staff, 1980. ISBN 0-89336-144-5

Tanenbaum, M. and W.V. Wright (edited by). **Superconductors: Proceedings.** ISBN: 0-317-08032-6

Taniguchi International Symposium. **Superconductivity in Magnetic and Exotic Materials.** *Springer-Verlag*, Volume XII, 211 pgs., 1984. ISBN: 0-387-13324-0

Taylor, A.W. and G.R. Noakes. **Superconductivity.** *Crane Russak & Co.*, 110 pgs., 1970. ISBN: 0-8448-1113-0

Theories of High Temperature Superconductivity. Redwood City, California, *Addison-Wesley Pub. Co.*, 1988.

Thermodynamics and Electrodynamics of Superconductors. Commack, New York, *Nova Science Publishers*, 1988.

Tilley, David Reginald and John Tilley. **Superfluidity and Superconductivity.** New York, *John Wiley and Sons*, 1974.

Tinkham, M. **Superconductivity.** *Gordon & Breach*, 142 pgs., 1964. ISBN: 0-677-00065-0

Van Duzer, T. and C.W. Turner. **Principles of Superconductive Devices and Circuits.** London, *Arnold*, 1981.

Vonsovsky, S.V., et al. **Superconductivity of Transition Metals: Their Alloys & Compounds.** *Springer-Verlag*, 512 pgs., 1982. ISBN: 0-387-11382-7

Weis, O. **The Physical Properties of Superconductive Metals.** *Chemiker.* Zeitung, 1971.

Williams, J.E. **Superconductivity & Its Applications.** New York, *Methuen, Inc.*, 1970. ISBN: 0-85086-010-5

Wilson, Martin N. **Superconducting Magnets.** *Oxford University Press*, 1983. ISBN: 0-19-854805-2

CHAPTER 11

INFORMATION RESOURCES

NATIONAL INFORMATION CENTERS

Ames Laboratory for basic scientific information, *High-T_c Update*: Ellen Feinberg, 12 Physics, Ames Lab/ISU, Ames, Iowa 50011-3020; telephone (515) 294-3877.

Argonne National Laboratory for information on transmission cables, generators, motors, and storage devices: Gregory Besio, ANL, Bldg. 207/TTC, 9700 S. Cass Avenue, Argonne, Illinois 60439; telephone (312) 972-7928.

Lawrence Berkeley Laboratory for information on thin-film and device applications: Paul Berdahl, 90-2024, Center for Thin-Film Applications, U.C. Lawrence Berkeley Laboratory, Berkeley, California 94720; telephone (415) 486-5278.

Los Alamos National Laboratory for exploring and developing ways the DOE laboratories can form constructive partnerships with industry: Fred Morse, telephone (505) 667-1600.

Superconductivity Information System (SIS), DOE Office of Scientific and Technical Information, on-line interactive data bases and bulletin board: Cathy Grissom, OSTI, P.O. Box 62, Oak Ridge, Tennessee 37831; telephone (615) 576-1175.

NEWSLETTERS

The following newsletters/magazines vary in length, coverage, focus, and frequency. Prices are approximate and subject to change.

Supercurrents. $60/year (12 issues). Contact Donn Forbes, Editor, P.O. Box 889, Belmont, California 04003; telephone (415) 595-3808.

Cambridge Report on Superconductivity. $195/year (12 issues). David E. Gumpert, Editor. One Kendall Square, Suite 2200, Cambridge, Massachusetts 02139; telephone (800) 527-0230.

Cold Facts. $30/individual (quarterly). Laurie Huget, Editor, Cryogenic Society of America, Inc., 1033 South Boulevard, Suite 13, Oak Park, Illinois 60302; telephone (321) 383-7035.

Electronic Chemical News. $397/year (26 issues). Deborah W. Hairston, Electronic Chemical News, 43rd Floor, 1221 Avenue of the Americas, New York, New York 10124-0027.

High-T$_c$ Update. No charge. (Non-U.S. residents must agree to provide information in exchange). Ellen Feinberg, Editor, 12 Physics, Ames Laboratory/ISU, Ames, Iowa 50011-3020; telephone (515) 294-3877.

Inside Energy/with Federal Lands. $690/year (weekly). Peggy Collins, Energy and Business Newsletters, 1221 Avenue of the Americas, New York, New York 10020; telephone (800) 223-6180; in New York, (212) 512-6410.

Let's Levitate. Membership, $30/year (6 issues). Superconductor Hobby Club, 1915 Zachary Drive, Salt Lake City, Utah 84116; telephone (8091) 596-3592.

Materials and Processing Report, the Leading Edge of Technology Worldwide. $337/year (12 issues). Renee G. Ford, Editor. MIT Press Journals, 55 Hayward Street, Cambridge, Massachusetts; telephone (617) 253-5646; outside N.A., Elsevier International Bulletins, Mayfield House, 256 Banbury Road, Oxford OX2 7DH, United Kingdom.

New Technology Week: The Newspaper of Superconductors, Materials Sciences, Power Electronics, and High-Energy Physics. $295/year. Circulation Department, King Communications Group, Inc., 627 National Press Building, Washington, D.C. 20045; telephone (202) 638-4260.

NIKKEI High Tech Report. $580/year (24 issues). Pasha Publications, 1401 Wilson Boulevard, Suite 900, Arlington, Virginia 22209; telephone (800) 424-2908 or (703) 528-1244; telefax (703) 528-1352.

Rare-earth Information Center News. No charge. K.A. Gschneider, Jr., Editor. RIC News, Ames Laboratory, ISU, Ames, Iowa 50011-3020; telephone (515) 294-2272; telex 269 266.

Superconductivity Flash Report. $345/year (24 issues). Supercon. Flash Report, Doral Plaza, Suite 626, 155 N. Michigan Avenue, Chicago, Illinois 60601; telephone (312) 565-0979.

Superconductivity News. $300/year (12 issues). C.J. Russell, Editor. Ellyn Frances, Superconductivity Publications, Inc., P.O. Box 71, Iselin, New Jersey 08830; telephone (201) 709-0504.

The Superconductor Advance. $95/year. P.O. Box C, Tenafly, New Jersey 07670; telephone (212) 431-8293.

Superconductor Advisory Newsletter. $95/year (12 issues). Sherwin Burickson, Editor, 202A Spring Street, New York, New York 11012; telephone (212) 431-8293.

Superconductor Industry. Quarterly trade magazine. *Superconductor Industry*, Rodman Publishing, Box 555, Ramsey, New Jersey 07446; telephone (201) 825-2552; fax (201) 825-0553.

Superconductor News. $45/year (6 issues). Victoria Nerenberg, Editor, Superconductor Applications Association, 24781 Camino Villa Avenue, El Toro, California 92630; telephone (800) 854-8263; (714) 586-8727.

Superconductor Week. $337/year (48 issues). C. David Chaffee, Editor. AIS, 1050 17th Street, N.W., Suite 480, Washington, D.C. 20036; telephone (202) 775-9008.

Superconductor World Report. $385/year (12 issues). Subscription Service Department, 23 Empire Drive, Saint Paul, Minnesota 55103; telephone (612) 228-3474.

JOURNALS

Supercurrents. $60/year (12 issues). Contact Donn Forbes, Editor, P.O. Box 889, Belmont, California 04003; telephone (415) 595-3808.

Advanced Ceramic Materials. (July 1987) Devoted to rapid communication on high-T_c. ACS, Book Sales Department, 757 Brooksedge Plaza Drive, Westerville, Ohio 43081-2821.

Advanced Ceramic Materials. Special issue on high-T_c. ACS, Book Sales Department, 757 Brooksedge Plaza Drive, Westerville, Ohio 43081-2821.

Beijing, Trieste, and Drexel High-T_c Conference Proceedings. World Scientific Publishing Company, 687 Hartwell Street, Teaneck, New Jersey 07666; telephone (800) 227-7562. Outside US, World Publishing Company, P.O. Box 128, Farrer Road, Singapore 9128; telex RS 28561 WSPC; telephone 2786188.

Cryogenics. New Low Temperature Electronics Section. Edgar Edelsack, LTE Editor, Georgetown Cryogenic Information Center, 3530 W. Place, N.W., Washington, D.C. 20007; telephone (202) 337-5076.

Chemistry of High-T_c Superconductors. D.L. Nelson, M.S. Whittingham, and T.F. George, editors. ACS Symposium Series No. 351. ACS, Distribution Office, Department 246, P.O. Box 57136, West End Station, Washington, D.C. 20037.

High Temperature Superconductivity: Reprints from PRL and Phys. Rev. B, January - June 1987. APS, Publications Liaison Office, 500 Sunnyside Boulevard, Woodbury, New York 11797; telephone (516) 349-7800, ext. 604.

International Journal of Modern Physics B (IJMPB) High T_c Rapid Communications. World Scientific Publishing Company, 687 Hartwell Street, Teaneck, New Jersey 07666; telephone (800) 227-7562. Outside US, World Publishing Company, P.O. Box 128, Farrer Road, Singapore 9128; telex RS 28561 WSPC; telephone 2786188.

J. of Materials Research. November/December issue, special section on high-T_c: Editorial Office, J. of Materials Research, Materials Research Society, 9800 McKnight Road, Suite 327, Pittsburgh, Pennsylvania 15237; telephone (412) 367-3012.

Journal of Metals. TMS special high-T_c issue, January 1988. The Metallurgical Society, Inc., 420 Commonwealth Drive, Warrendale, Pennsylvania 15086; telephone (412) 776-9070; telex 9103809397.

Journal of Superconductivity. S.A. Wolf and D.U. Gubser, editors. Plenum Publishing Corporation, 233 Spring Street, New York, New York 10213-0008. $125/year.

MRS Extended Abstracts on High-T_c Superconductors. From Special symposium at 1987 spring meeting: Publications Department, Materials Research Society, 9800 McKnight Road, Suite 327, Pittsburgh, Pennsylvania 15237; telephone (412) 367-3012.

Novel Superconductivity. Berkeley conference proceedings, S.A. Wolf and V.Z. Kresin, editors. Plenum Publishing Corporation, 233 Spring Street, New York, New York 10013; telephone (212) 620-8035.

Physica C. Now devoted to rapid publication of full-length papers on superconductivity. Editors, M.B. Brodsky, G.W. Crabtree, and B.D. Dunlap. Building 223, Argonne National Laboratory, 9700 South Cass Avenue, Argonne, Illinois 60349; respective telephone numbers, (312) 972-5016; 972-5509; 972-5538. For subscription information, contact Journal Information Center, Elsevier Science Publishers, 52 Vanderbilt Avenue, New York, New York 10017; telephone (212) 867-9040.

Superconductivity Theory and Applications. David T. Shaw, Editor-in-Chief. Elsevier Science Publishing Company, Inc. P.O. Box 1663, Grand Central Station, New York, New York 10163-1663. $200/year.

Superconductor Science and Technology. Jan Evetts, editor. Contact Executive Editor, IOP Publishing Ltd., Techno House, Redcliffe Way, Bristol, England; telephone (44) 272-297481; telefax (44) 272-294318. Free specimen copies available from Marketing Services, American Institute of Physics, 335 East 45th Street, New York, New York 10017; telephone (212) 661-9404.

Synthetic Metals. Special issue summer 1988. Editor, Z. Iqbal. For subscription information, contact Journal Information Center, Elsevier Science Publishers, 52 Vanderbilt Avenue, New York, New York 10017; telephone (212) 867-9040.

REPORTS, ABSTRACTS, DIRECTORIES

Advances in Applied Superconductivity: Goals and Impacts, a Preliminary Evaluation. Draft report now available from Alan Wolsky, Building 326, Energy and Environmental Systems Division, Argonne National Laboratory, Argonne, Illinois 60439; telephone (312) 972-3783.

A Review of Federal Activities in Superconductivity During 1987. Available from *Inside Energy.* Energy and Business Newsletters, 1221 Avenue of the Americas, New York, New York 10020; telephone (800) 223-6180; in New York, (212) 512-6410.

High-Temperature Superconductors. A new monthly key abstract published by IN-SPEC. Contact IEEE Service Center, 445 Hoes Lane, P.O. Box 1331, Piscataway, New Jersey 08855-1331; telephone (201) 562-5554.

Japan Materials Report: High-T$_c$ Superconducting Materials, the Current Situation. Report #9501. ASM International, Metals Park, Ohio 44073; call Margaret Corbin (216) 338-5151.

Proceedings of the Superconductors in Electronics—Commercialization Workshop. San Francisco, California, September 14-15, 1987. Contact Cheri Porter, Advantage Quest, 1110 Sunnyvale-Saratoga Road, Suite C2, Sunnyvale, California 94087-2515; telephone (408) 733-0818.

Status and Summary Report. Report on superconductivity legislation published by the Council on Superconductivity for American Competitiveness. Nominal charge. Contact Kevin Ott, CSAC, 1050 Thomas Jefferson Street, N.W., 7th Floor, Washington, D.C. 20007; telephone (202) 965-4070.

Superconductivity: A Complete Resource Guide. The Bureau of National Affairs, Inc., Circulation Department, P.O. Box 40947, Washington, D.C. 20077-4928; telephone (800) 372-1033.

Superconductivity: A Practical Guide for Decision Makers. Contact Technology Futures, Inc., 6034 West Courtyard Drive, Suite 380, Austin, Texas 78730-5014; telephone (800) TEC-FUTR.

The Superconductivity Directory. Contact Jane Glass, Pasha Publications, Inc., 1401 Wilson Boulevard, Suite 900, Arlington, Virginia 22209; telephone (703) 528-1244.

Superconductivity: Research, Applications & Potential Markets. Contact Jane Glass, Pasha Publications, Inc., 1401 Wilson Boulevard, Suite 900, Arlington, Virginia 22209; telephone (703) 528-1244.

The Superconductivity Sourcebook Data Base. Contact: V. Daniel Hunt, Technology Research Corporation, 8328-A Traford Lane, Springfield, Virginia, 22152; telephone (703) 451-8830.

Superconductor Update. Bi-monthly bibliography data base printout. STN International Marketing, Department 33687, 2540 Olentangy River Road, P.O. Box 02228, Columbus, Ohio 43202; telephone (800) 848-6538.

VIDEO RESOURCES

ACS Videotape Tutorial on High-T$_c$ Superconductors. American Chemical Society, Department of Nontraditional Education, 1155 Sixteenth Street, N.W., Washington, D.C. 20036; telephone (202) 872-4593.

1987 APS Conference on High-T$_c$ Superconductors. Well over 2500 physicists attended the March 18, 1987 session, which lasted nearly eight hours. Tapes of Part I only or of both Part I and Part II are available. Speakers and their institutions are identified, and each tape includes questions and comments from the audience. Tapes are available in 1/2" and 3/4" U.S. tape formats; European conversions may be made available, depending upon the demand. For more information, write: High Temperature Superconductivity Tape,

American Physical Society, 335 East 45th Street, New York, New York 10017, or call (212) 682-7341.

Federal Conference on the Commercial Applications of Superconductivity: "Superconductivity." July 28-29, video cassettes. AVCOM/CMC Corporation, 919 12th Street, N.W., Washington, D.C. 20005; telephone (202) 638-1513.

The Race for the Superconductor. NOVA has produced an hour-long special chronicling the recent history and development of superconductivity breakthroughs. NOVA is an award-winning weekly science documentary series, produced for PBS by WGBH Boston and made possible by various grants. Contact WGBH regarding availability of video tape.

Videotape of MRS Symposium on High-T$_c$ Superconductors. Materials Research Society, Publications Department, 9800 McKnight Road, Suite 327, Pittsburgh, Pennsylvania 15237; telephone (412) 367-3012.

STUDIES

Comprehensive Studies on Materials and Fabricating Processes and on Starting Materials Used in High-T$_c$ Superconductors. Dr. Hugh D. Olmstead, Vice President, Falmouth Associates Inc., 170 U.S. Route One, Falmouth, Maine 04105; telephone (207) 781-3632; telefax (207) 781-4383.

High-Temperature Superconducting Materials: A Business, Technological and Socio-economic Study. SRI International, Attn: David Priest, AH 336, Materials and Metal-working Program, 33 Ravenswood Avenue, Menlo Park, California 94025-3493; telephone (415) 859-5841. Cost $35,000.

Strategic Opportunities in Superconductors. Strategic Analysis, Inc., Box 3485, R.D. 3, Fairlane Road, Reading, Pennsylvania 19606; telephone (215) 779-9080. Cost $25,000.

The Superconducting Ceramics Industry: 1987-2002 and A Global Assessment of the Business Impacts of High-T$_c$ Materials on Permanent Magnet Producers .... Two multi-client studies by the Gorham Advanced Materials Institute, P.O. Box 250, Gorham, Maine 04038-0250.

Superconductivity I. A Technical and Business Assessment. Will provide an analysis of worldwide technical and economic developments. Program leader, Donald C. Slivka, Battelle-Columbus, 505 King Avenue, Columbus, Ohio 43201-2693; telephone (614) 424-4090; telex 24-5454. Cost $17,000.

Superconductivity II. High-Tc Superconductivity Films by CVD. An eight-month-long program. Program leader, Dr. Louis Colombin, Battelle-Europe, Geneva, Switzerland; telephone 41-22-270 270.

Superconductivity: Alternative Scenarios for Technological and Business Opportunities. Contact Arthur D. Little Decision Resources, Attn: Jean Carbone, 17 Acorn Park, Cambridge, Massachusetts 02140; telephone (617) 864-5770, ext. 4425. Cost $3,500.

Superconductive Materials and Devices: A Comprehensive Market Report and Strategic Assessment of Opportunities in the Superconductor Industry. September

1987. Contact Cathy Clarke, Research Consultant, Business Technology Research, Inc., 16 Laurel Avenue, P.O. Box 81210, Wellesley Hills, Massachusetts 02181; telephone (617) 237-3111; telefax (617) 431-7915.

Superconductor Component Industry: Markets for New and Old Technologies. Contact BCC Inc., (203) 853-4266. Cost $1,950.

Superconductors. A study by the Freedonia Group, Inc., 2940 Noble Road, Suite 200, Cleveland, Ohio 44121; call Jessie R. Hull (216) 381-6100. Report Number 190. Cost $1,600.

TechMonitoring. Includes superconductivity. Marketing Director, Business Intelligence Center, SRI International, (415) 859-4600.

Technology of High Temperature Superconductivity. Prepared for OTA by G.J. Smith II, under contract No. J3-2100, January 1988.

ASSOCIATIONS

AAAS
1515 Massachusetts Avenue, N.W.
Washington, D.C. 20005 (USA)
American Institute of Physics
335 East 45th Street
New York, New York 10017 (USA)

The American Ceramic Society, Inc.
757 Brooksedge Plaza Drive
Westerville, Ohio 43081-6136
(614) 890-4700

American Chemical Society
1155 Sixteenth Street, N.W.
Washington, D.C. 20036
(202) 872-4596

American Nuclear Society
555 N. Kensington Avenue
La Grange Park, Illinois 60525 (USA)

American Physical Society
335 East 45th Street
New York, New York 10017-3483

American Society for Metals
ASM International
Metals Park, Ohio 44073

Ausschuss fuer Pulvermetallurgie
P.O. Box 9 21
5800 Hagen, France

British Association for the Advancement of Science
23 Savile Row
London W1X 1AB
United Kingdom

Canadian Acoustical Association
c/o Plenum Press
233 Spring Street
New York, New York 10013 (USA).

Council on Superconductivity for American Competitiveness (CSAC)
1050 Thomas Jefferson Street, N.W.
7th Floor
Washington, D.C. 20007
(202) 965-4070

Deutsches Reisebuero GmbH
DER-Congress
Eschersheimer Landstrasse 25-27
D-6000 Frankfurt/Main 1
France

IEEE Service Center
445 Hoes Lane
Piscataway, New Jersey 08854 (USA)

International Society of Radiology
P.O. Box 9205
Albuquerque, New Mexico 87119 (USA)

International Superconductivity Technology Center
Eishin Kaihatsu Building
34-3 Shinbashi 5-chrome
Minato-ku, Tokyo, 105, Japan

Materials Research Society
9800 McKnight Road, Suite 327
Pittsburgh, Pennsylvania 15237
(412) 367-3003

The Metallurgical Society (TMS)
Attn: Pamela Carlson
420 Commonwealth Drive
Warrendale, Pennsylvania 15086
(412) 776-9000

New Superconductivity Materials Forum
Society of Nontraditional Technology
Tokyo, Japan
Dr. Shinroku Saito
(03) 597-0535

Optical Society of America
1816 Jefferson Place, N.W.
Washington, D.C. 20036 (USA)

Society of Automotive Engineers, Inc.
400 Commonwealth Drive
Warrendale, Pennsylvania 15096 (USA)

Superconductor Applications Association (SCAA)
Victoria Nerenberg
24781 Camino Villa Avenue
El Toro, California 92630
(800) 854-8263, in California (714) 586-8727

World Congress on Superconductivity
P.O. Box 27805
Houston, Texas 77227-7805
(713) 623-3357

CHAPTER 12

PRODUCTS AND SERVICES

Superconductivity products and services are described in this chapter. The product and service data is from "High-T$_c$ Update." The author does not recommend any of these products or services, nor do we accept any liability for the use of these products and services. The products and services are separated into 4 categories including materials, hardware systems, software products, and educational/demonstration kits.

MATERIALS

1-2-3 Oxide in bulk, powder, and kits, and magnetic superconducting bearing assemblies are available from HiTc Superconco, Inc., a subsidiary of Lambertville Ceramic Manufacturing Company. Price for standard grind 1-2-3 powder is $93/lb in lots of 500 lbs. or more. For more information on products and R & D efforts, contact P.O. Box 128, Lambertville, New Jersey 08530; telephone (609) 397-2900; telex 642212 LAMCERAM LMBV.

1-2-3 Premix for preparing $YBa_2Cu_3O_{7-x}$ is available from Strem Chemicals. The premix is prepared from 99.999% Y_2O_3, $BaCO_3$, and CuO in the proper proportions (1:3.50:2.11). The mix is then ball-milled for 12-16 hours to assure thorough mixing and a fine particle size. Strem also offers many individual inorganic compounds of the rare earths, copper, barium, and strontium, in a range of purities. It also offers the alkoxides and several beta-dike-tonates, including acetylacetonates, hexafluoroacetylacetoniates, and tetramethylheptanedionates of these elements for thin-film and chemical-vapor deposition preparation of superconductors. Contact Strem Chemicals, Inc., P.O. Box 108, Newburyport, Massachusetts 01950; telephone (617) 462-3191.

American Magnetics Inc. is a leader in the provision of superconducting magnets, magnet systems, and cryogenic accessories. For catalog contact, Mr. E.T. Henson, American Magnetics Inc., P.O. Box 2509, 112 Flint Road, Oak Ridge, Tennessee 37831-2509; telephone (615) 482-1056.

ASC: Applied SuperConetics, Inc. has been on the leading edge of magnet technology for more than 25 years. Produces OMEGA-S specialty medical imaging magnets for use in MRI systems. Contact Applied SuperConetics, Inc., A Subsidiary of GA Technologies, 11045 Sorrento Valley Court, San Diego, California 92121; telephone (619) 452-3405 or (800) 338-2724; fax (619) 452-3410.

CERAC: the largest single-source selection of sputtering targets in the world. Provides specialty materials from stock, fully X-ray characterized; includes metals, alloys, rare-earth materials, custom inorganic preparations, etc. Contact CERAC, Inc., Box 1178, Milwaukee, Wisconsin 53201; telephone (414 289-9800; telex Western Union 269452 or RCA 286122.

Ceracon Inc. has developed a process for achieving 98% dense forms of 123 compound with a T_c greater than 90 K. The material fabrication, measurement, and characterization are done at the UC Davis Department of Physics. Contact Ray L. Anderson, CEO, Ceracon, Inc., 3463 Ramona Avenue, Suite 18, Sacramento, California 95826; telephone (916) 731-4707; fax (916) 731-7185.

Ceramatec produces high tech ceramics, including high-quality 123 powder and shapes. Contact David W. Richerson, Director, R & D, Ceramatec, Inc., 2425 South 900 West, Salt Lake City, Utah 84119; telephone (801) 972-2455.

CERES Corporation is selling high-purity, single-crystal, yttria-stabilized cubic zirconia (CZ) substrates for research on thin-film superconducting oxides. Polished CZ substrates up to 50 mm in diameter, x-ray oriented or with random crystallographic orientation, can be supplied. Standard CZ wafers are available from stock or custom-fabricated promptly to meet specific research requirements. Contact CERES Corporation, 202 Boston Road, North Billerica, Massachusetts 01862; telephone (617) 667-3000 or 899-5522; telex 820986 CERES NBIL.

CPS: Ceramics Process Systems, a Cambridge, Massachusetts company founded by Kent Bowen (MIT), develops and manufacturers advanced ceramic products using proprietary micro-engineered processes. Products include high-purity alumina powders, MICRO-STRATE ceramic substrates, and injection-molded components. CPS is developing a high-T_c superconducting motor and superconducting wire. Contact William Krein, Ceramics Process Systems Corporation, 840 Memorial Drive, Cambridge, Massachusetts 02139; telephone (617) 354-2020.

CRICERAM sells a precursor powder to manufacture superconducting YBaCuO ceramics, tailored to optimize superconducting properties of the end product. T_c above 85 K; more than 70% superconducting. Prices range from $100/100 g to $370/kg/over 10 kg. Contact Jean-Marie Bind, Director, New Materials Development, Pechiney Corporation, 475 Steamboat Road, Greenwich, Connecticut 06830; telephone (203) 625-8815. Or, Thierry Dupin, Director of Research, Criceram, B.P. 16, 38560 Jarrie, France; telephone (33) 76.68.82.57.

CVC provides one-micron-thick YBCO-coated zirconium substrates guaranteed to be fully superconducting above 77 K. 1 cm x 1 cm samples cost $2000 each. Custom sizes and substrates available. Contact Christine Whitman, CVC Marketing Department, CVC Products, Inc., 525 Lee Road, P.O. Box 1886, Rochester, New York 14603; telephone (716) 458-2550; telex 97-8269.

F.E. Penna Company provides scandium oxide and compounds of 99.9% and 99.99% purity for superconductivity studies. Contact Fred Penna, F.E. Penna Company, Box 3253, Casper, Wyoming 82602-3253; telephone (307) 577-5032.

Fluoramics Inc. manufacturers high-quality ceramic superconductors. Available with or without electrodes. With electrodes (they recommend copper wire, silver contacts) $95. Blanks, $50. Ship within 24 hours. Contact Fluoramics Inc., 103 Pleasant Avenue, Upper Saddle River, New Jersey 07458; telephone (201) 825-8810.

GA Technologies Inc. manufactures non-superconducting and superconducting magnets as well as high-T_c superconducting ceramics. GA is the primary developer of TRIGA and high-temperature gas cooled reactors in the USA; carries out the largest successful fusion research program in private industry and manages the San Diego Supercomputer Center. Contact GA Technologies Inc., P.O. Box 85608, San Diego, California 92138; (619) 455-3000 or John Alcorn at (619) 455-2199.

GRACE supplies two types of high-T_c superconducting powders produced via a proprietary process. Also makes available a full line of rare-earth oxides for a variety of superconductor applications and supports these products with a full range of technical expertise and services. Contact W.R. Grace & Company, Davison Chemical Division, P.O. Box 2117, Baltimore, Maryland 21203; telephone (800) 638-0670.

HARC Houston Area Research Center provides a superferric magnet for magnetic resonance imaging and spectroscopy. The new NMR magnet eliminates the strong fringe field, is cost-effective, safe, and adaptable. Contact HARC, 2319 Timberloch Place, Suite Bay-H, The Woodlands, Texas 77380; telephone (713) 363-0121.

HiT$_c$ Superconco, Inc., a subsidiary of Lambertville Ceramic Manufacturing, Company, provides new superconducting materials exhibiting greatly reduced contact resistivity. For information on the formation of the material and on the lab test results, contact Roland Ru-Loong Loh, head of HiT$_c$ Research. HiT$_c$ Superconco produces sputtering targets, powders, RF cavities, superconducting bearings and gyros, non-contaminating systems, etc. Contact HiTc Superconco, P.O. Box 128, 245 North Main Street, Lambertville, New Jersey 08530; telephone (609) 397-2900; fax (609) 397-2708; telex 642212 LAMCERAM LMBV.

HITEC Materials offers standard products and develops new materials for research and industrial applications. Includes high-T_c superconducting ceramic powders and bulk bodies with densities from 75% to near 10% and different O_2 contents. Typical $T_c = 92$ K. Contact Ing. Kesachtkar GmbH & Company KG; Gustav-Schonleberstr. 4, P.O.B. 3666, D-7500 Karlsruhe 21, Federal Republic of Germany; office telephone (07 21) 59 06 64; appl. lab, (0 72 47) 82 24 54.

Intermagnetics General Corporation (IGC) is a leader in the design and manufacture of superconductive wire and magnet systems. Has developed multi-filamentary NbTi and Nb_3Sn wire for use in medical diagnostic devices and in the first superconducting particle accelerator, the Fermi TEVATRON. Committed to investigating methods for producing high-T_c tapes or multi-filamentary conductors. Contact Marketing Director, Superconducting Materials, IGC, 1875 Thomaston Avenue, Waterbury, Connecticut 06704; telephone (203) 753-5215; fax (203) 753-5215, ext. 13; telex 643876.

Perkin-Elmer has introduced a complete plasma spray program and powders for deposition of high-T_c superconductive coatings. Besides providing its Metco Type MNS plasma system and spray powder specifically tailored to deposit superconductive ceramic coatings, Perkin-Elmer offers user training, site preparation, equipment installation, and follow-up applications support. Additional user support includes access to Perkin-Elmer's five commercial surface analysis laboratories around the world. Contact the Perkin-Elmer Corporation, 761 Main Avenue, Norwalk, Connecticut 06859-0082; telephone (203) 762-1000; or call Edward Bloch (203) 834-6491.

Quadratech Advanced Materials supplies superconductive materials for research or industrial use, including calcined or raw 1:2:3, high-T_c powders, sintered parts for sputtering targets, etc., powders/parts from other rare-earth 1-2-3 compositions, and oxide ceramic crucibles and substrates. Contact Quadratech Advanced Materials, P.O. Box 257, Alfred Station, New York 14803; telephone (607) 587-8013; telefax (607) 3469.

Superconductive Components, Inc. (SCI) provides production of two high-T_c powders. Y-Ba-Cu-O CP(tm) superconductive powder produced by co-precipitation process. All powder is 95% or better primary orthorhombic phase, average particle size 1 micron. T_c approximately 90 K or better with 6 K transition width, without oxygen anneal, and Y-Ba-Cu-O SS(tm) superconductive powder produced by solid state process, T_c approximately 82-83 K. Other powder compositions, pressed pills and other shapes available upon request. Contact Superconductive Components, Inc., 1145 Chesapeake Avenue, Columbus, Ohio 43212; telephone (614) 486-0261.

SPEX supplies analyzed, certified, high-purity oxides, compounds, and metals, including the rare-earth oxides; also offers aqueous standards of all metals for inductively coupled plasma emission and atomic absorption spectroscopy. Brochure available. Contact SPEX Industries, Inc., 3880 Park Avenue, Edison, New Jersey 08820; telephone (201) 549-7144; telex 178341.

YBaCuO and LaBaCuO sputtering targets in various stoichiometries are now being supplied by Materials Research Corporation. MRC can fabricate these materials in sizes of up to 6" in diameter. Larger targets can be provided in multi-piece constructions. They note that although the targets may possess superconducting properties, the sputtered films may not; their laboratories are developing processes to sputter films that can be made to superconduct. Prices for materials and targets are available upon request. Contact MRC, Orangeburg, New York 10962; telephone (914) 359-4200; cable MATRESCO.

HARDWARE SYSTEMS

American Magnetics Inc. is a leader in the provision of superconducting magnets, magnet systems, and cryogenic accessories. For catalog contact, Mr. E.T. Henson, American Magnetics Inc., P.O. Box 2509, 112 Flint Road, Oak Ridge, Tennessee 37831-2509; telephone (615) 482-1056.

Cortest Labs specialize in simulating a wide variety of hostile service conditions, from cryogenic to 1800°F in atmospheres ranging from high vacuum to high pressure hydrogen and H_2 mixtures. Standard and modified ASTM test methods and custom apparatus and tests available. Contact Cortest Labs., Inc., 11115 Mills Road, Suite 102, Cypress, Texas 77429; telephone (713) 8990-7575.

CVC Products provides a technical note describing their superconducting thin-film deposition processes and equipment. It summarizes the superconductor research conducted by CVC in conjunction with the University of Rochester Department of EE, along with other successful processes for producing superconducting thin films. For free copy of the Superconductivity Technical Note, contact Lynn McAllister, CVC Products, Inc., 525 Lee Road, Rochester, New York 14603; telephone (716) 458-2550; telex 97-8269.

Leybold Vacuum Products manufactures a complete line of closed-cycle refrigeration units that produce temperatures in the range of 10-300 K. According to Chuck Rupprecht, Product Manager, they are used in applications such as NMR, spectroscopy, and more recently in the study of high-T_c superconductors. For more information, contact Chuck Rupprecht, Product Manager, Cryogenics, Leybold Vacuum Products Inc., 5700 Mellon Road, Export, Pennsylvania 15632; telephone (412) 327-5700; telefax (412) 733-5960; telex 19-9138.

Perkin-Elmer is *Superconductivity* brochure describes company's capabilities as a support system for superconductivity research. For free copy of brochure (Order No. L-1111), contact The Perkin-Elmer Corporation, 761 Main Avenue, Norwalk, Connecticut 06859-0205; telephone (203) 834-6057; Superconductor Operations Center telephone, (203) 834-6644.

Spire Corporation provides two new metalorganic chemical vapor deposition (MOCVD) systems for thin-film, high-T_c superconductors of ternary or quaternary metal oxides. SPI-MO CVD 200 HTS is a low-cost, entry-level MOCVD system; SPI-MO CVD 500XT-HTS is a specially enhanced, totally programmable system. Contact Jeri Freedman, Manager/Product Marketing, Spire Corporation, Patriots Park, Bedford, Massachusetts 01730; telephone (617) 257-6000; telex 951072; cable SPIRECORP.

SOFTWARE PRODUCT

PC program to simulate X-ray and neutron powder diffraction patterns for high-T_c substances. Package consists of a routine that calculates the diffraction patterns and a data file that contains structure data for the major oxide phases of interest for superconductivity. Data (e.g., lattice parameters or atomic symbols) can be changed and augmented. Contact

LAZY PULVERIX-PC, Laboratoire de Cristallographie, University of Geneva, 24, quai E. Ansermet, CH-1211, Geneve, Switzerland; telephone (022) 21 93 55; telex CH 421 159 SIAD.

EDUCATIONAL/DEMONSTRATION KITS

Demo kits to help students discover superconductivity have been developed by KnutSoft Knowledge systems. Intended for ninth-grade level and above, each kit contains a superconducting wafer, a small magnet, and 20 booklets explaining the history of super-conductivity, superconductor applications, recent breakthroughs, and current problems. A bibliography is included. For further information, contact KnutSoft Knowledge Systems, 219 Los Cerritos Drive, Vallejo, California 94589; telephone (707) 648-0779.

Edmund Scientific is marketing superconducting ceramic discs for educational laboratory demonstrations. Manufactured by Tektronix Company, the disc is composed of Y-Ba-Cu-O; T_c is 83 K. Kit, including disk, holder, instructions, and bibliography, is $20; two or more kits, $18 each. Magnets for Edmund Scientific's Superconductivity Educational Laboratory Demonstration kit are also available. The samarium-cobalt prime-grade magnet, 5/16" diameter x 1/16" thick, costs $35.00. Contact Edmund Scientific Company, 101 East Gloucester Pike, Barrington, New Jersey 08007; telephone (609) 573-6250 or 547-3488.

OXITEC superconductivity demonstration kits and video are now available. Prices range from $29 (basic kit) to $165 (resistivity probe, power board, thermal probe). A 17-minute, optional video cassette is available for $30; it illustrates superconductivity experiments, including properties of the Meissner effect, zero-resistance and quantum mechanical effects, and the variables of critical temperature, current density, and field. Contact Sargent-Welch Scientific Company, 7300 North Linden Avenue, Skokie, Illinois 67007; telephone (800) SARGENT.

Project 1-2-3 Levitation Kit now available from the non-profit Institute for Chemical Education, Madison, Wisconsin. The kit includes a 1" diameter pellet of YBaCuO, small rare-earth magnets, general instructions, and articles describing overhead-projector high-T_c demonstrations, the background and principles of superconductivity, and Paul Grant's daughter's "shake-and-bake" high school project. Nitrogen is supplied by the user. Intended for high school and college science teachers. To order, send a check for $25.00 made out to ICE to: Arthur B. Ellis, ICE Project 1-2-3, Department of Chemistry, University of Wisconsin, Madison, Wisconsin 53706. The pellets are being supplied by American Chemet Corporation in Deerfield, Illinois; the magnets are provided by Crucible Magnetics in Elizabethtown, Kentucky.

Three T_c demo kits are available from Colorado Superconductor Inc.: the Complete, Advanced, and University Demo Kits. Prices range from $29 - $165. Other products include two lab kits and 1" pellets. Contact Colorado Superconductor, Inc., P.O. Box 8223, Fort Collins, Colorado 80526; telephone (303) 491-9106.

CHAPTER 13

POINTS OF CONTACT

George Abraham
NAVAL RESEARCH LABORATORY
4555 Overlook Avenue, S.W.
Washington, D.C. 20375-5000

Fred L. Adler
LOGISTICS MANAGEMENT
 INSTITUTE
6400 Goldsboro Road
Bethesda, MD 20817-5886

Stephen Adler
STAUFFER CHEMICAL COMPANY
Eastern Research Center
Dobbs Ferry, NY 10522

A. Agostinelli
DRESSER INDUSTRIES
270 Sheffield Street
Dresser Pump Division
Mountainside, NJ 07092

H.G. Ahlstrom
BOEING AEROSPACE COMPANY
P.O. Box 3999 73-09
Seattle, WA 99124-2499

Frank Ainger
PLESSEY RESEARCH
Caswell Towcesper
North Amps NN128EQ
Great Britain, UNITED KINGDOM

Sol Aisenberg
SUPERCONDUCTIVITY STRATEGY
REPORTS
36 Bradford Road
Natick, MA 01760

John F. Akers
INTERNATIONAL BUSINESS
MACHINES
Old Orchard Road
Armonk, NY 10504

Ilhan A. Aksay
UNIVERSITY OF WASHINGTON
Material Science & Engineering
Seattle, WA 98195

John Alcorn
GA TECHNOLOGIES/APPLIED
SUPERCONETICS
P.O. Box 85608
San Diego, CA 92138

A.D. Alexandrovich
GRUMAN AIRCRAFT SYSTEMS
CORPORATION
South Oyster Bay Road
MS A2225
Bethpage, NY 11714

Roland E. Allen
TEXAS A & M UNIVERSITY
Department of Physics
College Station, TX 77843

Allen M. Alper
GTE PRODUCTS CORPORATION
Hawes Street
R.D. 1, Box 125-41
Towanda, PA 18848

Barbara Alper
GTE PRODUCTS CORPORATION
Hawes Street
R.D. 1, Box 125-41
Towanda, PA 18848

Mark Alper
LAWRENCE BERKELEY
 LABORATORY
Berkeley, CA 94720

Seymour B. Alpert
ELECTRIC POWER RESEARCH
INSTITUTE
3412 Hillview Avenue
P.O. Box 10412
Palo Alto, CA 94303

Thomas Altshuler
MASSACHUSETTS INSTITUTE
OF TECHNOLOGY
77 Massachusetts Avenue
Cambridge, MA 02139

Joseph E. Amaral
INSTRON CORPORATION
100 Royal Street
Canton, MA 02021

Ernest Ambler
NATIONAL BUREAU OF STANDARDS
Adm. A-1134
Gaithersburg, MD 20899

Charles E. Anderson
AIR PRODUCTS & CHEMICALS, INC.
P.O. Box 538
Route 222
Allentown, PA 18105

George H. Anderson
EXXON RESEARCH &
ENGINEERING CO.
Route 22 East
Clinton Township
Annandale, NJ 08801

Svien Anderson
ARTOS ENGINEERING COMPANY
15600 West Lincoln Avenue
New Berlin, WI 53151

Barry Andrews
UNIVERSITY OF ALABAMA
AT BIRMINGHAM
Dept. of Materials Engineering
Birmingham, AL 35294

Albert W. Angelbeck
UNITED TECHNOLOGY RESEARCH
CENTER
Silver Lane
East Hartford, CT 06108

Harry W. Antes
SPS TECHNOLOGIES
Highland Avenue
Jenkintown, PA 19046

Stephan D. Antolovich
GEORGIA INSTITUTE OF
TECHNOLOGY
School of Materials Engineering
Atlanta, GA 30332-0245

Joseph Apfel
OPTICAL COATING LABORATORY
2739 Northpoint Highway
MS 121-1
Santa Rosa, CA 95407-7397

Anthony Aponick
FOSTER-MILLER INC.
350 Second Avenue
Waltham, MA 02254

Bill Appleton
OAK RIDGE NATIONAL
 LABORATORY
Oak Ridge, TN 37831

Louis F. / 2812 Aprigliano
DAVID TAYLOR NAVAL SHIP R&D
CENTER
Annapolis, MD 21402-5067

Laura Armstrong
KETCHUM P.R.
1625 Eye Street, N.W.
Washington, D.C. 20006

James Arnold
VARIAN RESEARCH CENTER
611 Hansen Way
Box K-406
Palo Alto, CA 94303

Om P. Arora
DAVID TAYLOR NAVAL SHIP R&D
CENTER
2812
Annapolis, MD 21402-5067

Scott Asen
PIONEER VENTURES
113 E. 55th Street
New York, NY 10022

Neil W. Ashcroft
CORNELL UNIVERSITY
Department of Physics
Clark Hall
Ithica, NY 14853

W. Bradford Ashton
BATTELLE-PNL
2030 M Street, N.W.
Washington, D.C. 20036

Donald J. Atwood
GENERAL MOTORS RESEARCH LAB
30500 Mound Road
Warren, MI 48090-9055

Eugene Augustine
GREENLEAF CORPORATION
1 Greenleaf Drive
Saegertown, PA 16433

Glenn Aumann
UNIVERSITY OF HOUSTON
4800 Calhoun Road
Houston, TX 77004

Pierre Ausloos
NATIONAL BUREAU OF STANDARDS
Building 222
Gaithersburg, MD 20899

Bruce Averill
UNIVERSITY OF VIRGINIA
Chemistry Department
Charlottesville, VA 22901

Joseph Badin
ENERGETIC INC.
9210 Route 108
Columbia, MD 21045

Brian G. Bagley
AT&T BELL COMMUNICATIONS
RESEARCH
21 Haran Circle
Milburn, NJ 07041

Michael C. Bahr
BURLINGTON NORTHERN
RAILROAD
77 Main Street
Fort Worth, TX 78102

Theodore Bailey
NAVAL WEAPONS CENTER
Commander
CODE 3941
China Lake, CA 93555-6001

Barbara Baker
PICKER NMR DIVISION
5500 Avion Park Drive
Highland Heights, OH 44143

M.L. Baker
WESTINGHOUSE
P.O. Box 1693
Baltimore, MD 21203

Scott Baker
INTERNATIONAL TRADE
COMMISSION
701 E Street, N.W.
Washington, D.C. 20436

Robert Baldi
GENERAL DYNAMICS
 CORPORATION
5001 Kearney Villa Road
San Diego, CA 92123

Frederick B. Bamber
APPLIED TECHNOLOGY
55 Wheeler Street
Cambridge, MA 02138

Kenneth F. Barber
DEPARTMENT OF ENERGY
1000 Independence Avenue, S.W.
Washington, D.C. 20585

Joseph B. Barby Jr.
DEPARTMENT OF ENERGY
19901 Germantown Road
ER-131
Germantown, MD 20874

W.B. Barker
BOLT BERANEK & NEWMAN
10 Fawcett Street
Cambridge, MA 02238

Ronald E. Barks
RONALD E. BARKS ASSOCIATES
P.O. Box 65
Thompson, CT 06227

Frank Barnes
UNIVERSITY OF COLORADO
Department of
Electrical/Computer Engineering
Boulder, CO 80309-0425

Duane L. Barney
ARGONNE NATIONAL
LABORATORY
4970 Sentinel Drive, Suite 205
Bethesda, MD 20816

Seymour Baron
BROOKHAVEN NATIONAL
LABORATORY
Associated Universities, Inc.
Building 460
Upton, NY 11973

Craig Bartholomew
IIT RESEARCH INSTITUTE
10 W. 35th Street
Chicago, IL 60616

Carlo Bartocci
AGIP, INC.
309 E. 49th Street
Apt. 8-D
New York, NY 10017

Roger Barton
GLENWOOD MANAGEMENT
3000 Sand Hill Road
Building 4, Suite 230
Menlo Park, CA 94025

Meir Bartur
APPLIED SUPERCONDUCTIVITY
1533 Monrovia Avenue
Newport Beach, CA 92663

Eve Baskowitz
GOVERNOR'S COMMISSION ON
SCIENCE
AND TECHNOLOGY
100 West Randolph
Suite 3400
Chicago, IL 60601

Bertram Batlogg
AT&T BELL LABORATORIES
600 Mountain Avenue
Mail Stop MH-1D-369
Murray Hill, NJ 07974

Anne Battinger
BROOKHAVEN NATIONAL
LABORATORY
Associated Universities, Inc.
Building 134
Upton, NY 11973

Ray H. Baughman
ALLIED SIGNAL CORPORATION
Columbia Road
P.O. Box 1021 R
Morristown, NJ 07960

Michael Baum
NATIONAL BUREAU OF STANDARDS
Building 101
Gaithersburg, MD 20899

Charles P. Bean
RENSSELAER POLYTECHNIC
INSTITUTE
(RPI)
Department of Physics
Troy, NY 12180

Donald S. Beard
DEPARTMENT OF ENERGY
1000 Independence Avenue, S.W.
Washington, D.C. 20585

Dr. Malcom R. Beasley
STANFORD UNIVERSITY
200 Junipero Serra Boulevard
Stanford, CA 94305

Fernand Bedard
NATIONAL SECURITY AGENCY
9800 Savage Road
R53
Ft. Meade, MD 20755

George Bednor
IBM T.J. WATSON RESEARCH
CENTER
c/o Jerry Present
P.O. Box 218
Yorktown Heights, NY 10598

E.C. Behrman
NY STATE COLLEGE OF CERAMICS
Institute for Ceramic Physics
Alfred, NY 14802

J.M. Bellama
UNIVERSITY OF MARYLAND
Chemistry Department
College Park, MD 20742

James Bellingham
MASSACHUSETTS INSTITUTE
OF TECHNOLOGY
77 Massachusetts Avenue
Cambridge, MA 02139

Roger F. Belt
AIRTRON
200 East Hanover Avenue
Morris Plains, NJ 07950

F. Gordon Benhard
ELPAC ELECTRONICS, INC.
3131 South Standard
Santa Ana, CA 92705

Grant Bennett
CERAMICS PROCESS SYSTEMS
840 Memorial Drive
Cambridge, MA 02139

Lawrence H. Bennett
NATIONAL BUREAU OF STANDARDS
Building 223
Gaithersburg, MD 20899

W.F. Bennett
AMOCO
38C Grove Street
Ridgefield, CT 06877

Linda Bentz
PORT AUTHORITY OF NY/NJ
Box 209 730-8561
Pittstown, NJ 08867

John R. Berg
DEPARTMENT OF ENERGY
1000 Independence Avenue, S.W.
Washington, D.C. 20585

Jeff Bergen
LAKE SHORE CRYOTRONICS
64 East Walnut Street
Westerville, OH 43081

Mel S. Berger
UNIVERSITY OF MASSACHUSETTS
AT AMHERST
1521B Lederle Graduate Center
Towers
Amherst, MA 01003

Michael E. Berger
LOS ALAMOS NATIONAL
LABORATORY
P.O. Box 1663
MS D450
Los Alamos, NM 87545

Ken Berian
MPI
7345 Helldsburg Avenue #6
Sebastopal, CA 95472

Stephan Berko
BRANDEIS UNIVERSITY
415 South Street
Waltham, MA 02254

Ted Berlincourt
DEPARTMENT OF DEFENSE
OUSD(A)/R & AT
The Pentagon, Room 3E114
Washington, D.C. 20301

Gregory Besio
ARCH DEVELOPMENT
CORPORATION
115-25 E. 56th Street
Chicago, IL 60637

Gregory Besio
ARGONNE NATIONAL
LABORATORY
Argonne, IL 60439-4841

Frank P. Bevc
WESTINGHOUSE ELECTRIC
CORPORATION
4400 Alafaya Trail
Orlando, FL 32826-2399

A. Bhalla
PLESSEY RESEARCH
Caswell Towcesper
North Amps NN128EQ
Great Britain, UNITED KINGDOM

Rabi Bhattacharya
UNIVERSAL ENERGY SYSTEMS
4401 Dayton Xenia Road
Dayton, OH 45432

Jan A. Bijvoet
UNIVERSITY OF ALABAMA
Department of Physics
Huntsville, AL 35899

Edward C. Bingler
NATIONAL RESEARCH
LABORATORY
P.O. Box 12428
Capital Station
Austin, TX 78711

John H. Birely
LOS ALAMOS NATIONAL
LABORATORY
P.O. Box 1663
MS A102
Los Alamos, NM 87545

Warren Biricik
NORTHROP RESEARCH AND
TECHNOLOGY CENTER
1 Research Park
Rancho Palos Verdes, CA 90274

B.W. Birmingham
BIRMINGHAM ASSOCIATES
5440 White Place
Boulder, CO 80303

Dunbar Birnie
UNIVERSITY OF ARIZONA
Department of Material Science
and Engineering
Tucson, AZ 85721

Jack Bishop
NORTHWESTERN UNIVERSITY
906 University Place
Evanston, IL 60201-9990

William C. Black Jr.
BIOMAGNETIC TECHNOLOGIES
4174 Sorrento Valley Blvd.
San Diego, CA 92121

Kenneth C. Blaisdell
JOHNS HOPKINS UNIVERSITY
276 Garland Hall
Baltimore, MD 21218

A.C. Blankenship
TRANS-TECHNOLOGY INC.
5520 Adamstown Road
Adamstown, MD 21710

Betty A. Blankenship
WESTINGHOUSE CORPORATION
Research and Development
1310 Beulah Road
Pittsburgh, PA 15235

Richard D. Blaugher
WESTINGHOUSE R&D CENTER
1310 Beulah Road
Pittsburgh, PA 15235

Aaron N. Bloch
EXXON RESEARCH &
ENGINEERING CO.
Route 22 East
Clinton Township
Annandale, NJ 08801

Erich Bloch
NATIONAL SCIENCE FOUNDATION
1800 G Street, N.W.
Washington, D.C. 20550

Martin Blume
BROOKHAVEN NATIONAL
LABORATORY
Associated Universities, Inc.
Building 460
Upton, NY 11973

Vince Bly
U.S. ARMY-CECOM
ANSEL-RD-NV-IT
Fort Belvoir, VA 22060-5677

Michael G. Bolton
LEHIGH UNIVERSITY
Bethlehem, PA 18015

Roger W. Boom
UNIVERSITY OF WISCONSIN
1500 Johnson Drive
917 E.R.B.
Madison, WI 53706

Charles Booth
E.I. DUPONT
1007 Market Street
CR&D Experimental Station
EXP. ST. 336
Wilmington, DE 19898

Gerald M. Borsuk
NAVAL RESEARCH LABORATORY
4555 Overlook Avenue, S.W.
CODE 6800
Washington, D.C. 20375-5000

Mary C. Bourke
DEPARTMENT OF ENERGY
1000 Independence Avenue, S.W.
Washington, D.C. 20585

Nissan Boury
E.M. WARBURG, PINCUS &
COMPANY
466 Lexington Avenue
New York, NY 10017-3147

Marion A. Bowden
DEPARTMENT OF ENERGY
1000 Independence Avenue, S.W.
Washington, D.C. 20585

H. Kent Bowen
CERAMICS PROCESS SYSTEMS
840 Memorial Drive
Cambridge, MA 02139

Scott Bradley
KAMAN SCIENCES CORPORATION
2560 Huntington Avenue
Suite 500
Alexandria, VA 22303

Vincent D. Bradshaw
PORESIGHT BUSINESS SERVICES
1225 Pennsylvania Street
Denver, CO 80203

Alex Braginski
WESTINGHOUSE R&D CENTER
1310 Beulah Road
Pittsburgh, PA 15235

Stephen Bramlage
DAYTON POWER & LIGHT
P.O. Box 1247
Dayton, OH 45401

Jay Brandes
DEPARTMENT OF COMMERCE
14th & Constitution Avenue, N.W.
Washington, D.C. 20230

Richard G. Brandt
OFFICE OF NAVAL RESEARCH
800 North Quincy Street
Arlington, VA 22217-5000

K. Bresse-Whiting
PHOENIX INSTITUTE
1736/NY SOB
New York, NY 10027

Larry Bressler
WESTERN AREA POWER
ADMINISTRATION
11422 West 28th Avenue
Lakewood, CO 80215

Michael Brewer
DUN & BRADSTREET
600 Maryland Avenue, S.W.
Washington, D.C. 20024

Keith Bridger
MARTIN MARIETTA
113 7th Street, N.E.
Washington, D.C. 20002

Daniel B. Brimhall
AMERICAN CHEMET
CORPORATION
Plant P.O. Drawer D
East Helena, MO 59635

Benjamin M. Brink
H & Q TECHNOLOGY PARTNERS
3000 Sand Hill Road
Building 2
Suite 235
Menlo Park, CA 94025

William Brinkman
AT&T BELL LABORATORIES
600 Mountain Avenue
Mail Stop MH-1D-369
Murray Hill, NJ 07974

M.B. Brodsky
ARGONNE NATIONAL
LABORATORY
9700 S. Cass Avenue
Materials Science Division
Argonne, IL 60439

Phil Brody
HARRY DIAMOND LABS
RT-RA
2800 Powder Mill Road
Adelphi, MD 20783

John J. Brogan
DEPARTMENT OF ENERGY
1000 Independence Avenue, S.W.
Washington, D.C. 20585

Duncan Brown
ATM INC.
520-B Danbury Road
New Milford, CT 06776

Jerry Brown
ANSER
1215 Jefferson Davis Highway
Arlington, VA 22202

Kent Brown
MASSACHUSETTS INSTITUTE
OF TECHNOLOGY
77 Massachusetts Avenue
Cambridge, MA 02139

Paul J. Brown
DEPARTMENT OF ENERGY
1000 Independence Avenue, S.W.
Washington, D.C. 20585

Wayne S. Brown
W.S. BROWN, INC.
908 E. South Temple
Salt Lake City, UT 84102

Charles Brukl
FONAR CORPORATION
110 Marcus Drive
Melville, NY 11747

Herbert Bryan
PATENT & TRADEMARK OFFICE
2021 Jefferson Davis Highway
Arlington, VA 22206

William L. Buchanan
OFFICE OF SCIENTIFIC
TECHNOLOGY INFORMATION
Department of Energy
P.O. Box 62
Oak Ridge, TN 37830

John H. Bucher
LUKENS STEEL COMPANY
Coatesville, PA 19320

Joseph Budnick
UNIVERSITY OF CONNECTICUT
Physics U-46
Storrs, CT 06268

Professor Robert A. Buhrman
CORNELL UNIVERSITY
Applied Engineering Physics Dept.
Clark Hall
Ithaca, NY 14853

Amos G. Bullard
CAROLINA POWER & LIGHT
P.O. Box 1551
Raleigh, NC 27602

Sidney Bulter
LEHIGH UNIVERSITY
Bethlehem, PA 18015

Johst Burk
STAUFFER CHEMICAL COMPANY
Eastern Research Center
Dobbs Ferry, NY 10522

Joe Burke
DEPARTMENT OF COMMERCE
International Trade Administration
14th & Constitution Ave., N.W.
Washington, D.C. 20230

John Burke
LAMBERTVILLE CERAMIC
MANUFACTURING
COMPANY
Lambertville, NJ 08530

Sibley C. Burnett
ADVANCED CRYO MAGNETICS
P.O. Box 210132
San Diego, CA 92121

Calvin G. Burnham
HOUSTON LIGHTING & POWER
P.O. Box 1700
Houston, TX 77001

Sylvia Burns
SRA TECHNOLOGIES, INC
c/o Department of Energy
1000 Independence Avenue, S.W.
CE-43-1
Washington, D.C. 20585

Daniel F. Burton
COUNCIL ON COMPETITIVENESS
1331 Pennsylvania Avenue, N.W.
Suite 900
Washington, D.C. 20004

Wayne G. Burwell
UNITED TECHNOLOGY RESEARCH
CENTER
Silver Lane
East Hartford, CT 06108

Mary S. Butigan
EF HUTTON & COMPANY
31 West 52nd Street
New York, NY 10019

John C. Butler
UNIVERSITY OF HOUSTON
4800 Calhoun Road
214 SR I
Houston, TX 77004

Nathan Butler
BETEC, INC.
1952 Gallows Road
Vienna, VA 22180

Audrey Buyrn
U.S. CONGRESS
Office of Technology Assessment
600 Pennsylvania Avenue, S.E.
Washington, D.C. 20510

Thomas Byrer
BATTELLE COLUMBUS LABS
505 King Avenue
Columbus, OH 43201-2693

Ellena Byrne
TII INDUSTRIES, INC.
1375 Akron Street
Copiague, NY 11726

Charles E. Byvik
NASA LANGLEY RESEARCH
CENTER
MS/468
Hampton, VA 23665

Robert Calaway
RESOURCE MANAGEMENT
INTERNATIONAL INC.
P.O. Box 9503
McLean, VA 22102

Edward Caldwell
SIEMENS
2191 Laurelwood Road
Santa Clara, CA 95054

Thomas L. Callcott
UNIVERSITY OF TENNESSEE
4218 Hiawatha Drive
Knoxville, TN 37919

Steve Callis
EASTMAN KODAK CORPORATION
1300 N. 17th Street
Arlington, VA 22209

I.E. (Ricky) Campisi
SE. UNIV. RES. ASSOC., INC.
12070 Jefferson Avenue
Newport News, VA 23606

Dr. A.D. Caplin
IMPERIAL COLLEGE OF
SCIENCE & TECHNOLOGY
Solid State Physics Group
Prince Consort Road
LONDON SW7, ENGLAND

Dr. Donald W. Capone II
ARGONNE NATIONAL
LABORATORY
9700 South Cass Avenue
Argonne, Illinois 60439-4841

Linda Capuano
GLENWOOD MANAGEMENT
3000 Sand Hill Road
Building 4, Suite 230
Menlo Park, CA 94025

R.P. Caren
LOCKHEED CORPORATION
4500 Park Granda Boulevard
Calabasas, CA 91399

Charlie Carey
SANDERS ASSOCIATES
95 Canal Street
Nashua, NH 03060

Joseph D. Cargioli
GENERAL ELECTRIC CORPORATE
R&D
P.O. Box 8
Schenectady, NY 12301

Donald M. Carlton
RADIAN CORPORATION
P.O. Box 201088
Austin, TX 78720-1088

J.A. Carpenter
MARTIN MARIETTA
P.O. Box X
Building 45 15
Oak Ridge, TN 37831

Paul H. Carr
ROME AIR DEVELOPMENT CENTER
RADC/EEAC
Hanscomb AFB, MA 01731

Steve H. Carr
NORTHWESTERN UNIVERSITY
906 University Place
Evanston, IL 60201-9990

Richard A. Carrigan
FERMILAB
P.O. Box 500
MS 316
Batavia, IL 60510

Stephen Carson
SHEARSON LEHMAN BROTHERS
One Sansome Street
38th Floor
San Francisco, CA 94104

Brent Carter
GEORGIA INSTITUTE OF
 TECHNOLOGY
School of Materials Engineering
Atlanta, GA 30332-0245

Robert Carter
PIEZO ELECTRIC PRODUCTS
186 Massachusetts Avenue
Cambridge, MA 02139

Richard B. Cass
HiTc SUPERCON COMPANY
P.O. Box 128
Lambertville, NJ 08530

Michael Cassidy
OXFORD INSTRUMENTS
34 Alfred Circle
Bedford, MA 01730

G. Richard Cataldi
ASSOCIATION OF AMERICAN
RAILROADS
50 F Street, N.W.
Washington, D.C. 20001

Richard Cataldi
ASSOCIATION OF AMERICAN
RAILROADS
50 F Street, N.W.

Robert M. Catchings
HOWARD UNIVERSITY
2355 6th Street, N.W.
Washington, D.C. 20059

Charles J. Cauchy
TELLUREX
1225 Woodman
Traverse City, MI 49684

Robert J. Cava
AT&T BELL LABORATORIES
600 Mountain Avenue
Murry Hill, New Jersey 07974

John M. Cawley
TII INDUSTRIES, INC.
1375 Akron Street
Copiague, NY 11726

Gerald P. Ceaser
STANDARD OIL ENGINEERED
MATERIALS COMPANY
4440 Warrensville Road
Cleveland, OH 44128

Walter S. Cebulak
ALUMINUM COMPANY OF AMERICA
Alcoa Technical Center
Alcoa Center, PA 15069-0001

Bert Chamberland
UNIVERSITY OF CONNECTICUT
Chemistry U-60
Storrs, CT 06268

Dudley Chance
IBM
Old Orchard Road
Armonk, New York 10504

Grish Chandra
DOW CORNING
Mail Stop C41C00
2200 Salzbury Road
Midland, MI 48686-0994

Ching-Sung Chang
GENERAL RESEARCH
CORPORATION
635 Discovery Drive
Huntsville, AL 35806

Jack C. Chang
EASTMAN KODAK CORPORATION
343 State Street
Building 82
Rochester, NY 14650

E.W. (Chap) Chappell
SCIENCE & ENGINEERING
ASSOCIATES, INC.
1420 King Street
Suite 400
Alexandria, VA 22314

Damon Chappie
BUREAU OF NATIONAL AFFAIRS
1231 25th Street, N.W.
Washington, D.C. 20037

Richard H. Chastain
SOUTHERN COMPANY SERVICE
P.O. Box 2625
Birmingham, AL 35202

Praveen Chaudhari
IBM CORPORATION
Watson Research Center
P.O. Box 218
Yorktown Heights, NY 10598

Monte S. Chawla
SYSTEM PLANNING CORPORATION
1500 Wilson Blvd.
Arlington, VA 22209-2454

Gregory Chen
EATON CORPORATION
4201 North 27th Street
Milwaukee, WI 53215

Jar-Mo Chen
MARTIN MARIETTA LABORATORIES
1450 South Rolling Road
Baltimore, MD 21227

H.N. Cheng
HERCULES INC.
RESEARCH CENTER
Wilmington, DE 19894

H. Wendel Cherry
HUMANA, INC.
500 West Main Street
Louisville, KY 40202

Albert A. Chesnes
DEPARTMENT OF ENERGY
1000 Independence Avenue, S.W.
Washington, D.C. 20585

Arthur W. Chester
MOBIL R&D CORPORATION
Paulsboro Research Laboratory
Paulsboro, NJ 08066

Albert Chiang
PITNEY BOWES
Lot 2
Zinn Road
Danbury, CT 06811

Roger Chiarodo
DEPARTMENT OF COMMERCE
International Trade Administration
14th & Constitution Ave., N.W.
Washington, D.C. 20230

C.L. Chien
JOHNS HOPKINS UNIVERSITY
Department of Physics & Astronomy
34th & Charles Street
Baltimore, MD 21218

Wai-Yim Ching
UNIVERSITY OF MISSOURI
1110 East 48th Street
Physics Department
Kansas City, MO 64110

John R. Chirichiello
NATIONAL SCIENCE FOUNDATION
1800 G Street, N.W.
Washington, D.C. 20550

Dr. Uma Chowdhry
E.I. DUPONT
Central R&D Department
Wilmington, DE 19898

Albert T. Christensen
GENERAL ELECTRIC
1331 Pennsylvania Avenue, N.W.
Suite 895
Washington, D.C. 20004

Clayton M. Christensen
CERAMICS PROCESS SYSTEMS
840 Memorial Drive
Cambridge, MA 02139

Linda E. Christenson
COORS CERAMICS/ADOLPH
COORS CO.
601 Pennsylvania Avenue, N.W.
Suite 500
Washington, D.C. 20004

M. Christiansen
TEXAS TECH UNIVERSITY
Dept. Electrical Engineering
Lubbock, TX 79409-4439

Paul Christian
EASTMAN KODAK CORPORATION
343 State Street
Research Lab
Rochester, NY 14650

Paul C.W. Chu
UNIVERSITY OF HOUSTON
4800 Calhoun Road
Houston, Texas 77004

Kukjin Chun
WASHINGTON STATE UNIVERSITY
Department of Physics
Pullman, WA 99164-2210

R.J. Churchill
AMERICAN RESEARCH
CORPORATION
P.O. Box 3406
Radford, VA 24143

Dr. Michael J. Cima
MASSACHUSETTS INSTITUTE
OF TECH.
Dept. Materials Science &
Engineering
Cambridge, MA 02138

Michael Cima
MASSACHUSETTS INSTITUTE
OF TECHNOLOGY
Ceramics Processing Laboratory
Room 12011
Cambridge, MA 02139

Lewis T. Claiborne
TEXAS INSTRUMENTS
P.O. Box 655936
MS 145
Dallas, TX 75265

Alan F. Clark
NATIONAL BUREAU OF STANDARDS
325 Broadway
Boulder, CO 80303

David Clark
UNIVERSITY OF FLORIDA
Department of
Material Science & Engineering
136 Mae Building
Gainesville, FL 32611

Judy A. Clark
MONOLITHIC SUPERCONDUCTORS
P.O. Box 1654
Lake Oswego, OR 97035

John Clarke
UNIVERSITY OF
CALIFORNIA-BERKELEY
Lawrence Berkeley Laboratory
Berkeley, CA 94720

John Clement
NATIONAL ACADEMY OF SCIENCES
2101 Constitution Avenue, N.W.
Washington, D.C. 20418

William L. Clinton
NATIONAL SCIENCE FOUNDATION
1800 G Street, N.W.
Washington, D.C. 20550

Albert M. Clogston
LOS ALAMOS NATIONAL
LABORATORY
P.O. Box 1663
MS K765
Los Alamos, NM 87545

James Clum
SUNY, BINGHAMTON
Vestal Parkway East
Binghamton, NY 13901

Bill Coblenz
NORTON COMPANY
Goddard Road
Northborough, MA 01532-1545

C. Norman Cochran
ALCOA TECHNICAL CENTER
Alcoa Center, PA 15069

Bruno L. Codispoti
OHIO EDISON COMPANY
76 S. Main Street
Akron, OH 44308-1890

Dr. Morrell H. Cohen
EXXON RESEARCH & ENGINEERING
Route 22, East
Clinton Township
Annandale, New Jersey 08801

Hirsh Cohen
IBM CORPORATION
Watson Research Center
P.O. Box 218
Yorktown Heights, NY 10598
(914) 945-3000

Marvin L. Cohen
UNIVERSITY OF
CALIFORNIA-BERKELEY
Department of Physics
Berkeley, CA 94720

Daniel R. Cohn
MIT PLASMA FUSION CENTER
167 Albany Street
NW 16-140
Cambridge, MA 02139

Barry Cole
HONEYWELL, INC.
10701 Lyndale Avenue South
Bloomington, MN 55420

Henry Cole
MASSACHUSETTS INSTITUTE OF
TECHNOLOGY
77 Massachusetts Avenue
Cambridge, MA 02139

James Coleman
DEPARTMENT OF ENERGY
1000 Independence Avenue, S.W.
Washington, D.C. 20585

Richard Coleman
RVC INSTRUMENTS CORPORATION
529 14th Street, N.W.
Washington, D.C. 20045

Myron A. Coler
COLER ENGINEERING COMPANY
One Washington Square
New York, NY 10012

Ward Collin
DOW CORNING
Mail Stop C41C00
2200 Salzbury Road
Midland, MI 48686-0994

E.W. Collings
BATTELLE COLUMBUS LABS
505 King Avenue
Columbus, OH 43201-2693

Joey Colwell
J.M. HUBER CORPORATION
Kaolin Division
P.O. Box 30833
Wrens, Georgia 30833

Denis Connolly
NASA
2100 Brookpark Road
Cleveland, OH 44135

John Connolly
NATIONAL SCIENCE FOUNDATION
1800 G Street, N.W.
Washington, D.C. 20550

David E. Conrad
GARRETT CORPORATION
1625 Eye Street, N.W.
Suite 520
Washington, D.C. 20006

Thomas N. Cooley
NATIONAL SCIENCE FOUNDATION
1800 G Street, N.W.
Washington, D.C. 20550

Bill Costerhuis
OFFICE OF SCIENCE &
TECHNOLOGY POLICY
5002 New Executive Office Building
Washington, D.C. 20508

Greg Covax
GLENWOOD MANAGEMENT
3000 Sand Hill Road
Building 4, Suite 230
Menlo Park, CA 94025

N. Walter Cox
GEORGIA INSTITUTE OF
 TECHNOLOGY
School of Materials Engineering
Atlanta, GA 30332-0245

Bill Craig
BECHTEL NATIONAL, INC.
P.O. Box 6935
San Francisco, CA 94119

Charles H. Craig
DEPARTMENT OF DEFENSE
400 Army-Navy Drive
Suite 300
Arlington, VA 22202

R.W. Craig
LTV
P.O. Box 655907
Dallas, TX 75265-5907

Chris Crane
EER, INC.
1801 Alexander Bell Drive
Reston, VA 22192

Robert W. Crane
PEREGRINE VENTURES
20833 Stevens Creek Boulevard
Cupertino, CA 95014

David L. Cross
3M
3M Center
St. Paul, MN 55144-1000

L. Eric Cross
PENNSYLVANIA STATE
UNIVERSITY
202 Materials Research Laboratory
University Park, PA 16802

Jack Crow
TEMPLE UNIVERSITY
Physics Department
Philadelphia, PA 19122

Lowell Crow
LASER GENETICS CORPORATION
2362-E Cume Drive
San Jose, CA 95131

John C. Crowley
ASSOCIATION OF AMERICAN
UNIVERSITIES
One Dupont Circle, N.W.
Washington, D.C. 20036

Floyd L. Culler
ELECTRIC POWER RESEARCH
INSTITUTE
3412 Hillview Avenue
P.O. Box 10412
Palo Alto, CA 94303

William H. Culver
OPTELECOM INC.
15930 Luanne Drive
Gaithersburg, MD 20877

George B. Cvijanovich
AMP INCORPORATED
P.O. Box 3608
Harrisburg, PA 17105-3608

George A. Cypher
INTERNATIONAL
COPPER RESEARCH
708 Third Avenue
27th Floor
New York, NY 10017

Bhadrik Dalal
AMDAHL CORPORATION
1250 East Arques
Sunnyvale, CA 94086

Ford A. Daley
DYNATECH SCIENTIFIC
99 Erie Street
Cambridge, MA 02139

Ed Daniels
ARGONNE NATIONAL
LABORATORY
9700 S. Cass Avenue
Argonne, IL 60439

Joseph C. Danko
UNIVERSITY OF TENNESSEE
College of Engineering
53 Turner House
Knoxville, TN 37996-2000

S.J. Dapkunas
NATIONAL BUREAU OF STANDARDS
Building 223
Gaithersburg, MD 20899

Michael R. Darby
U.S. TREASURY DEPARTMENT
15th & Pennsylvania Avenues, N.W.
Room 3454
Washington, D.C. 20220

Amit DasGupta
DEPARTMENT OF ENERGY
9800 S. Cass Avenue
Building 362
Argonne, IL 60439

E. David Daugherty
TENNESSEE VALLEY AUTHORITY
3N 79A Missionary Ridge Place
Chattanoga, TN 37402-2801

James M. Daughton
HONEYWELL INCORPORATED
12001 State Highway 55
Plymouth, MN 55441

L. Craig Davis
FORD MOTOR COMPANY
Scientific Research Laboratory
Dearborn, MI 58121-2053

Alberta Dawson
MIT PLASMA FUSION CENTER
167 Albany Street
Cambridge, MA 02139

John F. Day III
STRATEGIES UNLIMITED
201 San Antonio Circle
Room 201
Mt. View, CA 94040

Theo De Winter
BOSTON UNIVERSITY
2 Glengarry
Winchester, MA 01890

Marcello A. DeGiorgis
PIRELLI ENTERPRISES
CORPORATION
800 Rahway Avenue
Union, NJ 07083

Adriaam M. DeGraaf
NATIONAL SCIENCE FOUNDATION
1800 G Street, N.W.
Washington, D.C. 20550

Mark R. DeGuire
CASE WESTERN RESERVE
UNIVERSITY
10900 Euclid Avenue
Cleveland, OH 44106

Lutgard C. DeJonghe
LAWRENCE BERKELEY
LABORATORY
One Cyclotron Road
Berkeley, CA 94720

Mike Deal
STANFORD UNIVERSITY
200 Junipero Serra Boulevard
CIS 113
Stanford, CA 94305

Satyen Deb
SOLAR ENERGY RESEARCH
INSTITUTE
1617 Cole Boulevard
Golden, CO 80401

Dr. Jagadish Debsikdar
IDAHO NATIONAL ENGINEERING
LAB
INEL Research Center
Box 1625
Idaho Falls, ID 83415

Rudolf Decher
NASA
Marshall Space Flight Center
E901
Marshall SFC, AL 35812

C. David Decker
GTE LABORATORIES, INC.
40 Sylvan Road
Waltham, MA 02254

James Decker
DEPARTMENT OF ENERGY
1000 Independence Avenue, S.W.
Washington, D.C. 20585

Clare Delmar
OFFICE OF TECHNOLOGY
 ASSESSMENT
U.S. Congress
Washington, D.C. 20510

Lance Delong
UNIVERSITY OF KENTUCKY
Physics Department
Physics 0510
Lexington, KY 40506

Anthony J. Demaria
UNITED TECHNOLOGY RESEARCH
CENTER
Silver Lane
MS 77
East Hartford, CT 06108

Reid W. Dennis
INSTITUTIONAL VENTURE
 PARTNERS
3000 Sand Hill Road
Suite 290
Menlo Park, CA 94025

Fred Denny
EDISON ELECTRIC INSTITUTE
1111 19th Street, N.W.
Washington, D.C. 20036-3691

Victor Der
DEPARTMENT OF ENERGY
1000 Independence Avenue, S.W.
Washington, D.C. 20585

Jack Devine
ELECTRIC POWER RESEARCH
INSTITUTE
3412 Hillview Avenue
P.O. Box 10412
Palo Alto, CA 94303

Ralph Devries
UNIVERSITY OF WYOMING
P.O. Box 3355
Physics Department
Laramie, WY 82071

Nicholas DiGiacomo
THE KEYWORTH COMPANY
3050 K Street, N.W.
Suite 360
Washington, D.C. 20007

Robert C. Dickeson
STATE OF COLORADO
136 State Capital
Denver, CO 80203

John Dickman
UNIVERSITY OF CINCINNATI
ML #30
Cincinnati, OH 45221

Edward Dickson
ADVANTAGE QUEST, INC.
1110 Sunnyvale-Saratoga Road
Sunnyvale, CA 94087-0818

John Dimmock
BOLLING AIR FORCE BASE
Washington, D.C. 20331

Arthur Diness
OFFICE OF NAVAL RESEARCH
800 N. Quincy Street
Arlington, VA 22217-5000

David A. Dingee
DEPARTMENT OF ENERGY
1000 Independence Avenue, S.W.
Washington, D.C. 20585

David Ditmars
NATIONAL BUREAU OF STANDARDS
Materials Building
Gaithersburg, MD 20899

Harvey L. Dixon
PALMER SERVICE CORPORATION
300 Unicorn Park Drive
Woburn, MA 01801

David Douglass
UNIVERSITY OF ROCHESTER
Physics Department
Rochester, NY 14627

Eugenia Douglass
UNIVERSITY OF ROCHESTER
Physics Department
Rochester, NY 14627

Grant A. Dove
MICROELECTRONICS AND
COMPUTER TECHNOLOGY
3500 W. Balcones Center Drive
Austin, TX 78759

David Dragnich
BATTELLE NORTHWEST LABS
2033 Trevino Court
P.O. Box 999
Richland, WA 99352-9785

Mark J. Dreiling
PHILLIPS PETROLEUM COMPANY
122 AL
Bartlesville, OK 74004

Dennis Drew
UNIVERSITY OF MARYLAND
Physics Department
College Park, MD 20742

A.J. Driscoll
BOLLING AIR FORCE BASE
AFOSR/CC
Washington, D.C. 20331

Frank Druding
FORD AEROSPACE &
COMMUNICATIONS
3939 Fabian Way
Palo Alto, CA 94303

Timonthy Dues
AIR FORCE SYSTEM COMMAND
HQAFSC/DLAA
Andrews AFB, MD 20334-5000

Edward J. Dulis
CRUCIBLE MATERIALS
CORPORATION
P.O. Box 88
Pittsburgh, PA 15230

Bradford A. Dulmaine
CARPENTER TECHNOLOGY
 CORPORATION
P.O. Box 662
Reading, PA 19603

R.F. Dundervill
BDM CORPORATION
7915 Jones Branch Drive
McLean, VA 22102

Edgar Durbin
OD&E/CSG/SED
Washington, D.C. 20505

Thomas J. Dwyer
CORNING
Sullivan Park
FR025
Corning, NY 14831

Alexander P. Dyer
AIR PRODUCTS & CHEMICALS, INC.
P.O. Box 538
Route 222
Allentown, PA 18105

Robert C. Dynes
AT&T BELL LABORATORIES
600 Mountain Avenue
Mail Stop MH-1D-369
Murray Hill, NJ 07974

Robert Eagaly
BIOMAGNETIC TECHNOLOGIES
4174 Sorrento Valley Blvd.
San Diego, CA 92121

Robert J. Eagan
SANDIA NATIONAL LABORATORIES
P.O. Box 5800
Department 1840/1150
Albuquerque, NM 87185-5800

James Early
NATIONAL BUREAU OF STANDARDS
Administrative A-1002
Gaithersburg, MD 20899

Russell Eaton III
DEPARTMENT OF ENERGY
1000 Independence Avenue, S.W.
Washington, D.C. 20585

Chris Ebel
NORTON COMPANY
Goddard Road
Northborough, MA 01532-1545

James Eberhardt
DEPARTMENT OF ENERGY
1000 Independence Avenue, S.W.
Washington, D.C. 20585

H. Eckhardt
ALLIED SIGNAL CORPORATION
Columbia Road
P.O. Box 1021 R
Morristown, NJ 07960

Edgar Edelsack
GEORGETOWN CRYOSENIC
INFORMATION
CENTER
3530 W Place, N.W.
Washington, D.C. 20007

Dayton D. Eden
LTV MISSILES & ELECTRIC GROUP
P.O. Box 650003
MS-TH-85
Dallas, TX 75265-0003

David Edgerly
NATIONAL BUREAU OF STANDARDS
Building 101
Gaithersburg, MD 20899

Ester Edgerton
MASSACHUSETTS INSTITUTE OF
TECH.
Dept. Materials Science & Eng.
Cambridge, MA 02138

Harold E. Edgerton
MASSACHUSETTS INSTITUTE OF
TECH.
Dept. Materials Science & Eng.
Electronic Materials
Cambridge, MA 02138

James S. Edmonds
ELECTRIC POWER RESEARCH
INSTITUTE
3412 Hillview Avenue
P.O. Box 10412
Palo Alto, CA 94303

Dr. Charles C. Edwards
SCRIPPS CLINIC AND
RESEARCH FOUNDATION
10666 North Torrey Pines Road
La Jolla, CA 92037

Kenneth R. Efferson
AMERICAN MAGNETICS INC.
112 Flint Road
P.O. Box 2509
Oak Ridge, TN 37831-2509

P.M. Eisenberger
EXXON RESEARCH & ENGINEERING
CO.
Route 22 East
Clinton Township
Annandale, NJ 08801

Jack Ekin
NATIONAL BUREAU OF STANDARDS
325 Broadway
Mail Stop 724.05
Boulder, CO 80303

Charles Elbaum
BROWN UNIVERSITY
Box 1862
Providence, RI 02912

J. David Ellett Jr.
PENNIE & EDMONDS
1155 Avenue of the Americas
New York, NY 10036

Don Elliott
GENERAL ELECTRIC
Medical Systems Group
P.O. Box 414
Milwaukee, WI 53201

Frank Elliott
NAVY NEWS AND UNDERSEA
TECHNOLOGY
1401 Wilson Boulevard
Suite 910
Arlington, VA 22209

Dr. Victor J. Emery
BROOKHAVEN NATIONAL
 LABORATORY
Building 134
Upton, New York 11973

Tom England
INDUSTRIAL DRIVES
201 Rock Road
Radford, VA 34141

Ed Engler
IBM ALMADEN RESEARCH CENTER
650 Harry Road
Mail Stop K32/803(D)
San Jose, CA 95120-6099

Mike Epstein
BATTELLE COLUMBUS LABS
505 King Avenue
Columbus, OH 43201-2693

Donald E. Erb
DEPARTMENT OF ENERGY
1000 Independence Avenue, S.W.
Washington, D.C. 20585

Craig Ernsberger
CTS CORPORATION
905 North West Boulevard
Elkhart, IN 46514

Dennis K. Evans
U.S. ARMY
1800 Jefferson Park Avenue
#69
Charlottesville, VA 22903

E.A. Evans
HANFORD ENGINEERING
Westinghouse
Richland, WA 99352

Michael Evans
CASTLE TECHNOLOGY
52 Dragon Court
Woburn, MA 01801

Tony Evans
UNIVERSITY OF CALIFORNIA
AT SANTA BARBARA
College of Engineering
Santa Barbara, CA 93106

Tony Evensen
3M
3M Center
St. Paul, MN 55144-1000

Greg Eyring
U.S. CONGRESS
Office of Technology Assessment
600 Pennsylvania Avenue, S.E.
Washington, D.C. 20510

Yehia Eyssa
UNIVERSITY OF WISCONSIN
1500 Johnson Drive
Physics Department
Madison, WI 53706

John Fabian
ANSER
1215 Jefferson Davis Highway
Arlington, VA 22202

William M. Fairbank
STANFORD UNIVERSITY
200 Junipero Serra Boulevard
Physics Department
Stanford, CA 94305

Michael Fallon
GLENWOOD MANAGEMENT
3000 Sand Hill Road
Building 4, Suite 230
Menlo Park, CA 94025

Sadeg M. Faris
HYPRES INC.
500 Executive Boulevard
Elmsford, NY 10523

Andrew D. Farrell
IIT RESEARCH INSTITUTE
1825 K Street
Suite 1101
Washington, D.C. 20006

Joseph Farrell
TELEDYNE BROWN ENGINEERING
Cummings Research Park
Huntsville, AL 35807

J.S. Faulkner
FLORIDA ATLANTIC UNIVERSITY
Physics Department
Boca Raton, FL 33431

Patricia Feafock
V. GARBER INTERNATIONAL
ASSOCIATION
1215 Jefferson Davis Highway
Suite 1105
Arlington, VA 22202

Harold Federow
BOEING COMMERCIAL
AIRPLANE CO.
Seattle, WA 98124

Larry Fehrenbacker
TA&T INC.
2002 Huntwood Drive
Cambrills, MD 21054

Ellen O. Feinberg
AMES LABORATORY
Iowa State University
12 Physics
Ames, IA 50011

Alan Feinerman
THERMO ELECTRON
101 1st Avenue
P.O. Box 9046
Waltham, MA 02254-9046

Gene D. Feit
HARRIS CORPORATION
P.O. Box 94000
Melbourne, FL 32902

Albert Feldman
NATIONAL BUREAU OF STANDARDS
Materials Building
Gaithersburg, MD 20899

Saul Feldman
APPLIED SUPERCONDUCTIVITY
1533 Monrovia Avenue
Newport Beach, CA 92663

Justo Fernandez
SUPERCON INCORPORATED
830 Boston Turnpike
Shrewsbury, MA 01545

Raymond R. Fessler
BASIC INDUSTRY RESEARCH LAB
2020 Ridge Avenue
2nd Floor
Evanston, IL 60201

H. Fickenscher
AT&T BELL LABORATORIES
260 Cherry Hill Road
Parsippany, NJ 07054

Chase Fielding
PEABODY BARNES
1500 Massachusetts Avenue
Washington, D.C. 20005

Paul Filios
DEFENSE NUCLEAR AGENCY
7187 Fairfield Court
Alexandria, VA 22306

John Fillo
SUNY, BINGHAMTON
Vestal Parkway East
Binghamton, NY 13901

Douglas Finnemore
AMES LABORATORY
Ames, IO 50011

Robert G. Finney
ELECTRONIC ASSOCIATES, INC.
185 Monmouth Parkway
W. Long Branch, NJ 07764

Richard Fischer
GARRETT AIRESEARCH
2525 West 190th Street
Torrance, CA 90509

Patrick Fitch
GOULD ELECTRONICS
1755 Jefferson Davis Highway
Arlington, VA 22203

Donna R. Fitzpatrick
DEPARTMENT OF ENERGY
1000 Independence Avenue, S.W.
Washington, D.C. 20585

Terry Fleener
BALL AEROSPACE SYSTEMS
P.O. Box 1062
Boulder, CO 80306

Paul Fletcher
TERMIFLEX CORPORATION
316 Daniel Webster Highway
Merrimack, NH 03054

W. Wendell Fletcher
U.S. CONGRESS
Office of Technology Assessment
600 Pennsylvania Avenue, S.E.
Washington, D.C. 20510

Paul E. Fleury
AT&T BELL LABORATORIES
600 Mountain Avenue
Mail Stop MH-1D-369
Murray Hill, NJ 07974

Carl Flick
WESTINGHOUSE ELECTRIC
 CORPORATION
4400 Alafaya Trail
GRSD, MC-100
Orlando, FL 32826-2399

Jonathan A. Flint
BURR, EGAN, DELEAGE &
 COMPANY
One Post Office Square
Boston, MA 02109

Eric B. Forsyth
BROOKHAVEN NATIONAL
LABORATORY
Accelerator Development Dept.
Upton, NY 11973

Joan S. Fortune
FRONTENAC VENTURE COMPANY
208 South LaSalle Street
Chicago, IL 60604

William Fox
NAVAL RESEARCH LABORATORY
4555 Overlook Avenue, S.W.
Code 6101
Washington, D.C. 20375-5000

Arthur G. Foyt
UNITED TECHNOLOGY RESEARCH
CENTER
Silver Lane
East Hartford, CT 06108

Frank Fradin
ARGONNE NATIONAL
 LABORATORY
9700 S. Cass Avenue
Materials Science Division
Argonne, IL 60439

David H. Fradkin
WAYNE STATE UNIVERSITY
Physics & Astronomy Department
Detroit, MI 48202

Ellyn Frances
SUPERCONDUCTIVITY NEWS
65 Jackson Drive
Suite 2000
Cranford, NJ 07016

Joseph Frank
NATIONAL INSTITUTES OF HEALTH
Magnetic Resonance Imaging
Research Clinical Center
Building 10
Bethesda, MD 20892

Donald C. Fraser
C.S. DRAPER LABORATORY
555 Technology Square
Cambridge, MA 02139

Stephen W. Freiman
NATIONAL BUREAU OF STANDARDS
Building 223
Gaithersburg, MD 20899

Bruce Friedman
DAVID TAYLOR NAVAL SHIP R&D
CENTER
2812
Annapolis, MD 21402-5067

Ken Friedman
DEPARTMENT OF ENERGY
1000 Independence Avenue, S.W.
Washington, D.C. 20585

Milo E. Friesen
RANSBURG CORPORATION
3939 West 56th Street
Indanapolis, IN 46254

Paul J. Fromme
EASTMAN KODAK CORPORATION
343 State Street
Rochester, NY 14650

Robert Frosch
GENERAL MOTORS RESEARCH LAB
30200 Mound Road
Warren, MI 48090-9055

Brian R.T. Frost
ARCH DEVELOPMENT
CORPORATION
115-25 E. 56th Street
Chicago, IL 60637

Robert Fry
MICRON METALS
8728 S. Littlecloud Road
Sandy, UT 84092

Samuel Fuller
DIGITAL EQUIPMENT
CORPORATION
146 Main Street
Maynard, MA 01754

Don Fuqua
AEROSPACE INDUSTRIES
ASSOCIATION
OF AMERICA
1725 DeSales Street, N.W.
Washington, D.C. 20036

Rosendo Fuquen
TIMKEN COMPANY
1835 Dueber Avenue, S.W.
Canton, OH 44706

Theodore H. Gabelle
STANFORD UNIVERSITY
Stanford, CA 94305

William J. Gallagher
IBM CORPORATION
Watson Research Center
P.O. Box 218
Yorktown Heights, NY 10598

R. Ganapathy
J.T. BAKER CHEMICAL COMPANY
3510 LaFayette Drive
Bethlehem, PA 18017

John Garder
OREGON STATE UNIVERSITY
Corvallis, OR 97331-4003

E.M. Gardiner
PRELIMINARY DESIGN
P.O. Box 3999
Seattle, WA 98124

John A. Gardner Jr.
OREGON STATE UNIVERSITY
Corvallis, OR 97331-4003

Angelo Gattozzi
TYLER POWER SYSTEMS
8684 Tyler Boulevard
Mentor, OH 44060

Laurie Gavrin
U.S. CONGRESS
Office of Technology Assessment
600 Pennsylvania Avenue, S.E.
Washington, D.C. 20510

Theodore H. Geballe
STANFORD UNIVERSITY
Applied Physics
Stanford, CA 94305

Allan Gelb
MARLOW INDUSTRIES, INC.
10451 Vista Park Road
Dallas, TX 75238

Thomas F. George
STATE UNIVERSITY OF NEW YORK
BUFFALO
411 Capen Hall
North Campus
Buffalo, NY 14260

Bill Georgen
SANTA FE CORPORATION
4900 Seminar Road
Suite 500
Alexandria, VA 22311

Reza Ghafurian
CON EDISON
4 Irving Place
Room 1428
New York, NY 10003

B.B. Ghate
AT&T BELL LABORATORIES
555 Union Boulevard
Allentown, PA 18103

Ali H. Ghovanlou
THE MITRE CORPORATION
7525 Colshire Drive
McLean, VA 22102

Philip T. Gianos
INTERWEST PARTNERS
3000 Sand Hill Road
Building 3
Suite 255
Menlo Park, CA 94025

Carolyn Gibbard
CARATEC, INC.
P.O. Box 1585
Wayne, NJ 07470

H. Frank Gibbard
POWER CONVERSION, INC
495 Boulevard
Elmwood Park, NJ 07407

Chet F. Giermak
ERIEZ MAGNETICS
23rd & Asbury Road
P.O. Box 10608
Erie, PA 16514

Robert F. Giese
ARGONNE NATIONAL
LABORATORY
9700 S. Cass Avenue
Materials Science Division
Argonne, IL 60439

Bill C. Giessen
BARNETT INSTITUTE
Northeastern University
Boston, MA 02115

John Gilman
LAWRENCE BERKELEY
LABORATORY
One Cyclotron Road
Berkeley, CA 94720

Charles Gilmore
GEORGE WASHINGTON
UNIVERSITY
School of Engineering and
Applied Science
Washington, D.C. 20052

Donald M. Ginsberg
UNIVERSITY OF ILLINOIS
1110 West Green Street
Urbana, IL 61801

William Glaberson
NATIONAL SCIENCE FOUNDATION
1800 G Street, N.W.
Washington, D.C. 20550

William A. Glass
BATTELLE NORTHWEST LABS
2033 Trevino Court
P.O. Box 999
Richland, WA 99352-9785

Tim F. Glennon
SUNDSTRAND CORPORATION
4747 Harrison Avenue
P.O. Box 7002
Rockford, IL 61125

Maurice Glicksman
BROWN UNIVERSITY
Box 1862
Providence, RI 02912

Rolfe E. Glover III
UNIVERSITY OF MARYLAND
Physics Department
College Park, MD 20742

Nev Gokcen
BUREAU OF MINES
P.O. Box 70
Albany, OR 97321

Gerald Goldberg
U.S. PATENT & TRADEMARK OFFICE
Crystal Plaza, Bldg 4, Rm 9D19
Washington, D.C. 20231

Dr. Allen Goldman
UNIVERSITY OF MINNESOTA
Department of Physics
116 Church Street
Minneapolis, MN 55454

Edwin Goldwasser
LAWRENCE BERKELEY
LABORATORY
Central Design Group for the SSC
Building 90—4040
Berkeley, CA 94720

Jeremy A. Good
CRYOGENIC CONSULTANTS
Metrostore Building
231 The Vale
London W3 7QS, ENGLAND

Mary Good
ALLIED-SIGNAL, CORPORATION
50 E. Algonquin Road
Box 5018
Des Planes, IL 60017-5016

J.B. Goodenough
UNIVERSITY OF TEXAS
Engineering Division
Austin, TX 78712

David M. Goodman
NJ COMMISSION ON SCIENCE &
TECHNOLOGY
122 West State Street
CN 832
Trenton, NJ 08625

R.B. Goodrich
LOUISIANA STATE UNIVERSITY
Department of Physics
Baton Rouge, LA 70803

Ken Gordon
NATIONAL BUREAU OF STANDARDS
Building 101
Gaithersburg, MD 20899

Ron Gordon
CERAMATEC, INCORPORATED
2425 South 900 West
Salt Lake City, UT 84119

Ronald S. Gordon
CERAMATEC, INCORPORATED
2425 South 900 West
Salt Lake City, UT 84119

Scott Gordon
DIGITAL EQUIPMENT
CORPORATION
77 Reed Road
Hudson, MA 01749

William R. Graham
OFFICE OF THE SCIENCE ADVISOR
360 Old Executive Office Building
Washington, D.C. 20506

Paul M. Grant
IBM ALMADEN RESEARCH CENTER
650 Harry Road
Mail Stop K32/803(D)
San Jose, CA 95120-6099

Robert Grant
INTERNATIONAL ENERGY
ASSOCIATION
LTD
3211 Germantown Road
Fairfax, VA 22030

Reau Graves Jr.
GA TECHNOLOGIES/APPLIED
SUPERCONETICS
P.O. Box 85608
San Diego, CA 92138

Allen G. Gray
ASM INTERNATIONAL
4301 Esteswood Drive
Nashville, TN 37215

Paul E. Gray
MASSACHUSETTS INSTITUTE
OF TECHNOLOGY
77 Massachusetts Avenue
Cambridge, MA 02139

James C. Greene
COMMITTEE ON SCIENCE, SPACE,
AND TECHNOLOGY
2321 Rayburn HOB
Washington, D.C. 20515

Matthew Greenfield
ABS VENTURES L.P.
135 East Baltimore Street
Baltimore, MD 21202

James M. Greenleaf
GREENLEAF CORPORATION
1 Greenleaf Drive
Saegertown, PA 16433

Charryl L. Greenwood
ENGELHARD CORPORATION
Menlo Park, CN40
Edison, NJ 08818

Eric Gregory
SUPERCON INCORPORATED
830 Boston Turnpike
Shrewsbury, MA 01545

William D. Gregory
DIOTEC CORPORATION
P.O. Box 187
Potsdam, NY 13676

Paul J. Gripshover
SOUTHWALL TECHNOLOGIES INC.
1029 Corporation Way
Palo Alto, CA 94303

Michael I. Grove
GROVE & ASSOCIATES
996 Riverside Drive
22
Burbank, CA 91506

W. Andrew Grubbs
EMERGING GROWTH PARTNERS
400 East Pratt Street
Suite 610
Baltimore, MD 21202

Harold L. Grubin
SCIENTIFIC RESEARCH
ASSOCIATES, INC.
P.O. Box 1058
50 Nye Road
Glastonbury, CT 06033

Thomas W. Grudkowski
UNITED TECHNOLOGY RESEARCH
CENTER
Silver Lane
East Hartford, CT 06108

John A. Gruver
WESTINGHOUSE
1801 K Street, N.W.
Washington, D.C. 20006

Jacob Gubbay
C.S. DRAPER LABORATORY
555 Technology Square
Cambridge, MA 02139

Don Gubser
NAVAL RESEARCH LABORATORY
4555 Overlook Avenue, S.W.
CODE 4630
Washington, D.C. 20375-5000

Arthur H. Guenther
AIR FORCE WEAPONS LAB
AFWL/CA
Kirtland AFB, NM 87117-6008

Peter Guernsey
GUERNSEY COATING LABS
4464 McGrath Street
Unit 106
Ventura, CA 93003

Robert P. Guertin
TUFTS UNIVERSITY
120 Packard Avenue
Medford, MA 02155

Terry Gulden
GA TECHNOLOGIES/APPLIED
SUPERCONETICS
P.O. Box 85608
San Diego, CA 92138

James L. Gumnick
OAK RIDGE ASSOCIATED
UNIVERSITY
P.O. Box 117
Oak Ridge, TN 37831-0117

Indra Gupta
INLAND STEEL COMPANY
Research Labs
3001 E. Columbus Drive
East Chicago, IL 46312

William M. Gust II
BROVENTURE CAPITAL
MANAGEMENT
16 West Madison Street
Baltimore, MD 21201

W.C. Guyker
ALLEGHENY POWER SYSTEM
800 Cabin Hill Drive
Greensburg, PA 15601

Jaime Guzman
NEWMONT MINING CORPORATION
200 Park Avenue
36th Floor
New York, NY 10166

Robert A. Hageman
KIDDER PEABODY & COMPANY
10 Hanover Square
New York, NY 10005

Ross Haghighat
FOSTER-MILLER INC.
350 Second Avenue
Waltham, MA 02254

Gerald M. Haines
HYPRES INC.
500 Executive Boulevard
Elmsford, NY 10523

John D. Hale
KERR-MCGEE CORPORATION
P.O. Box 25861
Oklahoma City, OK 73125

John Halloran
CERAMICS PROCESS SYSTEMS
840 Memorial Drive
Cambridge, MA 02139

Jack Halter
GENERAL SIGNAL CORPORATION
2 High Ridge Park
P.O. Box 10010
Stanford, CT 06904

Lawrence D. Hamlin
SOUTHERN CALIFORNIA EDISON
2244 Walnut Grove Avenue
P.O. Box 800
Rosemead, CA 91770

Barry Hammond
DAVID TAYLOR NAVAL SHIP R&D
CENTER
CODE 2812
Annapolis, MD 21402-5067

George S. Hammond
ALLIED SIGNAL CORPORATION
Columbia Road
P.O. Box 1021 R
Morristown, NJ 07960

Robert Hammond
LOS ALAMOS NATIONAL
LABORATORY
P.O. Box 1663
MS D429
Los Alamos, NM 87545

Damian Hampshire
APPLIED SUPERCONDUCTIVITY
CENTER
1533 Monrovia Avenue
Newport Beach, CA 92663

Charles Hamrin Jr.
UNIVERSITY OF KENTUCKY
Physics Department
Chem. Eng. 0046
Lexington, KY 40506

Fred Haney
3I VENTURES
450 Newport Center Drive
Suite 250
Newport Beach, CA 92660

Eric Hanson
HYPRES INC.
500 Executive Boulevard
Elmsford, NY 10523

John Hanson
M.R.J. INC.
10455 White Granite Drive
Suite 200
Oakton, VA 22124

Robert Hard
CABOT CORPORATION
Billerica Technical Center
Concord Road
Billerica, MA 01821

John Harding
DYNATECH SCIENTIFIC
99 Erie Street
Cambridge, MA 02139

Thomas C. Hardy
AFESC/RDC
Tyndall AFB
Panama City, FL 32403

Neal T. Hare
POWELL INDUSTRIES INC.
P.O. Box 12818
Houston, TX 77217

O. Harling
MASSACHUSETTS INSTITUTE
OF TECHNOLOGY
138 Albany Street
Cambridge, MA 02139

Frederic Harman
MORGAN STANLEY VENTURE
GROUP
1251 Avenue of the Americas
28th Floor
New York, NY 10020

Martin Harmer
LEHIGH UNIVERSITY
Bethlehem, PA 18015

Mark Harmor
ICI AMERICAS
1600 M Street, N.W.
Suite 702
Washington, D.C. 20036

Roy V. Harrington
FERRO CORPORATION
7500 East Pleasant Valley Road
Independence, OH 44131

Leonard A. Harris
NASA
600 Independence Avenue, S.W.
OAST
Washington, D.C. 20546

Richard Harris
DEPARTMENT OF COMMERCE
2645 Briarwood Drive
Boulder, CO 80303

Timothy R. Hart
STEVENS INSTITUTE OF
TECHNOLOGY
Castle Point Station
Materials & Metallurgy
Hoboken, NJ 07030

R. Hartke
AEROSPACE INDUSTRIES
ASSOCIATION
OF AMERICA
1725 DeSales Street, N.W.
Washington, D.C. 20036

William V. Hassenzahl
LAWRENCE BERKELEY
LABORATORY
One Cyclotron Road
Berkeley, CA 94720

Daryl G. Hatano
SEMICONDUCTOR INDUSTRY
ASSOCIATION
10201 Torre Avenue
Suite 275
Cupertino, CA 95014

Oscar Hauptman
HARVARD UNIVERSITY
Soldiers Field
Morgan Hall 7
Boston, MA 02163

W. Lance Haworth
NATIONAL SCIENCE FOUNDATION
1800 G Street, N.W.
Washington, D.C. 20550

Arthur F. Hebard
AT&T BELL LABORATORIES
600 Mountain Avenue
Murray Hills, NJ 07974

Siegfried S. Hecker
LOS ALAMOS NATIONAL
LABORATORY
P.O. Box 1663
MS A100
Los Alamos, NM 87545

A. Zeev Hed
SUPERCONDUCTIVITY STRATEGY
RPTS.
36 Bradford Road
Natick, MA 01760

Bob Heistand
DOW CHEMICAL NEW ENGLAND
LAB
P.O. Box 400
Wayland, MA 01778

David Helland
MASSACHUSETTS INSTITUTE
OF TECHNOLOGY
National Magnetic Laboratory
NW14-3030
Cambridge, MA 02139

B. Hello
ROCKWELL INTERNATIONAL
1745 Jefferson Davis Highway
Arlington, VA 22202

Helmut W. Hellwig
NATIONAL BUREAU OF STANDARDS
Administrative A-1134
Gaithersburg, MD 20899

Harold E. Helms
GENERAL MOTORS CORPORATION
P.O. Box 420
M.S. T-18
Indianapolis, IN 46206-0420

Patrick Hemenger
U.S. AIR FORCE
303 Orton Road
Yellow Springs, OH 45387

Greg Henderson
OFFICE OF SCIENCE &
TECHNOLOGY
POLICY
5002 New Executive Office Building
Washington, D.C. 20508

Bryan Hendrix
COLUMBIA UNIVERSITY
918 Mudd Building
New York, NY 10027

George C. Hennessy
DAVID SARNOFF RESEARCH
CENTER
Princeton, NJ 08543-5300

Donald Hense
PRAIRIE VIEW A&M UNIVERSITY
Department of Physics
Prairie View, TX 77445

Stan Herman
J.M. HUBER CORPORATION
333 Thornall Street
Edison, NJ 08818

Allen Hermann
UNIVERSITY OF ARKANSAS
2801 University Avenue
Little Rock, AR 72204

John Herrington
DEPARTMENT OF ENERGY
1000 Independence Avenue, S.W.
Washington, D.C. 20585

Robert Hershey
SCIENCE MANAGEMENT
CORPORATION
2100 M Street, N.W.
Washington, D.C. 20037

William F. Herwig
UNION CARBIDE CORPORATION
8615 Bradgate Road
Alexandria, VA 22308

Douglas Herz
AT&T NETWORK SYSTEMS
Crawfords Corner Road
4J613
Holmdel, NJ 07733

Charles Herzfeld
AETNA, JACOBS, RAMO
375 Park Avenue
New York, NY 10152

Claudio Herzfeld
CEMCOM CORPORATION
10123 Senate Drive
Lanham, MD 20851

Robert V. Hess
NASA/LEWIS RESEARCH CENTER
LGRC
Hampton, VA 23605

Nicholas P. Heymann
DREXEL BURNHAM LAMBERT
60 Broad Street
New York, NY 10004

Donald L. Hildenbrand
SRI INTERNATIONAL
333 Ravenswood Avenue
Menlo Park, CA 94025-8493

Norm Hill
GEORGIA INSTITUTE OF
TECHNOLOGY
School of Materials Engineering
Atlanta, GA 30332-0245

Eugene Hirschkoff
BIOMAGNETIC TECHNOLOGIES
4174 Sorrento Valley Blvd.
San Diego, CA 92121

James Ho
WICHITA STATE UNIVERSITY
Department of Physics
Wichita, KS 67201

Gary Hoffer
KAMAN CORPORATION
P.O. Box 7463
Colorado Springs, CO 90833

Allan Hoffman
NATIONAL ACADEMY OF SCIENCES
2101 Constitution Avenue, N.W.
Washington, D.C. 20418

Tom Hoke
OKLAHOMA GAS & ELECTRIC
COMPANY
P.O. Box 321
Oklahoma City, OK 73101

L. Holdeman
COMSAT LABORATORIES
22300 Comsat Drive
Clarksburg, MD 20871

Charles J. Holloman
TRANS-LUS CORPORATION
110 Richards Avenue
Norwalk, CT 06854

Gunnar Holmdahl
A.S.E.A. INCORPORATED
2975 Westchester Avenue
Purchase, NY 10577

Harry Holmgren
S.U.R.A.
1776 Massachusetts Avenue, N.W.
Washington, D.C. 20036

Richard Holt
DEPARTMENT OF ENERGY
1000 Independence Avenue, S.W.
Washington, D.C. 20585

Kirby C. Holte
SOUTHERN CALIFORNIA EDISON
2244 Walnut Grove Avenue
P.O. Box 800
Rosemead, CA 91770

William C. Holton
SEMICONDUCTOR RESEARCH
CORPORATION
P.O. Box 12053
Research Triangle Park, NC 27709

John Holzrichter
LAWRENCE LIVERMORE NATIONAL
LAB
Livermore, CA 94550

Seung Hong
OXFORD SUPERCONDUCTING
TECHNOLOGY
600 Milik Street
Carteret, NJ 07008

Seungok Hong
OXFORD SUPERCONDUCTING
TECHNOLOGY
600 Milik Street
Carteret, NJ 07008

Richard H. Hopkins
WESTINGHOUSE R&D CENTER
1310 Beulah Road
Pittsburgh, PA 15235

Peiherng Hor
UNIVERSITY OF HOUSTON
4800 Calhoun Road
Physics Department
Houston, TX 77004

Tom Horlacher
EMERSON ELECTRIC CORPORATION
8000 W. Foorissant
St. Louis, MO 63136

Stuart B. Horn
U.S. ARMY
AMSEL-RD-NV-IT
Fort Belvoir, VA 22060-5677

William R. Horst
NCR CORPORATION
1700 South Patterson Boulevard
Dayton, OH 45479

Bob Horstmeyer
S-TRON
101 Twin Dolphin Drive
Redwood City, CA 94065

Gerald L. Houze
ALLEGHENY LUDLUM
CORPORATION
Technical Center
Brackenridge, PA 15014

Charles Howard
NORDSON CORPORATION
28601 Clemens Road
Westlake, OH 44145

Deh Hsiung
NATIONAL SCIENCE FOUNDATION
1800 G Street, N.W.
Washington, D.C. 20550

William A. Hubbard
MOBAY CORPORATION
5601 Eastern Avenue
Baltimore, MD 21224

Stephen J. Hudgens
ENERGY CONVERSION DEVICES
1675 West Maple Road
Troy, MI 48084

Francis Hughes
AMERICAN SUPERCONDUCTOR
21 Erie Street
Cambridge, MA 02139

John Hughes-Blanks
SCIENCE & ENGINEERING
ASSOCIATES, INC.
6301 Indian School Road, N.E.
Suite 700
Albuquerque, NM 87110

John K. Hulm
WESTINGHOUSE R&D CENTER
1310 Beulah Road
Pittsburgh, PA 15235

Patrick Hunt
COMMUNICATION WORKERS OF
AMERICA
1925 K Street, N.W.
Washington, D.C. 20006

V. Daniel Hunt
TECHNOLOGY RESEARCH
CORPORATION
Springfield Professional Park
8328-A Traford Lane
Springfield, VA 22152

Paul G. Huray
UNIVERSITY OF TENNESSEE
College of Engineering
53 Turner House
Knoxville, TN 37996-2000

John P. Hurrell
THE AEROSPACE CORPORATION
P.O. Box 92957
Los Angeles, CA 90009

Jim Ingram
IBM CORPORATION
1801 K Street, N.W.
Washington, D.C. 20006

John Ings
AIRTRON
200 East Hanover Avenue
Morris Plains, NJ 07950

Louis C. Inniello
DEPARTMENT OF ENERGY
1000 Independence Avenue, S.W.
Washington, D.C. 20585

Tsuyoshi Inque
KOBE DEVELOPMENT
CORPORATION
P.O. Box 13608
Research Triangle Park, NC 27709

Zafar Iqbal
ALLIED SIGNAL CORPORATION
Columbia Road
P.O. Box 1021 R
Morristown, NJ 07960

Dick Iverson
AMERICAN ELECTRONICS
ASSOCIATION
1612 K Street, N.W.
Suite 1100
Washington, D.C. 20006

E.S. Iverson
FORD AEROSPACE
1235 Jefferson Davis Highway
Suite 1300
Arlington, VA 22202

Daniel G. Jablonski
SUPERCOMPUTING RESEARCH
CENTER
4380 Forbes Boulevard
Lanham, MD 20706

Michael Jack
HYPRES INC.
500 Executive Boulevard
Elmsford, NY 10523

Gilbert S. Jackson
DEPARTMENT OF ENERGY
1000 Independence Avenue, S.W.
Washington, D.C. 20585

Thomas R. Jacobsen
BURLINGTON NORTHERN
RAILROAD
3700 Continental Plaza
Forth Worth, TX 76102

Harold Jaffe
DEPARTMENT OF ENERGY
1000 Independence Avenue, S.W.
Washington, D.C. 20585

Joe Jahoda
ASTRON CORPORATION
470 Spring Park Place
Suite 100
Herndon, VA 22070

Adish Jain
SERVO CORPORATION OF AMERICA
111 New South Road
Hicksville, NY 11802

Babu L. Jain
FLORIDA STATE UNIVERSITY
Department of Physics
Tallahassee, FL 32307

Robert A. Jake
AMERICAN MAGENTICS INC.
112 Flint Road
P.O. Box 2509
Oak Ridge, TN 37831-2509

Stephan O. James
BIOMAGNETIC TECHNOLOGIES
4174 Sorrento Valley Blvd.
San Diego, CA 92121

Chris Jamison
RANSBURG CORPORATION
3939 West 56th Street
Indanapolis, IN 46254

John Janecek
WILKES COLLEGE
Department of Engineering
Wilkes-Barre, PA 18766

Gary M. Janney
HUGHES AIRCRAFT COMPANY
P.O. Box 902
Mail Station E55-G206
El Segundo, CA 90245

Stephen C. Jardin
PRINCETON UNIVERSITY
P.O. Box 451
Princeton, NJ 08544

Mike Jaremchuk
MARTIN MARIETTA LABORATORIES
1450 South Rolling Road
Baltimore, MD 21227

John W. Jarve
MENLO VENTURES
3000 Sand Hill Road
Building 4
Suite 100
Menlo Park, CA 94025

Benny E. Jay
ELECTROSOURCE, INC.
6500 Tracor Lane
Austin, TX 78725

Y.C. Jean
UNIVERSITY OF MISSOURI
1110 East 48th Street
Physics Department
Kansas City, MO 64110

Michael Jeffries
GENERAL ELECTRIC
P.O. Box 414
Milwaukee, WI 53201

Jane Liu Jernow
NATIONAL ACADEMY OF SCIENCES
2101 Constitution Avenue, N.W.
Washington, D.C. 20418

J.D. Joannopoulos
MASSACHUSETTS INSTITUTE
OF TECHNOLOGY
77 Massachusetts Avenue
Cambridge, MA 02139

Donald Johnson
I.A.P. RESEARCH
2763 Culver Avenue
Dayton, OH 45429

Herbert H. Johnson
CORNELL UNIVERSITY
Bard Hall
Ithaca, NY 14853

Larry Johnson
ARGONNE NATIONAL
LABORATORY
9700 S. Cass Avenue
Materials Science Division
Argonne, IL 60439

Mark J. Johnson
AT&T NETWORK SYSTEMS
Room 4H-602
Crawfords Corner Road
Holmdel, NJ 07733

Milo Johnson
TEXAS INSTRUMENTS
P.O. Box 655936
MS 145
Dallas, TX 75265

Robert A. Johnson
GENERAL DYNAMICS SPACE
SYSTEMS
DIVISION
P.O. Box 85990
San Diego, CA 92138

Robert Johnson
GENERAL DYNAMICS
CORPORATION
5001 Kearney Villa Road
San Diego, CA 92123

Walter Johnson
LAMSON & SESSIONS COMPANY
Bank One
Cleveland, OH 44122

Beverly Jones
GOVERNOR'S OFFICE OF S&T
150 East Kellogg Boulevard
St. Paul, MN 55101

H. Graham Jones
NEW YORK STATE S & T OFFICE
314 Loudonville Road
Loudonville, NY 12211

Howard E. Jordan
RELIANCE ELECTRIC COMPANY
24800 Tunesten Road
Cleveland, OH 44117

Dr. Earl C. Joseph
ANTICIPATORY SCIENCES, INC.
345 E. Sixth Street
Suite 700
St. Paul, MN 55101

Bobby R. Junker
OFFICE OF NAVAL RESEARCH
800 N. Quincy Street
Arlington, VA 22217-5000

Alan Kadin
UNIVERSITY OF ROCHESTER
Electrical Engineering Dept.
Rochester, NY 14627

Robert Kaiser
"ARGOS" ASSOCIATES, INC.
12 Glengarry
Winchester, MA 01890

Fritz R. Kalhammer
ELECTRIC POWER RESEARCH
INSTITUTE
3412 Hillview Avenue
P.O. Box 10412
Palo Alto, CA 94303

Gretchen Kalonji
MASSACHUSETTS INSTITUTE
OF TECHNOLOGY
77 Massachusetts Avenue
Cambridge, MA 02139

Dr. Paul G. Kaminski
H & Q TECHNOLOGY PARTNERS
5242 Lyngate Court
Burke, VA 22015

Raymond G. Kammer
NATIONAL BUREAU OF STANDARDS
Administrative A-1134
Gaithersburg, MD 20899

Robert Kamper
NATIONAL BUREAU OF STANDARDS
325 South Broadway
Boulder, CO

C.R. Kannewurf
NORTHWESTERN UNIVERSITY
906 University Place
Evanston, IL 60201-9990

Y.H. Kao
SUNY — STONY BROOK
Department of Physics
Stony Brook, NY 11794

Vijay K. Kapur
INSET
8635 Aviation Boulevard
Inglewood, CA 90301

Justin Karp
EDISON ELECTRIC INSTITUTE
1111 19th Street, N.W.
Washington, D.C. 20036-3691

Andi Kasarsky
SRA TECHNOLOGIES, INC
c/o Department of Energy
CE-43.1
1000 Independence Avenue, S.W.
Washington, D.C. 20585

Gerhard Kasper
AMERICAN AIR LIQUIDE
5230 S. East Avenue
Countryside, IL 60525

Stephen Katz
KATZ ASSOCIATES
55 East End Avenue
New York, NY 10028

Charlie Kaufmann
AMERICAN CYANAMID COMPANY
One Cyanamid Plaza
Wayne, NJ 07470

Elton N. Kaufmann
LAWRENCE LIVERMORE
LABORATORY
P.O. Box 808
L-350
Livermore, CA 94550

K. Kay
CORETECH
1735 New York Avenue, N.W.
Suite 500
Washington, D.C. 20006

Bernard Kear
RUTGERS STATE UNIVERSITY
OF NEW JERSEY
P.O. Box 909
Piscataway, NJ 08854

Don W. Keefer
IDAHO NATIONAL ENGINEERING
LAB
INEL Research Center
Box 1625
Idaho Falls, ID 83415

Phil Keif
DEPARTMENT OF ENERGY
1000 Independence Avenue, S.W.
Washington, D.C. 20585

G. Anthony Keig
G.A. KEIG CONSULTANTS
28 Saunders Lane
Ridgefield, CT 06877

Ronald L. Kelley
STRATEGIC PARTNERS
9101 Cricklewood Court
Vienna, VA 22180

Don Kelly
COMMERCE CABINET
Capitol Plaza Tower
24th Floor
Frankfort, KY 40601

Jeffrey G. Kenkel
W.J. SCHAFER ASSOCIATES
5100 Springfield Pike
Suite 311
Dayton, OH 45431

George Kenney
MASSACHUSETTS INSTITUTE
OF TECHNOLOGY
77 Massachusetts Avenue
Cambridge, MA 02139

John N. Kessler
KESSLER MARKETING
INTELLIGENCE
America's Cup Avenue
31 Bridge Street
Newport, RI 02840

William C. Kessler
U.S. AIR FORCE
APWAL/MLP
Wright-Patterson AFB, OH 45433-6533

Douglas A. Keszler
OREGON STATE UNIVERSITY
Corvallis, OR 97331-4003

Stephen C. Kiely
PRIME COMPUTER, INC.
500 Old Connecticut Path
Framingham, MA 01701

Young B. Kim
UNIVERSITY OF SOUTHERN
CALIFORNIA
1449 S Carmelina Avenue
Los Angeles, CA 90025

Gordon F. Kingsley
FIDELITY VENTURE ASSOCIATES
82 Devonshire Street
Boston, MA 02109

D.L. Kinser
VANDERBILT UNIVERSITY
Dept. of Materials Science
Box 1689, Station B
Nashville, TN 37235

Wiley P. Kirk
TEXAS A & M UNIVERSITY
Department of Physics
College Station, TX 77843

Mark Kirschner
B.O.C. GROUP TECHNICAL CENTER
100 Mountain Avenue
Murray Hill, NJ 07974

James Kirtley
MASSACHUSETTS INSTITUTE
OF TECHNOLOGY
77 Massachusetts Avenue
Cambridge, MA 02139

Professor Koichi Kitazawa
UNIVERSITY OF TOKYO
7-3-1 Hongo
Bunkyo-ku, Tokyo 113
Tokyo, Japan
81-03-812-2111

Thomas A. Kitchens
DEPARTMENT OF ENERGY
1000 Independence Avenue, S.W.
Washington, D.C. 20585

Dr. Steven A. Kivelson
SUNY - STONY BROOK
Stony Brook, NY 11794

Skip Klatt
AMERICAN CHEMET
P.O. Drawer D
East Helena, MT 59635

Barry M. Klein
NAVAL RESEARCH LABORATORY
4555 Overlook Avenue, S.W.
CODE 4680
Washington, D.C. 20375-5000

James D. Klein
EIC LABORATORIES
111 Downey Street
Norwood, MA 02062

Kenneth Klein
DEPARTMENT OF ENERGY
1000 Independence Avenue, S.W.
Washington, D.C. 20585

Robert Kleinberg
SCHLUMBERGER
Old Quarry Road
Ridgefield, CT 06877

Marvin Klinger
DEPARTMENT OF ENERGY
P.O. Box 3621
Portland, OR 97222

Barry Koepke
HONEYWELL, INC.
5121 Winnetka Avenue North
New Hope, MN 55428

Henry Kolm
EML RESEARCH INC./COMMAND
625 Putman Avenue
Cambridge, MA 02139

Curt Koman
ERNST & WHINNEY
2300 National City Center
Cleveland, OH 44114

R. Komanduri
R. KOMANDURI LTD.
1322 Ruffner Road
Schenectady, NY 12309

Ranga Komanduri
NATIONAL SCIENCE FOUNDATION
Materials Engineering & Processing
1800 G Street, N.W.
Room 1110
Washington, D.C. 20550

Greg Kovacs
STANFORD UNIVERSITY
Ginzton Lab
200 Junipero Serra Boulevard
Stanford, CA 94305

George Kovatch
DEPARTMENT OF
TRANSPORTATION
Kendall Square
Cambridge, MA 02142

Bruce Kramer
GEORGE WASHINGTON
UNIVERSITY
Department of Physics
Washington, D.C. 20008

Martha Krebs
LAWRENCE BERKELEY
LABORATORY
One Cyclotron Road
Berkeley, CA 94720

Vladimir Z. Kresin
LAWRENCE BERKELEY
LABORATORY
One Cyclotron Road
Berkeley, CA 94720

Henry Kressel
E.M. WARBURG, PINCUS &
COMPANY
466 Lexington Avenue
New York, NY 10017-3147

Harry Kroger
MICROELECTRONICS AND
COMPUTER TECHNOLOGY
3500 W. Balcones Center Drive
Austin, TX 78759

Lennard G. Kruger
LIBRARY OF CONGRESS
SPRD
Washington, D.C. 20540

M. Kuchnir
FERMILAB
P.O. Box 500
MS 316
Batavia, IL 60510

Carl Kukkonen
JET PROPULSION LABORATORY
Center for Space Technology
4800 Oak Grove Drive
Pasadena, CA 91109

M.A. Kuliasha
OAK RIDGE NATIONAL
LABORATORY
P.O. Box X, MS 201
Oak Ridge, TN 37831-6255

Sudhir Kulkarni
CERAMATEC
163 West 1700 South
Salt Lake City, UT 84115

Binod Kumar
UNIVERSITY OF DAYTON
RESEARCH INSTITUTE
300 College Park Drive
Dayton, OH 45469

Pradeep Kumar
UNIVERSITY OF FLORIDA
Department of Physics
215 Williamson Hall
Gainesville, FL 32611

Edward Kuroki
SHIMIZU TECHNOLOGY CENTER
OF AMERICA
One Kendall Square
Suite 2200
Cambridge, MA 02139

Lawrence M. Kushner
MITRE CORPORATION
7525 Colshire Drive
McLean, VA 22102-3481

Chris Kuyatt
NATIONAL BUREAU OF STANDARDS
Rad. Phy. Building
Gaithersburg, MD 20899

Stephen B. Kuznetsov
PSM TECHNOLOGIES INC.
3508 Woodbine Street
Chevy Chase, MD 20815

Erik W. Kvam
LEBOEUF, LAMB, LEIBY & MACRAE
520 Madison Avenue
New York, NY 10022

Thomas F. Lamb
NYNEX CORPORATION
1113 Westchester Avenue
White Planes, NY 10604

James Lambert
PACIFIC POWER & LIGHT CO.
920 SW 6th Avenue
Portland, OR 97204

Roger Lambertson
LOCKHEED MISSILES &
SPACE CO., INC
Suite 1100
1825 Eye Street, N.W.
Washington, D.C. 20006

Wayne R. Lambertson
LOCKHEED MISSILES & SPACE
COMPANY, INC.
1825 I Street, N.W.
Suite 1100
Washington, D.C. 20006

Horace Lander
NIPPON STEEL COMPANY
Materials Consultant
34 Bay View Drive
Swampscott, MA 01907

Hans Landsberg
RESOURCES FOR THE FUTURE
1616 P Street, N.W.
Washington, D.C. 20036

John Langan
STI
150 Aero Camino
Goleta, CA 93117

Kristine Langdon
GENERAL ACCOUNTING OFFICE
1825 K Street, N.W.
Suite 515
Washington, D.C. 20006

Rick Langman
COORS CERAMICS COMPANY
17550 West 32nd Avenue
Golden, CO 80401

Joseph E. Lannutti
FLORIDA STATE UNIVERSITY
SCRI/Physics
Tallahassee, FL 32306

David C. Larbalestier
UNIVERSITY OF WISCONSIN
Applied Superconductivity Ctr.
1500 Johnson Drive
Dept. Metallurgical Engineering
Madison, WI 53706

Ana Larkin
DEPARTMENT OF ENERGY
1000 Independence Avenue, S.W.
Washington, D.C. 20585

Arvid G. Larson
BOOZ, ALLEN & HAMILTON
4330 East West Highway
Bethesda, MD 20814-2655

Charles F. Larson
INDUSTRIAL RESEARCH INSTITUTE
100 Park Avenue
New York, NY 10017

John G. Larson
GENERAL MOTORS RESEARCH LAB
30200 Mound Road
Warren, MI 48090-9055

Gunnar Larsson
SWEDISH EMBASSY
600 New Hampshire Avenue, N.W.
Washington, D.C. 20037

William K. Lasala
COMPTEK RESEARCH
110 Broadway
Buffalo, NY 14203

Alan Lauder
E.I. DUPONT
1007 Market Street
CR&D Experimental Station
D6009
Wilmington, DE 19898

Larry Lauderdale
BABCOCK & WILCOX COMPANY
1735 I Street, N.W.
Suite 814
Washington, D.C. 20006

Michael J. Lavelle
JOHN CARROLL UNIVERSITY
North Park & Miramar Boulevards
Physics Department
University Heights, OH 44118

Stephen J. Lawrence
CONDUCTOR TECHNOLOGIES
1001 Connecticut Avenue, N.W.
Washington, D.C. 20036

W.G. Lawson
PIRELLI CABLE CORPORATION
800 Rahway Avenue
Union, NJ 07083

Steven Lazarus
ARCH DEVELOPMENT
 CORPORATION
115-25 E. 56th Street
Chicago, IL 60637

Fred Leavitt
OFFICE OF SCIENCE &
TECHNOLOGY POLICY
5002 New Executive Office Building
Washington, D.C. 20508

Burton Lederman
ELECTRO-KINETIC SYSTEMS
701 Chestnut Street
Trainer, PA 19013

J.D. Lee
E.I. DUPONT
1007 Market Street
CR&D Experimental Station
D6009
Wilmington, DE 19898

Swenam R. Lee
DOE/PITTSBURG ENGINEERING
RESEARCH CENTER
P.O. Box 10940
Pittsburgh, PA 15236

William Lee
ARTHUR D. LITTLE INC.
Acorn Park
Cambridge, MA 02140

T.F. Lemke
INCO ALLOYS INTERNATIONAL
Huntington, WV 25720

Douglas K. Lemon
BATTELLE NORTHWEST LABS
2033 Trevino Court
P.O. Box 999
Richland, WA 99352-9785

James T. Leong
CIA
Washington, D.C. 20505

Josh Lerner
GENERAL ACCOUNTING OFFICE
1825 K Street, N.W.
Suite 515
Washington, D.C. 20006

Chet Leroy
TELEDYNE
Wah Chang Albany Division
Box 460
Albany, OR 97321

Martin S. Leung
AEROSPACE CORPORATION
2350 E. El Segundo Boulevard
El Segundo, CA 90245-4691

Aaron Levine
DAVID SARNOFF RESEARCH
CENTER
Princeton, NJ 08543-5300

Leslie S. Levine
FUSION SYSTEMS CORPORATION
7600 Standish Place
Rockville, MD 20855

Mark Levinson
GTE LABORATORIES, INC.
40 Sylvan Road
Waltham, MA 02254

C.H. Li
LINTEL TECHNOLOGY
379 Elm Drive
Roslyn, NY 11576

Paul Liberman
PATENT & TRADEMARK OFFICE
2021 Jefferson Davis Highway
Arlington, VA 22206

D.H. Liebenberg
DEPARTMENT OF ENERGY
1000 Independence Avenue, S.W.
Washington, D.C. 20585

Judith S. Liebman
UNIVERSITY OF ILLINOIS
801 South Wright Street
107 Coble Hall
Champaign, IL 61820

Wolfgang K. Liebmann
UNIVERSITY OF VERMONT
428 Waterman Building
Burlington, VT 05405

Gerald L. Liedl
PURDUE UNIVERSITY
Materials Engineering
CMET
West LaFayette, IN 47907

Yeong Lin
DIGITAL EQUIPMENT
CORPORATION
10500 Ridgeview Court
Cupertino, CA 95014

Walter Linde
LAGUS CORPORATION
200 Park Avenue
New York, NY 10166

William B. Lindgren
QUANTUM DESIGN
11578 Sorrento Valley Road
San Diego, CA 92121

Lai Chang Ling
LITTON INDUSTRIES
1215 S. 52nd Street
Tempe, AZ 85281

J.D. Litster
MASSACHUSETTS INSTITUTE
OF TECHNOLOGY
77 Massachusetts Avenue
Cambridge, MA 02139

Gary Little
VENROCK ASSOCIATES
30 Rockefeller Plaza
New York, NY 10112

John A. Little
SERVO CORPORATION OF AMERICA
111 New South Road
Hicksville, NY 11802

William L. Little
STANFORD UNIVERSITY
200 Junipero Serra Boulevard
Stanford, CA 94305

Michael Littlejohn
ARMY RESEARCH OFFICE
Durham, NC 27701

Chuan S. Liu
UNIVERSITY OF MARYLAND
Physics Department
College Park, MD 20742

Jennifer Lobo
DOMAIN ASSOCIATES
One Palmer Square
Princeton, NJ 08542

E.M. Logothetis
FORD MOTOR COMPANY
Scientific Research Laboratory
Box 1899, Rm. S-3953
Dearborn, MI 58121-2053

Gordon T. Longerbeam
LAWRENCE LIVERMORE
LABORATORY
P.O. Box 808, L-795
Livermore, CA 94550

R.N. Longuemare
WESTINGHOUSE
P.O. Box 1693, MS 1107
Baltimore, MD 21203

Robert E. Lorenzini
REAL VENTURE CAPITAL
405 National Avenue
Mountain View, CA 94043

Liang-Fu Lou
NORTHROP RESEARCH AND
TECHNOLOGY CENTER
1 Research Park
Rancho Palos Verdes, CA 90274

Terry Loucks
AMERICAN SUPERCONDUCTOR
21 Erie Street
Cambridge, MA 02139

Austin Lowery
KAMAN SCIENCES CORPORATION
2560 Huntington Avenue
Suite 500
Alexandria, VA 22303

Robert J. Loyd
BECHTEL NATIONAL, INC.
P.O. Box 6935
San Francisco, CA 94119

Chin-Shun Lu
XINIX INCORPORATED
3500-B Thomas Road
Santa Clara, CA 95054

J. Lucak
KEITHLEY INSTRUMENT
28775 Aurora Road
Cleveland, OH 44139

Thomas S. Luhman
BOEING AEROSPACE COMPANY
P.O. Box 3999 73-09
Seattle, WA 99124-2499

Jerry R. Lundien
H.Q.D.A.
Attn: ASRD-TR
Washington, D.C. 20310-3558

James Luper
V. GARBER INTERNATIONAL
1215 Jefferson Davis Highway
Suite 1105
Arlington, VA 22202

W. Stuart Lyman
CIA
Washington, D.C. 20505

Lahmer Lynds
UNITED TECHNOLOGIES RESEARCH
CTR.
Silver Lane
East Hartford, CT 06108

Jeffrey Lynn
UNIVERSITY OF MARYLAND
Physics Department
College Park, MD 20742

R.K. MacCrone
RENSSELAER POLYTECHNIC
INSTITUTE
(RPI)
Department of Physics
Troy, NY 12180

Bruce A. MacDonald
NATIONAL SCIENCE FOUNDATION
1800 G Street, N.W.
Washington, D.C. 20550

Barry A. MacKinnon
RESONEX, INC.
720 Palomar Avenue
Sunnyvale, CA 96086

Allan J. MacLaren
TRACOR AEROSPACE INC.
P.O. Box 196
San Ramon, CA 94583

Arlene P. MacLin
NATIONAL RESEARCH COUNCIL
2101 Constitution Avenue, N.W.
Washington, D.C. 20418

Carl J. Magee
NASA LANGLEY RESEARCH
CENTER
835 Yorktown Road
Poquoson, VA 23662

Tai-Il Mah
UNIVERSAL ENERGY SYSTEMS
4401 Dayton Xenia Road
Dayton, OH 45432

Andrew Maisley
TRW SPACE & TECHNOLOGY
GROUP
1 Space Park
MS R1-1086
Redondo Beach, CA 90278

Alex Malozemoff
IBM CORPORATION
Watson Research Center
P.O. Box 218
Yorktown Heights, NY 10598

William Manly
CAMDEC
Route 1, Box 157A
Kingston, TN 37763

Bob Manning
SSI TECHNOLOGIES
P.O. Box 768
Creedmoor, NC 27522

Chuck Manto
GARTNER GROUP INC.
72 Cummings Point Road
P.O. Box 10212
Stamford, CT 06904

Brian Maple
UNIVERSITY OF CALIFORNIA
AT SAN DIEGO
Department of Physics, B019
La Jolla, CA 92093

William Marancik
OXFORD SUPERCONDUCTING
TECHNOLOGY
600 Milik Street
Carteret, NJ 07008

Harris Marcus
UNIVERSITY OF TEXAS
Engineering Division
Austin, TX 78712

Giorgio Margaritones
UNIVERSITY OF WISCONSIN
Synchrotron Radiation Center
Stroughton, WI 53589-3097

Faruq Marikar
HOECHST-CELANESE
P.O. Box 1000
Summit, NJ 07901

Raymond Marlow
MARLOW INDUSTRIES, INC.
10451 Vista Park Road
Dallas, TX 75238

Nicholas Maropis
U.T.I. CORPORATION
200 West 7th Avenue
Collegeville, PA 19426

Ogden Marsh
HUGHES RESEARCH
LABORATORIES
3011 Malibu Canyon Road
Mail Station RL63
Malibu, CA 90265

Edith W. Martin
BOEING ELECTRONICS COMPANY
P.O. Box 24969
M/S 7J-20
Seattle, WA 98124-6269

Cloyd E. Marvin
HARVEST VENTURES
10080 North Wolfe Road
Building SW3
#365
Cupertino, CA 95014

A.J. Masley
TRW SPACE & TECHNOLOGY
GROUP
1 Space Park
Redondo Beach, CA 90278

Wesley N. Mathews Jr.
GEORGETOWN UNIVERSITY
Department of Physics
Washington, D.C. 20057

Leonard F. Mattheiss
AT&T BELL LABORATORIES
600 Mountain Avenue
Mail Stop MH-1D-369
Murray Hill, NJ 07974

Dennis L. Matthies
S.R.I.
David Sarnoff Research Center
Princeton, NJ 08543-5300

Bob Maurer
CORNING
Sullivan Park
FR17
Corning, NY 14831

Donald Maurer
EMPI INCORPORATED
1275 Grey Fox Road
St. Paul, MN 55112

Donald B. Mausshardt
KAISER ENGINEERS, INC.
1117 North 19th Street
Arlington, VA 22209

Eugene McAllister
ECONOMIC POLICY COUNCIL
The White House
1600 Pennsylvania Avenue, N.W.
Washington, D.C. 20500

Susan McBride
HYPRES INC.
500 Executive Boulevard
Elmsford, NY 10523

David W. McCall
AT&T BELL LABORATORIES
600 Mountain Avenue
Mail Stop MH-1D-369
Murray Hill, NJ 07974

William McCallum
AMES LABORATORY
Iowa State University
12 Physics
Ames, IA 50011

Ben McConnell
OAK RIDGE NATIONAL
LABORATORY
P.O. Box X
Oak Ridge, TN 37831-6255

Robert McConnell
SOLAR ENERGY RESEARCH
INSTITUTE
1617 Cole Blvd.
Golden, CO 80401

Bob McCune
NATIONAL ELECTRICAL
MANUFACTURERS
ASSOCIATION
2101 L Street, N.W.
Washington, D.C. 20037

James E. McEvoy
COUNCIL FOR CHEMICAL
RESEARCH
One Bethlehem Plaza
Bethlehm, PA 18108

Paul J. McGinn
UNIVERSITY OF NOTRE DAME
P.O. Box E
Physics Department
Notre Dame, IN 46556

George W. McKinney III
AMERICAN SUPERCONDUCTOR
45 Milk Street
Boston, MA 02109

Malcolm McLaren
RUTGERS STATE UNIVERSITY
OF NEW JERSEY
P.O. Box 909
Piscataway, NJ 08854

Michael McNeil
NUCLEAR REGULATORY
COMMISSION
NL007
Washington, D.C. 20555

Tim McNulty
ARGONNE NATIONAL
LABORATORY
9700 S. Cass Avenue
Materials Science Division
Argonne, IL 60439

John McTague
FORD MOTOR COMPANY
Scientific Research Laboratory
Box 1899
Dearborn, MI 58121-2053

Gary L. McVay
PACIFIC NORTHWEST
LABORATORY
P.O. Box 999
Richland, WA 99352

Edward J. Mead
E.I. DUPONT
1007 Market Street
CR&D Experimental Station
D6028
Wilmington, DE 19898

Martha Mecartney
UNIVERSITY OF MINNESOTA
151 Amundson Hall
421 Washington Avenue, SE
Minneapolis, MN 55455

George Mechlin
UNIVERSITY OF PITTSBURGH
Fifth & Bigelow
Department of Physics
Pittsburgh, PA 15260

Robert Mehrabian
UNIVERSITY OF CALIFORNIA
AT SANTA BARBARA
College of Engineering
Santa Barbara, CA 93106

Carolyn Meinel
CARDINAL COMMUNICATIONS
Route 2
Box 120
Lovettsville, VA 22080

Michael Melich
NAVAL RESEARCH LABORATORY
4555 Overlook Avenue, S.W.
CODE 8305
Washington, D.C. 20375-5000

Robert Melin
AT&T BELL LABORATORIES
555 Union Boulevard
Allentown, PA 18103

Thomas E. Melloy
GENERAL ACCOUNTING OFFICE
441 G Street, N.W.
Washington, D.C. 20548

Jasper D. Memory
UNIVERSITY OF NORTH CAROLINA
P.O. Box 2688
General Administration
Chapel Hill, NC 27515-2688

Kevin Mengelt
BEST POWER TECHNOLOGY, INC.
P.O. Box 280
Necedah, WI 54646

Bruce Merrifield
DEPARTMENT OF COMMERCE
14th & Constitution Avenue, N.W.
Washington, D.C. 20230

Robert F. Merwin
ERIEZ MAGNETICS
23rd & Asbury Road
P.O. Box 10608
Erie, PA 16514

Sheldon Meth
SCIENCE APPLICATIONS
INTERNATIONAL
CORPORATION
1710 Goodridge Drive
McLean, VA 22102

Roger D. Meyer
DEPARTMENT OF ENERGY
1000 Independence Avenue, S.W.
Washington, D.C. 20585

Ronald F. Meyer
ARTOS ENGINEERING COMPANY
15600 West Lincoln Avenue
New Berlin, WI 53151

Gerald Michael
CORP FOR SCIENCE &
TECHNOLOGY
One North Capitol
Indianapolis, IN 46204

Michael Michaelis
HOUSTON AREA RESEARCH
CENTER
910 17th Street, N.W.
Suite 517
Washington, D.C. 20006

Herman C. Mihalich
RHONE-POULENC
P.O. Box 125
Monmouth Junction, NJ 08852

Nelson Milder
UNITED STATES CONGRESS
Committee on
Science, Space & Technology
Washington, D.C. 20515

Earl Mills
ASTRON CORPORATION
470 Spring Park Place
Suite 100
Herndon, VA 22070

Lawrence Minkoff
FONAR CORPORATION
110 Marcus Drive
Melville, NY 11747

William C. Mitchell
3M
3M Center
St. Paul, MN 55144-1000

Nalini Mitra
COLORADO SCHOOL OF MINES
Physics Department
Golden, CO 80401

Steven Mocarski
KLINE & COMPANY
330 Passaic Avenue
Fairfield, NJ 07006

David L. Moffat
CORNELL UNIVERSITY
Laboratory of Nuclear Studies
Ithaca, NY 14853-5001

Barry Moline
MERIDIAN CORPORATION
4300 King Street
Suite 400
Alexandria, VA 22302

John Moll
HEWLETT-PACKARD
4111 Old Trace Road
Palo Alto, CA 94306

Peter Mongeau
EML RESEARCH INC./COMMAND
625 Putman Avenue
Cambridge, MA 02139

Bobby L. Montague
CAROLINA POWER & LIGHT
P.O. Box 1551
Raleigh, NC 27602

N. Montanarelli
S.D.I.O.
SDIO/CA
Washington, D.C. 20301-7100

Paul Moon
DEKALB ELECTRONICS COMPANY
P.O. Box 272
Dekalb, IL 60115

John Moore
NATIONAL SCIENCE FOUNDATION
1800 G Street, N.W.
Washington, D.C. 20550

K. Moorjani
JOHNS HOPKINS UNIVERSITY
Applied Physics Laboratory
John Hopkins Road
Laurel, MD 20707

Natalie Moorman
SUPERCONDUTING CONSTITUENCY
1736 / NYSOB
New York, NY 10027

Terri Moreland
STATE OF ILLINOIS
444 North Capitol Street, N.W.
Washington, D.C. 20001

Joe Morrison
MOMBAY CORPORATION-BAYER
5801 Eastern Avenue
Baltimore, MD 21210

Richard E. Morrison
NATIONAL SCIENCE FOUNDATION
1800 G Street, N.W.
Washington, D.C. 20550

Fred A. Morse
LOS ALAMOS NATIONAL
LABORATORY
P.O. Box 1663, MS A114
Los Alamos, NM 87545

Jerome G. Morse
COLORADO SCHOOL OF MINES
Physics Department
Golden, CO 80401

Marvin K. Moss
OFFICE OF NAVAL RESEARCH
800 N. Quincy Street
Arlington, VA 22217-5000

D. Muelenberg
COMSAT LABORATORIES
22300 Comsat Drive
Clarksburg, MD 20871

Carl Alex Mueller
IBM T.J. WATSON RESEARCH
CENTER
c/o Jerry Present
P.O. Box 218
Yorktown Heights, NY 10598

Fred Mueller
LOS ALAMOS NATIONAL
LABORATORY
P.O. Box 1663
Los Alamos, NM 87545

C. Richard Mullen
INTERMAGNETICS GENERAL
Charles Industrial Park
P.O. Box 566
Guilderland, NY 12084

Bernard Multon
WHEELER INDUSTRIES
2611 Jefferson Davis Highway
Arlington, VA 22202

Daniel R. Mulville
NASA
600 Independence Avenue, S.W.
Washington, D.C. 20546

Narayan P. Murarka
IIT RESEARCH INSTITUTE
10 W. 35th Street
Chicago, IL 60616

David Murphree
INSTITUTE FOR TECHNOLOGY
DEVELOPMENT
700 State Street
Jackson, MS 39212

Donald Murphy
AT&T BELL LABORATORIES
600 Mountain Avenue
Mail Stop MH-1D-369
Murray Hill, NJ 07974

Joellyn Murphy
PUBLIC SERVICE COMPANY OF NM
Alvarado Square
Albuquerque, NM 87158

Michael Murphy
OXFORD SUPERCONDUCTIVITY
TECH.
600 Milik Street
Carteret, NJ 07008

L.E. Murr
OREGON GRADUATE CENTER
19600 NW. Von Newmann Drive
Beaverton, OR 97006

Lawrence E. Murr
MONOLITHIC SUPERCONDUCTORS
Box 1654
Lake Oswego, OR 97035

John Murray
UNIVERSITY OF ALASKA
Physics Department
Fairbanks, AK 99709

Peter Murray
WESTINGHOUSE
1801 K Street, N.W.
Washington, D.C. 20006

Bill Muston
TEXAS UTILITIES
400 North Olive Street
Dallas, TX 75201

Harold W. Myron
ARGONNE NATIONAL
LABORATORY
9700 S. Cass Avenue
Materials Science Division
Argonne, IL 60439

Thomas P. Nadolski
INTERMET, INC.
P.O. Box 3549
Reston, VA 22090

David J. Nagel
NRL
Code 4600
Washington, D.C. 20375

Alan Nagelberg
LANXIDE CORPORATION
1 Tralee Industrial Park
Newark, DE 19711

Steve Nagler
UNIVERSITY OF FLORIDA
Department of Physics
215 Williamson Hall
Gainesville, FL 32611

V. Narayanamurti
SANDIA NATIONAL LABORATORIES
P.O. Box 5800
Department 1840/1150
Albuquerque, NM 87185-5800

R.W. Nash
THE ENERGY BUREAU
331 Madison Avenue
New York, NY 10017

Donald Naugle
TEXAS A & M UNIVERSITY
Department of Physics
College Station, TX 77843

Stanley E. Nave
UNIVERSITY OF TENNESSEE
College of Engineering
53 Turner House
Knoxville, TN 37996-2000

Neale R. Neelameggham
AMEX MAGNESIUM
238 North 2200 West
Salt Lake City, UT 84116

Paul Nehring
DEKALB ELECTRONICS COMPANY
P.O. Box 272
Dekalb, IL 60115

Richard A. Neifeld
HARRY DIAMOND LABS
RT-RA
2800 Powder Mill Road
Adelphi, MD 20783

George A. Neil Jr.
CERAMICS PROCESS SYSTEMS
840 Memorial Drive
Cambridge, MA 02139

Richard D. Nelson
FORD AEROSPACE
Ford Road
Mail Code 2-25
Newport Beach, CA 92660

Stephen Nelson
CRAY RESEARCH
900 Lowater Road
Chippewa Falls, WI 54729

Burton Nemroff
GILBERT COMMONWEALTH
P.O. Box 1498
Reading, PA 19603

John J. Nepywoda
AMOCO CORPORATION
P.O. Box 87703
Chicago, IL 60680-0703

Victoria Nerenberg
SUPERCONDUCTIVITY
APPLICATION ASSOCIATION
24781 Camino villa Drive
El Toro, CA 92630

Herb Newborn
C.S. DRAPER LABORATORY
555 Technology Square
Cambridge, MA 02139

Charles M. Newstead
DEPARTMENT OF STATE
DES/NTS
Room 7820
Washington, D.C. 20520

Tg Nieh
LOCKHEED MISSILES & SPACE
COMPANY, INC.
3251 Hanover Street
Palo Alto, CA 94304-1187

Dale Niesz
AMERICAN CERAMIC SOCIETY
c/o Battelle
505 King Street
Columbus, OH 43201

Narain Ningorani
ELECTRIC POWER RESEARCH
INSTITUTE
3412 Hillview Avenue
P.O. Box 10412
Palo Alto, CA 94303

Martin Nisenoff
NAVAL RESEARCH LABORATORY
4555 Overlook Avenue, S.W.
CODE 6854
Washington, D.C. 20375-5000

Frank V. Nolfi Jr.
DEPARTMENT OF ENERGY
1000 Independence Avenue, S.W.
Washington, D.C. 20585

Daniel E. Nordell
NORTHERN STATES POWER
COMPANY
414 Nicollet Mall
Minneapolis, MN 55401

Clyde Northrup
THE PENTAGON
S.D.I.O.
Washington, D.C. 20301-7100

Lewis H. Nosanow
UNIVERSITY OF CALIFORNIA
Administration #155
Irvine, CA 92717

Robert M. Nowak
DOW CHEMICAL COMPANY
2030 Willard H. Dow Center
Midland, MI 48674

E.J. Nucci
ELECTRONIC INDUSTRIES
2001 Eye Street, N.W.
Washington, D.C. 20006

John N. Nustis
DEPARTMENT OF ENERGY
1000 Independence Avenue, S.W.
Washington, D.C. 20585

Mary O'Malley
PATTON, BOGGS, & BLOW
280 West Pratt Street
Suite 1100
Baltimore, MD 21201

R.P. Oberlin
AAI CORPORATION
P.O. Box 126
Hunt Valley, MD 21030

John M. Oblak
UNITED TECHNOLOGIES
1 Financial Plaza
Hartford, CT 06101

D. Oetterer
KALI-CHEMIE CORPORATION
41 West Putman Avenue
Greenwich, CT 06830

Greg Ogden
KAMAN SCIENCES CORPORATION
2560 Huntington Avenue
Suite 500
Alexandria, VA 22303

Tihiro Ohkawa
GA TECHNOLOGIES/APPLIED
SUPERCONETICS
P.O. Box 85608
San Diego, CA 92138

Kenneth Olen
FLORIDA POWER & LIGHT
P.O. Box 14000
Juno Beach, FL 33408-0420

Reinhard G. Olesch
DEPARTMENT OF ENERGY
1000 Independence Avenue, S.W.
Washington, D.C. 20585

Ray Olson
SUNDSTRAND CORPORATION
4747 Harrison Avenue
P.O. Box 7002
Rockford, IL 61125

Richard Opsahl
GRUMAN AIRCRAFT SYSTEMS
CORPORATION
South Oyster Bay Road
Bethpage, NY 11714

Neil Otto
SCIENCE APPLICATIONS
INTERNATIONAL CORPORATION
1701 E. Woodfield Road
Suite 819
Schamburg, IL 60173

Iris M. Ovshinsky
ENERGY CONVERSION DEVICES
1675 West Maple Road
Troy, MI 48084

Stanford Ovshinsky
ENERGY CONVERSION DEVICES
1675 West Maple Road
Troy, MI 48084

William A. Owczarski
PRATT & WITNEY
400 Main Street (165-35)
East Hartford, CT 06108

Robert E. Owen
COOPER INDUSTRIES, INC.
P.O. Box 100
Franksville, WI 53126-9526

David M. Palmer
COMBUSTION ENGINEERING
1000 Prospect Hill Road
Windsor, CT 06095

R.K. Pandey
TEXAS A & M UNIVERSITY
Department of Physics
College Station, TX 77843

Armand J. Panson
WESTINGHOUSE CORPORATION
Research and Development
1310 Beulah Road
Pittsburgh, PA 15235

D. Papaconstantopoulos
NAVAL RESEARCH LABORATORY
205 Apple Blossom Court
Vienna, VA 22180

Elaine Papadopoulos
UNC VENTURES INC.
195 State Street
#700
Boston, MA 02109

Rudolph Pariser
E.I. DUPONT
1007 Market Street
CR&D Experimental Station
Wilmington, DE 19898

John Park
CORDEC CORPORATION
8270-B Cinder Bed Road
Lorton, VA 22079

Y. Jin Park
AMAX R&D CENTER
5950 McIntyre Street
Golden, CO 80403

Dr. Stuart S.P. Parkin
IBM ALMADEN RESEARCH CENTER
650 Harry Road
San Jose, CA 95120-6099

Kenneth Partain
GA TECHNOLOGIES/APPLIED
SUPERCONETICS
P.O. Box 85608
San Diego, CA 92138

Dr. Gabriel Pasztor
SWISS INSTITUTE FOR
NUCLEAR RESEARCH
CH-5234
Villigen, SWITZERLAND

Bharat C. Patel
NJ DEPARTMENT OF COMMERCE
101 Commerce Street
Newark, NJ 07102-5102

Sushil Patel
NEW YORK, STATE UNIVERSITY OF
Electrical & Computer Engineering
330 Bonner Hall
Amhurst, NY 14260

Pamela Patrick
BEAR ANALYTICS
1615 M Street, N.W.
Suite 500
Washington, D.C. 20036

George A. Paulikas
THE AEROSPACE CORPORATION
P.O. Box 92957
Los Angeles, CA 90009

Ronald Paulson
EDO CORPORATION
44 Commerce Road
Stamford, CT 06904

Perkins C. Pedrick
LOGISTICS MANAGEMENT
INSTITUTE
6400 Goldsboro Road
Bethesda, MD 20817-5886

Sam Pellicari
GUERNSEY COATING LABS
4464 McGrath Street
Unit 106
Ventura, CA 93003

Dave Peltier
ASEA AUTOCLAVE
26825 Dare Road
Rockbridge, OH 43149

Glenn E. Penisten
SUPERCONDUCTOR
TECHNOLOGIES, INC.
2200 Sand Hill Road
Suite 250
Menlo Park, CA 94015

Gerald Perronne
ROCKETDYNE
6633 Canoga Avenue
Canoga Park, CA 91303

Tom Pestorius
DYNAMETRICS
6801 Whittier Avenue
#3048
McLean, VA 22101

Edward Peters
ARTHUR D. LITTLE INC.
Acorn Park
Cambridge, MA 02140

Dean Peterson
LOS ALAMOS NATIONAL
LABORATORY
P.O. Box 1663
MS G730
Los Alamos, NM 87545

Andrew Pettifor
ROCKWELL INTERNATIONAL
1049 Camino Dos Rios
Thousand Oaks, CA 91360

Bill Petuskey
ARIZONA STATE UNIVERSITY
Tempe, AZ 85287

N. Douglas Pewitt
SCIENCE APPLICATIONS
INTERNATIONAL CORPORATION
10260 Campus Point Drive
San Diego, CA 92121

George F. Pezoirtz
PALIDIN ASSOCIATES LIMITED
L'Enfant Plaza
Box 23166
Washington, D.C. 20026

Karl Pfitzer
MCDONNELL DOUGLAS
CORPORATION
5301 Bolsa
Huntington Beach, CA 92708

G.C. Phillips
T.W. BONNER NUCLEAR
LABORATORY
Rice University
P.O. Box 1892
Houston, TX 77251

Norman E. Phillips
LAWRENCE BERKELEY
LABORATORY
One Cyclotron Road
Berkeley, CA 94720

Richard Phillips
HUGHES AIRCRAFT COMPANY
P.O. Box 902
Mail Station E55-G206
El Segundo, CA 90245

Warren Pickett
NAVAL RESEARCH LABORATORY
Washington, D.C. 20375-5000

Henry Piehler
CARNEGIE-MELLON UNIVERSITY
4400 Fifth Avenue
Pittsburgh, PA 15213

James Pierce
CVI, INCORPORATED
P.O. Box 2138
Columbus, OH 43216

Albert Piesco
SINGER COMPANY
11800 Tech Road
Silver Spring, MD 20904

Barbara A. Piette
CHARLES RIVER VENTURES
67 Batterymarch Street
Boston, MA 02110

John Pincus
WEST END CONSULTANTS
817 West End Avenue
New York, NY 10025

David Pines
UNIVERSITY OF ILLINOIS
Loomis Laboratory of Physics
1110 W. Green Street
Urbana, IL 61801

R. Byron Pipes
UNIVERSITY OF DELAWARE
College of Engineering
137 Du Pont Hall
Newark, DE 19716

Teresa Piquette
ARCO
1333 New Hampshire Avenue, N.W.
Suite 1001
Washington, D.C. 20036

Bob Pirih
E G & G
40 William Street
Wellesley, MA 02181

Walter Podney
PHYSICAL DYNAMICS, INC
P.O. Box 1883
La Jolla, CA 92038

Dr. Roger B. Poeppel
ARGONNE NATIONAL
LABORATORY
9700 S. Cass Avenue
Materials Science Division
Argonne, IL 60439

Armand Poirier
SANDERS ASSOCIATES
95 Canal Street
MS NCA1-4393
Nashua, NH 03060

Peter Pollak
ALUMINUM ASSOCIATION
900 19th Street, N.W.
Washington, D.C. 20006

Tom Pollak
SANDERS ASSOCIATES
95 Canal Street
MS NCA1-4393
Nashua, NH 03060

John Pollard
U.S. ARMY-CECOM
CNVEO
Fort Belvoir, VA 22060-5677

Dietr Pommerrenig
EO PRODUCTS CORPORATION
23101 Moulton Parkway
Lagunda Hills, CA 92531

S.J. Poon
UNIVERSITY OF VIRGINIA
Physics Department
Charlottesville, VA 22901

Edward J.A. Pope
UNIVERSITY OF CALIFORNIA -
LOS ANGELES
405 Hilgard Avenue
Los Angeles, CA 90024-1405

Susan Pope
HARVARD UNIVERSITY
Soldiers Field
Morgan Hall 7
Boston, MA 02163

Herman Postma
OAK RIDGE NATIONAL
LABORATORY
P.O. Box X
Oak Ridge, TN 37831-6255

Julianne H. Prager
3M
3M Center
St. Paul, MN 55144-1000

Jerry Present
IBM T.J. WATSON RESEARCH
CENTER
P.O. Box 218
Yorktown Heights, NY 10598

Barbara Preston
SANTA FE CORPORATION
4900 Seminar Road
Suite 500
Alexandria, VA 22311

John Preston
MASSACHUSETTS INSTITUTE
OF TECHNOLOGY
77 Massachusetts Avenue
Cambridge, MA 02139

David Priest
SRI INTERNATIONAL
333 Ravenswood Avenue
Menlo Park, CA 94025-8493

S. Proffitt
USASDC
P.O. Box 1500
DASD-H-V
Huntsville, AL 35807-3801

Joseph M. Proud
GTE LABORATORIES, INC.
40 Sylvan Road
Waltham, MA 02254

Frederic Quan
CORNING
Sullivan Park
FR62
Corning, NY 14831

Michael L. Quick
ENGELHARD CORPORATION
Menlo Park
CN28
Edison, NJ 08818

N. Quick
APPLITECH OF INDIANA
8150 Zionsville Road
Indianapolis, IN 46200

S. Victor Radcliffe
NFC
2101 Connecticut Avenue, N.W.
Washington, D.C. 20008

Farhad Radpour
UNIVERSITY OF OKLAHOMA
202 West Boyd
Physics Department
Norman, OK 73019

Joseph Raguso
ALFRED UNIVERSITY
115 Filmore Street
Physics/Ceramics Department
Centerport, NY 11721

Richard Ralston
MASSACHUSETTS INSTITUTE OF
TECHNOLOGY LINCOLN
LABORATORY
244 Wood Street
Lexington, MA 02173

Venki Raman
AIR PRODUCTS & CHEMICALS, INC.
P.O. Box 538
Route 222
Allentown, PA 18105

R. Ranellone
NEWPORT NEWS SHIPBUILDING
22 Kenilworth Drive
Hampton, VA 23666

Bhakta B. Rath
NATIONAL RESEARCH
LABORATORY
Code 6000
4555 Overlook Avenue, S.W.
Washington, D.C. 20375-5000

Steve Rattien
NATIONAL RESEARCH COUNCIL
2101 Constitution Avenue, N.W.
Washington, D.C. 20418

Bernard A. Rausch
IIT RESEARCH INSTITUTE
10 W. 35th Street
Chicago, IL 60616

Doyle Rausch
NATIONAL STANDARD COMPANY
Research and Development
Niles, MI 49120

Siba P. Ray
ALUMINUM COMPANY OF AMERICA
Alcoa Technical Center
Alcoa Center, PA 15069-0001

Dennis Ready
OHIO STATE UNIVERSITY
Department of Physics
174 W. 18th Avenue
Columbus, OH 43210

Nuka Reddy
APPLIED POLYMER SYSTEMS INC.
P.O. Box 4040
College Point, NY 11356

Stanley Reible
MICRILOR
9 Lakeside Park N. Avenue
Wakefield, MA 01880

Joseph Reid
GTE LABORATORIES, INC.
40 Sylvan Road
Waltham, MA 02254

Jack Reilly
ELECTRO-KINETIC SYSTEMS
701 Chestnut Street
Trainer, PA 19013

Burton M. Rein
W.R. GRACE & COMPANY
7379 Route 32
Columbia, MD 21044

Kenneth Reinschmidt
STONE & WEBSTER ENGINEERING
CORP.
P.O. Box 2325
Boston, MA 02107

James V. Rett
PACIFIC POWER & LIGHT CO.
920 SW 6th Avenue
Portland, OR 97204

Dr. Kay Rhyne
DEFENSE ADVANCED RESEARCH
PROJECTS AGENCY
1400 Wilson Boulevard
Arlington, VA 22209

Roy W. Rice
W.R. GRACE & COMPANY
7379 Route 32
Columbia, MD 21044

Kenneth Richards
KERR-MCGEE CORPORATION
P.O. Box 25861
Oklahoma City, OK 73125

Al Riesen
TELEDYNE
Wah Chang Albany Division
Box 460
Albany, OR 97321

Brian Riley
COMBUSTION ENGINEERING
1000 Prospect Hill Road
Windsor, CT 06095

Kenneth W. Rind
OXFORD VENTURE CORPORATION
1266 Main Street
Stamford, CT 06902

P. Agustin Rios
GENERAL ELECTRIC COMPANY
Building 37
Room 311
Schenectady, NY 12345

Keith E. Ritala
COMINCO ELECTRONIC MATERIALS
15128 Euclid Avenue
Spokane, WA 99216

Kim Ritchie
AVX CORPORATION
17th Avenue South
Myrtle Beach, SC 29577

Lee W. Rivers
FEDERAL LAB CONSORTIUM
1825 K Street, N.W.
Suite 218
Washington, D.C. 20006

Alfred J. Roach
TII INDUSTRIES, INC.
1375 Akron Street
Copiague, NY 11726

Walt Robb
GENERAL ELECTRIC
Research and Development
P.O. Box 8
Schenectady
Schenectady, NY 12301

Raymond Roberge
HYDRO O.
1736/NYSOB
New York, NY 10027

Marga Roberts
FORD MOTOR COMPANY
Scientific Research Laboratory
Box 1899
Dearborn, MI 58121-2053

R.O. Roberts
GENERAL DYNAMICS
P.O. Box 748
Ft. Worth, TX 76101

Heyward G. Robinson
STANFORD UNIVERSITY
200 Junipero Serra Boulevard
Stanford, CA 94305

Mark M. Rochkind
NORTH AMERICAN PHILIPS
CORPORATION
345 Scarrborough Road
Briarcliff MNR, NY 10510

Louis H. Roddis
DETROIT EDISON COMPANY
330 Concord Street, 10FG
Charleston, SC 29401

C.C. Rolfe
DUKE POWER COMPANY
P.O. Box 33189
Charlotte, NC 28242

Michael Romei
SHELL OIL COMPANY
One Rockefeller Plaza
New York, NY 10020

Michael Root
DEPARTMENT OF THE ARMY
DAMO-SSL, 3B521
Pentagon
Washington, D.C. 20310-0420

Mike Roper
OFFICE OF SCIENCE & TECHNICAL
INFORMATION
U.S. Department of Energy
P.O. Box 62
Oak Ridge, TN 37831

Dr. Robert M. Rose
MIT & HARVARD MEDICAL SCHOOL
MIT Division of
Health Sciences and Technology
Materials Science & Engineering
Cambridge, MA 02138

Mark David A. Rosen
GRUMAN AIRCRAFT SYSTEMS
CORPORATION
South Oyster Bay Road
MS AD1-26
Bethpage, NY 11714

Stuart Rosenwasser
SPARTA, INC.
110413 Camino Delmar
Delmar, CA 92014

Peter A. Roshko
MOHR, DAVIDOW VENTURES
3000 Sand Hill Road
Building 4
#240
Menlo Park, CA 94025

Carl H. Rosner
INTERMAGENTICS GENERAL
CORPORATION
P.O. Box 566
New Karner Road
Guilderland, NY 12084

Charles J. Ross
KANSAS CITY POWER & LIGHT
P.O. Box 679
Kansas City, MO 64141

Ian Ross
AT&T BELL LABORATORIES
600 Mountain Avenue
Mail Stop MH-1D-369
Murray Hill, NJ 07974

Philip N. Ross Jr.
DEPARTMENT OF ENERGY
1000 Independence Avenue, S.W.
Washington, D.C. 20585

Steve M. Rothstein
CVC PRODUCTS, INC.
525 Lee Road
P.O. Box 1886
Rochester, NY 14603

Upendra Roy
DEPARTMENT OF COMMERCE
Patent & Trademark Office
2021 Jefferson Davis Highway
Arlington, VA 22202

T. James Rudd
EMHART CORPORATION
426 Colt Highway
Farmington, CT 06032

Dr. David A. Rudman
MASSACHUSETTS INSTITUTE OF
TECH.
Dept. Materials Science &
Engineering
Cambridge, MA 02138

C.J. Russell
SUPERCONDUCTIVITY NEWS
65 Jackson Drive
Suite 2000
Cranford, NJ 07016

Truman Rutt
AVX CORPORATION
17th Avenue South
Myrtle Beach, SC 29577

John Ruvalds
HARVARD UNIVERSITY
Lyman Laboratory of Physics
Cambridge, MA 02138

George P. Sabol
WESTINGHOUSE R&D CENTER
1310 Beulah Road
Pittsburgh, PA 15235

Alfred Sagarese
STRATEGIC ANALYSIS
Box 3485
RD3
Fairlane Road
Reading, PA 19606

George Sahady
NEW ENGLAND GOVERNMENT
CONFERENCE
76 Summer Street
Boston, MA 02110

Joseph F. Salgado
DEPARTMENT OF ENERGY
1000 Independence Avenue, S.W.
Washington, D.C. 20585

James Salsgiver
ALLEGHENY LUDLUM
CORPORATION
Technical Center
Brackenridge, PA 15014

Shyam Samanta
UNIVERSITY OF MICHIGAN
2250 GG Brown
Ann Arbor, MI 48109-2125

Dr. Nicholas P. Samios
BROOKHAVEN NATIONAL
LABORATORY
Associated Universities, Inc.
Building 134
Upton, NY 11973

William P. Samuels
UNION CARBIDE CORPORATION
39 Old Ridgebury Road
Danbury, CT 06817

Robert L. San Martin
DEPARTMENT OF ENERGY
1000 Independence Avenue, S.W.
Washington, D.C. 20585

Randall R. Sands
NAVAL AIR DEVELOPMENT CENTER
Code 6063
Warminster, PA 18974

Phillip A. Sanger
OXFORD SUPERCONDUCTING
TECHNOLOGY
600 Milik Street
Carteret, NJ 07008

S.G. Sankar
ADVANCED MATERIALS
CORPORATION
c/o Carnegie-Mellon University
4400 Fifth Avenue
Pittsburgh, PA 15213

Larry Satek
AMOCO CHEMICAL COMPANY
1316 Clifden Street
Wheaton, MD 60187

C.S. Saunders
I.C.I. AMERICAS INC.
P.O. Box 751
Wilmington, DE 19899

Sid Saunders
ICI AMERICAS
1600 M Street, N.W.
Suite 702
Washington, D.C. 20036

E.P. Scannel
AAI CORPORATION
P.O. Box 126
Hunt Valley, MD 21030

Dr. Harvey W. Schadler
GENERAL ELECTRIC CORPORATE
R&D
P.O. Box 8
Schenectady, NY 12301

Peter E. Schaub
POTOMAC ELECTRIC POWER
1900 Pennsylvania Avenue, N.W.
Washington, D.C. 20068

William W. Schertz
ARGONNE NATIONAL
LABORATORY
9700 S. Cass Avenue
Materials Science Division
Argonne, IL 60439

Michael Schilmoeller
SALT RIVER PROJECT
P.O. Box 52025
Phoenix, AZ 85072-2025

T.D. Schlabach
AT&T BELL LABORATORIES
600 Mountain Avenue
Mail Stop MH-1D-369
Murray Hill, NJ 07974

Ferenc Schmidt
AMTEK
Station Square TOO
Paoli, PA 19301

Roland W. Schmitt
GENERAL ELECTRIC CORPORATE
R&D
P.O. Box 8
Schenectady, NY 12301

Thomas R. Schneider
ELECTRIC POWER RESEARCH
INSTITUTE
3412 Hillview Avenue
P.O. Box 10412
Palo Alto, CA 94303

Richard Schoen
NATIONAL SCIENCE FOUNDATION
1800 G Street, N.W.
Washington, D.C. 20550

J. Clifford Schoep
GENERAL DYNAMICS
CORPORATION
7733 Forsyth Boulevard
St. Louis, MO 63105

Harlan Schone
COLLEGE OF WILLIAM & MARY
209 Kingswood Drive
Department of Physics
Williamsburg, VA 23185

Dr. James Schriber
SANDIA NATIONAL LABORATORIES
P.O. Box 5800
Department 1840/1150
Albuquerque, NM 87185-5800

Robert Schrieffer
UNIVERSITY OF CALIFORNIA
SANTA BARBARA
Institute for Theoretical Physics
Ellison Hall
Santa Barbara, CA 93106

Alan Schriesheim
ARGONNE NATIONAL
LABORATORY
9700 S. Cass Avenue
Materials Science Division
Argonne, IL 60439

Alan Schriesheim
ARGONNE NATIONAL
LABORATORY
9700 South Cass Avenue
Argonne, IL 60439

Klaus Schroder
SYRACUSE UNIVERSITY
320 Hinds Halls
Syracuse, NY 13244-1190

John Schultz
NAVAL AIR SYSTEMS COMMAND
CODE AIR-514 3F
Washington, D.C. 20361-5140

John C. Schumacher
CUSTOM ENGINEERED MATERIALS
4039 Avenida de la Plata
Oceanside, CA 92056

Richard T. Schumacher
VARIAN RESEARCH CENTER
611 Hansen Way
Box K-402
Palo Alto, CA 94303

Brian Schwartz
MIT FRANCIS BITTER NATIONAL
MAGNET LABORATORY
c/o TMAH Consultants
13 Shepart #5
Cambridge, MA 02138

Jo-Ann Schwartz
E.I. DUPONT
838 McComb Lane
Chadds Ford, PA 19317

Murray Schwartz
U.S. BUREAU OF MINES
Minerals and Materials Research
Washington, D.C. 20241

Fred Schwerer
ALCOA
218 Adams Street
Export, PA 15632

Richard L. Schwoebel
SANDIA NATIONAL LABORATORIES
P.O. Box 5800
Department 1840/1150
Albuquerque, NM 87185-5800

Patricia Seasock
V. GARBER INTERNATIONAL
ASSOC.
1215 Jefferson Davis Highway
Suite 1105
Arlington, VA 22202

David G. Seiler
NORTH TEXAS STATE UNIVERSITY
2128 Savannah Trail
Denton, TX 76205

Stanley Sekula
OAK RIDGE NATIONAL
LABORATORY
P.O. Box X
Oak Ridge, TN 37831

Jerry Selvaggi
ERIEZ MAGNETICS
23rd & Asbury Road
P.O. Box 10608
Erie, PA 16514

Philip Selwyn
OFFICE OF NAVAL RESEARCH
800 N. Quincy Street
Arlington, VA 22217-5000

Carl Setterstrom
HARRINGTON RESEARCH
COMPANY
436 West Main Street
Wyckoff, NJ 07481

Donald E. Shapero
NATIONAL RESEARCH COUNCIL
2101 Constitution Avenue, N.W.
Washington, D.C. 20418

Prem L. Sharma
RADIO MATERIALS CORPORATION
East Park Avenue
Attica, IN 47918

Shiv Sharma
UNIVERSITY OF HAWAII
Physics Department
Honolulu, HI 96822

Mary Sharp-Hayes
TEMPLE, BARKER, & SLOAN
33 Hayden Avenue
Lexington, MA 02173

Shambhu Shastry
GTE LABORATORIES, INC.
40 Sylvan Road
Waltham, MA 02254

David T. Shaw
STATE UNIVERSITY OF
NY - BUFFALO
330 Bonner Hall
Buffalo, NY 14260

Robert W. Shaw Jr.
ARET VENTURES INC.
6110 Executive Blvd., Suite 1040
Rockville, MD 20852

Thomas P. Sheahen
WESTERN TECHNOLOGY INC.
18708 Woodway Drive
Derwood, MD 20855

David P. Sheetz
DOW CHEMICAL COMPANY
2030 Willard H. Dow Center
Midland, MI 48674

Haskell Sheinberg
LOS ALAMOS NATIONAL
LABORATORY
P.O. Box 1663
MS G770
Los Alamos, NM 87545

Zhengzhi Sheng
UNIVERSITY OF ARKANSAS
2801 University Avenue
Little Rock, AR 72204

Kent P. Shepard
DREXEL BURNHAM LAMBERT
60 Broad Street
New York, NY 10004

William Shields
JANIS RESEARCH COMPANY
2 Jewel Drive
Wilmington, MA 01887

Daniel K. Shipp
NATIONAL ELECTRICAL
MANUFACTURERS
ASSOCIATION
2101 L Street, N.W.
Washington, D.C. 20037

David A. Shirley
LAWRENCE BERKELEY
LABORATORY
One Cyclotron Road
Berkeley, CA 94720

John Shoch
CONDUCTUS, INC.
2275 E. Bayshore Road
Palo Alto, CA 94303

Brian Shreve
ASTRON CORPORATION
470 Spring Park Place
Suite 100
Herndon, VA 22070

Roy Shultz
WHEELER INDUSTRIES
2611 Jefferson Davis Highway
Arlington, VA 22202

Edwin Shykind
DEPARTMENT OF COMMERCE
903 Burnt Crest Lane
Silver Spring, MD 20903

Seymour Siegel
UNIVERSITY OF CALIFORNIA -
LOS ANGELES
405 Hilgard Avenue
Los Angeles, CA 90024-1405

Richard Silberglitt
QUEST RESEARCH CORPORATION
1749 Old Meadow Road
McLean, VA 22102

Arnold H. Silver
TRW SPACE & TECHNOLOGY
GROUP
1 Space Park
Redondo Beach, CA 90278

Michael B. Simmonds
QUANTUM DESIGN
11578 Sorrento Valley Road
San Diego, CA 92121

Robert Simms
LITTON INDUSTRIES
1215 S. 52nd Street
Tempe, AZ 85281

R.W. Simon
TRW SPACE & TECHNOLOGY
GROUP
1 Space Park
Redondo Beach, CA 90278

S. Fred Singer
DEPARTMENT OF
TRANSPORTATION
400 Seventh Street, S.W.
Washington, D.C. 20590

Sidney Singer
SDIO/CA
2308 S. Rolfe Street
Arlington, VA 22202

Uday Sinha
SOUTHWIRE
145 Camp Drive
Carrollton, GA 30117

Earl Skelton
NAVAL RESEARCH LABORATORY
4555 Overlook Avenue, S.W.
Code 6483
Washington, D.C. 20375-5000

Arthur W. Sleight
E.I. DUPONT
1007 Market Street
CR&D Experimental Station
Wilmington, DE 19898

Robert S. Slott
SHELL DEVELOPMENT COMPANY
P.O. Box 2463
Houston, TX 77001

James Sluss
UNIVERSITY OF VIRGINIA
Thornton Hall
Charlottsville, VA 22901

Bruce Smart Jr.
DEPARTMENT OF COMMERCE
14th & Constitution Avenue, N.W.
Washington, D.C. 20230

Daniel G. Smedley
EATON CORPORATION
26201 Northwestern Highway
Southfield, MI 48037

Barry Smernoff
BJSA ENTERPRISES, INC.
P.O. Box 3136
Oakton, VA 22124

Dr. James L. Smith
LOS ALAMOS NATIONAL
LABORATORY
Center for Materials Science
P.O. Box 1663
Los Alamos, NM 87545

P. Smola
TERMIFLEX CORPORATION
316 Daniel Webster Highway
Merrimack, NH 03054

Donald M. Smyth
LEHIGH UNIVERSITY
Bethlehem, PA 18015

Milton Sneller
DATA DESIGN LAB
7925 Center Avenue
Rancho
Cucamonga, CA 91730

Dave Snoke
EMERSON ELECTRIC CORPORATION
8000 W. Foorissant
St. Louis, MO 63136

Joel A. Snow
DEPARTMENT OF ENERGY
1000 Independence Avenue, S.W.
Washington, D.C. 20585

William Snowden
DEPARTMENT OF DEFENSE
OUSD(A)/R & AT
The Pentagon, Room 3E114
Washington, D.C. 20301

Martin M. Sokoloski
NASA
600 Independence Avenue, S.W.
CODE RC
Washington, D.C. 20546

J.C. Solinsky
SCIENCE APPLICATIONS
INTERNATIONAL CORPORATION
10260 Campus Point Drive
San Diego, CA 92121

Ian W. Sorensen
ALLIED SIGNAL CORPORATION
AATC 900 West Maple Road
Troy, MI 48007-2602

Robert Soulen
NAVAL RESEARCH LABORATORY
Washington, D.C. 20375-5000

Stanley F. Spangenberger
DOW CHEMICAL COMPANY
2030 Willard H. Dow Center
Midland, MI 48674

Steven R. Speech
3M
3M Center
St. Paul, MN 55144-1000

S.T. Sprang
AAI CORPORATION
P.O. Box 126
Hunt Valley, MD 21030

Sriniva S. Sridhar
NORTHEASTERN UNIVERSITY
360 Huntington Avenue
Boston, MA 02115

Robert Srubas
SPECIALTY CABLE CORPORATION
P.O. Box 50
Walling Ford, CT 06492

Dr. Angelica M. Stacy
UNIVERSITY OF CALIFORNIA-
BERKELEY
Chemistry Department
Berkeley, CA 94720

Robert Standley
AMOCO CORPORATION
Amoco Research Center
P.O. Box 400
Naperville, IL 60568

Elizabeth Starbuck
MIDWEST TECHNICAL
DEVELOPMENT INSTITUTE
610 Conwed Tower
444 Cedar Street
St. Paul, MN 55101

Lawrence Starch
CONDUCTOR TECHNOLOGIES
1001 Connecticut Avenue, N.W.
Washington, D.C. 20036

Eugene E. Stark
FEDERAL LABORATORY
CONSORTIUM
H811 - Los Alamos National Lab
Los Alamos, NM 87545

Hermann Statz
RAYTHEON COMPANY
131 Spring Street
Lexington, MA 02173

John Stekly
INTERMAGENTICS GENERAL
CORPORATION
P.O. Box 566
New Karner Road
Guilderland, NY 12084

James Stephan
COORS BIOMEDICAL COMPANY
12860 W. Cedar Drive
Lakewood, CO 80004

Edward A. Stern
UNIVERSITY OF WASHINGTON
Material Science & Engineering
Physics FM-15
Seattle, WA 98195

Ernest Stern
MASSACHUSETTES INSTITUTE OF
TECHNOLOGY LINCOLN
LABORATORY
244 Wood Street
Lexington, MA 02173

Tom Stern
FILM TECH
7200 Ohms Lane
Edina, MN 55475

Ward Stevens
ATM INC.
520-B Danbury Road
New Milford, CT 06776

John C. Stiles
SINGER COMPANY
Kearfott Division
1150 McBride Avenue
Little Falls, NJ 07424

Ronald Stivers
GEORGE WASHINGTON
UNIVERSITY
Department of Physics
Washington, D.C. 20008

Frank Stodolsky
ARGONNE NATIONAL LABS
1611 N. Kent Street
No. 211
Arlington, VA 22209

Porter Stone
PIEZO ELECTRIC PRODUCTS
186 Massachusetts Avenue
Cambridge, MA 02139

Lawrence Storch
CONDUCTORS TECHNOLOGIES, INC.
21001 Connecticut Avenue, N.W.
Room 701
Washington, D.C. 20036

W. Stotesbery
MCC TECHNOLOGIES CENTER
3500 W. Balcones Center
Austin, TX 78759-6509

Jerry L. Straalsund
BATTELLE-PACIFIC NW
LABORATORY
2033 Trevino Court
P.O. Box 999
Richland, WA 99352-9785

P.A. Stranges
NORTHWEST TECHNICAL
INDUSTRIES, INC.
1840 18th Street, N.W.
Washington, D.C. 20009

Peter Stranges
NORTHWEST TECHNOLOGY
INDUSTRY
4 Chittenden Lane
Owings Mills, MD 21117

Robert Stratton
TEXAS INSTRUMENTS
P.O. Box 655936
MS 136
Dallas, TX 75265

Dr. Bruce Strauss
MASSACHUSETTS INSTITUTE OF
TECH.
Visiting Scientist
Cambridge, MA 02138

Donald M. Strayer
JET PROPULSION LABORATORY
4800 Oak Grove Drive
Pasadena, CA 91109

Alan J. Streb
DEPARTMENT OF ENERGY
1000 Independence Avenue, S.W.
Washington, D.C. 20585

Martin Strickley
BDM CORPORATION
7915 Jones Branch Drive
McLean, VA 22102

John Stringer
ELECTRIC POWER RESEARCH
INSTITUTE
3412 Hillview Avenue
P.O. Box 10412
Palo Alto, CA 94303

Joseph J. Stupak
SYNERTRON CORPORATION
12000 SW Garden Place
Portland, OR 97223

Jow-Lih Su
W.R. GRACE & COMPANY
7379 Route 32
Columbia, MD 21044

Cheryl R. Suchors
COOPERS & LYBRAND
National High Technology
Industries Program
One Post Office Square
Boston, MA 02109

Masaki Suenaga
BROOKHAVEN NATIONAL
LABORATORY
Building 460
Upton, NY 11973

Nam P. Suh
NATIONAL SCIENCE FOUNDATION
1800 G Street, N.W.
Washington, D.C. 20550

Harry Suhl
UNIVERSITY OF CALIFORNIA
AT SAN DIEGO
Department of Physics, B019
La Jolla, CA 92093

Tony Sun
CONDUCTUS
2275 E. Bayshore Road
Palo Alto, CA 94303

Robert Sundahl
ALLIED SIGNAL CORPORATION
Columbia Road
P.O. Box 1021 R
Morristown, NJ 07960

Dr. Steven A. Sunshine
AT&T BELL LABORATORIES
600 Mountain Avenue
Murry Hill, New Jersey 07974

Michael J. Superczynski
DAVID TAYLOR NAVAL SHIP R&D
CENTER
Code 2712
Annapolis, MD 21402-5067

David F. Sutter
DEPARTMENT OF ENERGY
1000 Independence Avenue, S.W.
Washington, D.C. 20585

John Swartz
LAKE SHORE CRYOTRONICS
64 East Walnut Street
Westerville, OH 43081

Paul S. Swartz
NEW YORK SCIENCE &
TECHNOLOGY
FOUNDATION
99 Washington Avenue
Albany, NY 12210

David A. Syracuse
AMERICAN PRECISION INDUSTRIES
270 Quaker Road
East Aurora, NY 14052

Bruce Taggart
BDM CORPORATION
7915 Jones Branch Drive
McLean, VA 22102

Dennis E. Talbert Jr.
PATENT & TRADEMARK OFFICE
2021 Jefferson Davis Highway
Arlington, VA 22206

T.J. Talley
TEXAS UTILITIES
400 North Olive Street
L.B. 81
Dallas, TX 75201

Robert S. Tamosaitis
E.I. DUPONT
1007 Market Street
CR&D Experimental Station
Wilmington, DE 19898

Shoji Tanaka
UNIVERSITY OF TOKYO
7-3-1 Hongo
Bunkyo-ku, Tokyo 113
Tokyo, Japan

David B. Tanner
UNIVERSITY OF FLORIDA
Department of Physics
215 Williamson Hall
Gainesville, FL 32611

James Tansey
NEW ENGLAND GOVERNMENT
CONFERENCE
76 Summer Street
Boston, MA 02110

Dr. Jean M. Tarascon
AT&T BELL COMMUNICATIONS
RESEARCH
21 Haran Circle
Milburn, NJ 07041

David Taschler
AIR PRODUCTS & CHEMICALS, INC.
P.O. Box 538
Route 222
Allentown, PA 18105

Sheridan Tatsumo
DATAQUEST/DUN & BRADSTREET
CORP.
1290 Ridder Park Drive
San Jose, CA 95131

Clyde Taylor
LAWRENCE BERKELEY
LABORATORY
One Cyclotron Road
Berkeley, CA 94720

Dag Tellefsen
GLENWOOD MANAGEMENT
3000 Sand Hill Road
Building 4
Suite 230
Menlo Park, CA 94025

V.J. Tennery
OAK RIDGE NATIONAL
LABORATORY
P.O. Box X
Oak Ridge, TN 37831-6255

Frank R. Tepe
UNIVERSITY OF CINCINNATI
ML #627
Cincinnati, OH 45221

Becky Terry
PALISADES INSTITUTE
3316 South 6th Street
Arlington, VA 22204

Daniel Tessler
FUSION SYSTEMS CORPORATION
7600 Standish Place
Rockville, MD 20855

Brian C. Thomas
PUGET SOUND POWER & LIGHT
P.O. Box 97034
Bellevue, WA 98009-9734

E.D. Thomas
THOMAS & BETTS
757 Cherry Valley Road
Princeton, NJ 08540

Floyd Thomas
GRUMAN AIRCRAFT SYSTEMS
CORPORATION
South Oyster Bay Road
Bethpage, NY 11714

Iran L. Thomas
DEPARTMENT OF ENERGY
1000 Independence Avenue, S.W.
Washington, D.C. 20585

Richard A. Thomas
BROOKHAVEN NATIONAL
LABORATORY
Upton, NY 11973

Robert J. Thomas
NATIONAL SCIENCE FOUNDATION
1800 G Street, N.W.
Washington, D.C. 20550

William A. Thomas
SUPERCONDUCTIVITY FLASH REP.
1400 N. State Parkway
#7B
Chicago, IL 60610

Jonathan Thompson
THE WHITE HOUSE
OFFICE OF SCIENCE &
TECHNOLOGY POLICY
Washington, D.C. 20500

Lewis B. Thompson
VACUUM BARRIER CORPORATION
Four Barten Lane
P.O. Box 529
Woburn, MA 01801-0529

Tommy Thompson
RIOTECH
P.O. Box 1369
Albuquerque, NM 87103

Ralph Thomson
AMERICAN ELECTRONICS
ASSOCIATION
1612 K Street, N.W.
Suite 1100
Washington, D.C. 20006

Clare Thornton
LABCOM
SLCET-D
Ft. Monmouth, NJ 07703-5302

G.F. Tice
FLUOR DANIEL
3333 Michelson Drive
(B4D)
Irvine, CA 92730

John K. Tien
COLUMBIA UNIVERSITY
918 Mudd Building
New York, NY 10027

Maury Tigner
CORNELL UNIVERSITY
Newman Laboratory
Ithica, NY 14853

Michael Tinkham
HARVARD UNIVERSITY
Physics Department
Cambridge, MA 02138

J.W. Tipping
I.C.I. AMERICAS INC.
P.O. Box 751
Wilmington, DE 19899

Erik Titland
BALTIMORE GAS & ELECTRIC
512 Idlewild Road
Bel Air, MD 21014

T.L. Tolbert
MONSANTO COMPANY
800 North Lindbergh
Saint Louis, MO 63167

John S. Toll
UNIVERSITY OF MARYLAND
Physics Department
Adelphi, MD 20783

Walter J. Tomasch
UNIVERSITY OF NOTRE DAME
P.O. Box E
Physics Department
Notre Dame, IN 46556

Louis Toth
NAVAL RESEARCH LABORATORY
Washington, D.C. 20375-5000

John Trani
GENERAL ELECTRIC
P.O. Box 414
Milwaukee, WI 53201

John Trefny
COLORADO SCHOOL OF MINES
Physics Department
Golden, CO 80401

Kris C. Tripathi
FOXBORO COMPANY
Foxboro, MA 02035

Anil Trivedi
ALLIED SIGNAL CORPORATION
Garrett AIresearch Division
Columbia Road
P.O. Box 1021 R
Morristown, NJ 07960

Joseph Trivisonno
NATIONAL SCIENCE FOUNDATION
1800 G Street, N.W.
Washington, D.C. 20550

Janett Truhatch
ARCH DEVELOPMENT
CORPORATION
115-25 E. 56th Street
Chicago, IL 60637

Ignatius Tsong
ARIZONA STATE UNIVERSITY
Tempe, AZ 85287

Raphael Tsu
N.C. A&T STATE UNIVERSITY
Dept. of Electrical Engineering
Greensboro, NC 27411

David Turner
LUCAS INDUSTRIES, INC.
5500 New King Street
Troy, MI 48098

Gilbert Ugiansky
CORTEST ENGINEERING SERVICES,
INC.
15200 Shady Grove Road #350
Rockville, MD 20850

Joe Underwood
NICHOLS RESEARCH CORPORATION
2340 Alamo, SE
Suite 105
Albuquerque, NM 87106

William D. Unger
MAYFIELD FUND
2200 Sand Hill Road
Menlo Park, CA 94025

Eugene Urban
NASA
Marshall Space Flight Center
E563
Marshall SFC, AL 35812

William F. Utlaut
DEPARTMENT OF COMMERCE
325 Broadway
Boulder, CO 80303

J. Van Den Sype
UNION CARBIDE CORPORATION
Tarrytown Technical Center
Tarrytown, NY 10591

Professor Theodore Van Duzer
UNIVERSITY OF CALIFORNIA
BERKELEY
Berkeley, CA 94720

Andre Van Rest
DEPARTMENT OF ENERGY
1000 Independence Avenue, S.W.
Washington, D.C. 20585

E.C. Van Reuth
TECHNOLOGY STRATEGIES INC.
10722 Shingle Oak Court
Burke, VA 22015

Lou Vance
ATOMIC ENERGY OF CANADA
White Shell Nuclear Research Lab
Pinawa, Manitoba CANADA RRE 1L0

Joe Vandermeulen
LEGISLATIVE SERVICE BUREAU
P.O. Box 30036
Lansing, MI 48909

John B. Vandersande
MASSACHUSETTS INSTITUTE
OF TECHNOLOGY
77 Massachusetts Avenue
Cambridge, MA 02139

Usha Vaseashta
VIRGINIA POLYTECHNIC INSTITUTE
305 W. Eakin Street
Blacksburg, VA 24060

Natalie Vassil
EDELSON TECHNOLOGY INC.
Park 80 West Plaza II
Saddle Brook, NJ 07662

Charles G. Vaughn
GENERAL RESEARCH
CORPORATION
635 Discovery Drive
Huntsville, AL 35806

Samuel L. Venneri
NASA
600 Independence Avenue, S.W.
CODE RM
Washington, D.C. 20546

Horacio Verdun
FIBERTEK
510-A Herndon Parkway
Herndon, VA 22070

John Vitko Jr.
SANDIA NATIONAL LABORATORIES
P.O. Box 969
Livermore, CA 94550

John Vohs
UNIVERSITY OF PENNSYLVANIA
Department of Physics
Philadelphia, PA 69702

F.L. Vook
SANDIA NATIONAL LABORATORIES
P.O. Box 5800
Albuquerque, NM 87185-5800

Richard Wagner
BABCOCK & WILCOX COMPANY
1735 I Street, N.W.
Suite 814
Washington, D.C. 20006

Richard Wahlers
ELECTRO SCIENCE LABORATORIES
416 E. Church Road
King of Prussia, PA 19406

John Walecka
BRENTWOOD ASSOCIATES
3000 Sand Hill Road
Building 2, Suite 210
Menlo Park, CA 94025

Carlton Walker
ARTHUR D. LITTLE INC.
Acorn Park
Cambridge, MA 02140

David Walker
GENERAL DYNAMICS
CORPORATION
5001 Kearney Villa Road
San Diego, CA 92123

William H. Walker, Jr.
S-TRON
101 Twin Dolphin Drive
Redwood City, CA 94065

Terry C. Wallace
LOS ALAMOS NATIONAL
LABORATORY
P.O. Box 1663
MST-DO/G 756
Los Alamos, NM 87545

Roger Walser
UNIVERSITY OF TEXAS
Engineering Division
Austin, TX 78712

John Walsh
UNION ELECTRIC COMPANY
1901 Gratiot Street
St. Louis, MO 83103

Martin Walt
LOCKHEED MISSILES & SPACE
COMPANY, INC.
3251 Hanover Street
Palo Alto, CA 94304-1187

Gully Walter
NATIONAL STRATEGIC MATERIALS
& MINES
Department of the Interior
Washington, D.C. 20240

Charles K. Waltt
GEORGIA INSTITUTE OF TECHNOL-
OGY
School of Materials Engineering
Atlanta, GA 30332-0245

Gerald D. Waltz
INDIANAPOLIS POWER & LIGHT
P.O. Box 15958
Indianapolis, IN 46206

James Wang
SUPERCON INCORPORATED
830 Boston Turnpike
Shrewsbury, MA 01545

Sou-Tien (Bert) Wang
WANG NMR, INC.
7074A Commerce Circle
Pleasanton, CA 94566

Yen-Chu Wang
HOWARD UNIVERSITY
2355 6th Street, N.W.
Washington, D.C. 20059

J. Washick
KEITHLEY INSTRUMENT
28775 Aurora Road
Cleveland, OH 44139

Edel Wasserman
E.I. DUPONT
1007 Market Street
CR&D Experimental Station
Wilmington, DE 19898

Simon Watkins
AMERICAN CYNAMID
1937 West Main Street
Stanford, CT 06904

Earl K. Weber
AMERICAN STEEL FOUNDRIES
3761 Canal Street
East Chicago, IN 46312

Richard W. Weeks
ARGONNE NATIONAL
LABORATORY
9700 S. Cass Avenue
Materials Science Division
Argonne, IL 60439

Felix Wehrli
GENERAL ELECTRIC
Medical Systems Group
P.O. Box 414
Milwaukee, WI 53201 53201

John M. Wehrung
INTERMET, INC.
P.O. Box 3549
Reston, VA 22090

R.J. Weimer
CORDEC CORPORATION
8270-B Cinder Bed Road
Lorton, VA 22079

Steven Weiner
AIR PRODUCTS & CHEMICALS
Route 222
Allentown, PA 18105

Roy Weinstein
UNIVERSITY OF HOUSTON
Physics Department
Houston, TX 77004

Harold Weinstock
BOLLING AIR FORCE BASE
AFOSR/NE
Washington, D.C. 20331

Stan I. Weiss
LOCKHEED MISSILES & SPACE
COMPANY, INC.
3251 Hanover Street
Palo Alto, CA 94304-1187

Richard C. Weissman
CHEM SYSTEMS, INC.
303 Broadway South
Tarrytown, NY 10591

David O. Welch
BROOKHAVEN NATIONAL
LABORATORY
Associated Universities, Inc.
Building 460
Upton, NY 11973

Dr. Nancy Welker
NATIONAL SECURITY AGENCY
Microelectronics R&D
9800 Savage Road
R53
Ft. Meade, MD 20755

William J. Welsh
UNIVERSITY OF MISSOURI
Department of Chemistry
St. Louis, MO 63121

Lowell E. Wenger
WAYNE STATE UNIVERSITY
Physics & Astronomy Department
Detroit, MI 48202

Jack Wernick
BELL COMMUNICATIONS
21 Haran Circle
Milburn, NJ 07041

James K. Wessel
DOW CORNING
12532 Fostoria Way
Gaithersburg, MD 20879

Bruce Wessels
NORTHWESTERN UNIVERSITY
134 17th Street
Wilmette, IL 60091

Barry Whalen
MICROELECTRONICS AND
COMPUTER TECHNOLOGY
3500 W. Balcones Center Drive
Austin, TX 78759

Professor Sung H. Whang
POLYTECHNIC UNIVERSITY
Dept. Metallurgy & Materials Sci.
333 Jay Street
Brooklyn, NY 11201

Clarence B. Whichard
VENTURE ECONOMICS
39 Evergreen Way
Medfield, MA 02052

Irvin White
NYS ENERGY R&D AUTHORITY
Two Rockefeller Plaza
Albany, NY 12223

Robert R. White
FERRO CORPORATION
2440 Virginia Avenue
D-1105
Washington, D.C. 20037

Robert M. White
NATIONAL ACADEMY OF
ENGINEERING
2101 Constitution Avenue, N.W.
Washington, D.C. 20418

James H. Whitley
AMP INCORPORATED
P.O. Box 3608
Harrisburg, PA 17105-3608

Charles Whitsett
MCDONNELL DOUGLAS
P.O. Box 516
Department 224
Building 110
St. Louis, MO 63166

Vern Whitt
TII INDUSTRIES, INC.
1375 Akron Street
Copiague, NY 11726

Marc Wigdor
NICHOLS RESEARCH CORPORATION
1764 Old Meadow Lane
Suite 150
McLean, VA 22102-4307

John P. Wiksow Jr.
VANDERBILT UNIVERSITY
Dept. of Physics and Astronomy
Box 1807, Station B
Nashville, TN 37235

William D. Wilkerson
PARKER HANNIFIN CORPORATION
18321 Jamboree Boulevard
Irvine, CA 92715

M.R. Willcott
NMR IMAGING
3502 Drummond
Houston, TX 77025

A.E. Williams
COMSAT LABORATORIES
22300 Comsat Drive
Clarksburg, MD 20871

Brown F. Williams
S.R.I.
David Sarnoff Research Center
Princeton, NJ 08543-5300

Don Williams
PACIFIC NORTHWEST
LABORATORIES
Richland, WA 99352

Ellen D. Williams
UNIVERSITY OF MARYLAND
Physics Department
College Park, MD 20742

Jim Williams
LOS ALAMOS NATIONAL
LABORATORY
Los Alamos, NM 87544, NM 87544

John Williams
MASSACHUSETTS INSTI.
TECHNOLOGY
Magnet Technology Division
National Magnet Laboratory
Cambridge, MA 02138

Ken D. Williamson Jr.
LOS ALAMOS NATIONAL
LABORATORY
P.O. Box 1663
MS F630
Los Alamos, NM 87545

Samuel J. Williamson
NEW YORK UNIVERSITY
4 Washington Place
Physics Department
New York, NY 10003

Wayne Willis
GA TECHNOLOGIES
1025 Connecticut Avenue, N.W.
Suite 704
Washington, D.C. 20036

Dow R. Wilson
GENERAL ELECTRIC
P.O. Box 414
Milwaukee, WI 53201

Fred Wilson
EUCLID PARTNERS
50 Rockefeller Plaza
New York, NY 10028

Jack Wilson
DATAQUEST
3 Walnut Avenue
Larkspur, CA 94930

Robert Wilson
NATIONAL CRITICAL MATERIALS
10674 Oakton Ridge Court
Oakton, VA 22124

Teck A. Wilson
TELEDYNE, INC.
1501 Wilson Boulevard
Arlington, VA 22209

Thomas A. Wilson
UNOCAL/MOLYCORP
1201 West 5th Street
Los Angeles, CA 90054

Roger Winkel
ROGERS CORPORATION
One Technology Drive
Rogers, CT 06263

Nick Winograd
PENNSYLVANIA STATE
UNIVERSITY
152 Davey Laboratory
University Park, PA 16802

Lester N. Winslow
NAVAL RESEARCH LABORATORY
4555 Overlook Avenue, S.W.
CODE 120
Washington, D.C. 20375-5000

Ed Winston
EDO WESTERN
2001 Jefferson Davis Highway
Arlington, VA 22202

Leslie Winter
G.E. AIRCRAFT GROUP
1 Newmann Way
MDM-87
Cincinnati, OH 45215-6301

F. Winterberg
UNIVERSITY OF NEVADA
Department of Physics
P.O. Box 60220
Reno, NV 89506

Rick Winterton
DOW CHEMICAL COMPANY
1776 Building
Midland, MI 48694

Steve Winzer
MARTIN MARIETTA LABORATORIES
1450 South Rolling Road
Baltimore, MD 21227

David G. Wirth
COORS CERAMICS COMPANY
17550 West 32nd Avenue
Golden, CO 80401

Gerald P. Wirtz
UNIVERSITY OF ILLINOIS
Department of Ceramic Engineering
203 Ceramics Building
105 S. Goodman Avenue
Urbana, IL 61801

Richard Withers
LINCOLN LABORATORY
P.O. Box 73
Lexington, MA 02173

Pete Woboril
ARTOS ENGINEERING COMPANY
15600 West Lincoln Avenue
New Berlin, WI 53151

David R. Wolcott
NYS ENERGY R&D AUTHORITY
Two Rockefeller Plaza
Albany, NY 12223

David Wolf
DRESSER INDUSTRIES
270 Sheffield Street
Dresser Pump Division
Mountainside, NJ 07092

Dr. Stuart A. Wolf
NAVAL RESEARCH LABORATORY
4555 Overlook Avenue, S.W.
CODE 4634
Washington, D.C. 20375-5000

Robert Wolf
CERN
CH/1221 LEP/MA
Geneva 23, SWITZERLAND

Alan M. Wolsky
ARGONNE NATIONAL
LABORATORY
9700 S. Cass Avenue
Materials Science Division
Argonne, IL 60439

Donald L. Wood
GENERAL ELECTRIC CORPORATE
R&D
P.O. Box 8
Schenectady, NY 12301

David M. Woodall
IDAHO NATIONAL ENERGY
LABORATORY
P.O. Box 1625
Physics Group
Idaho Falls, ID 83415

Herbert H. Woodson
UNIVERSITY OF TEXAS
Engineering Division
ECJ 10.324
Austin, TX 78712

William E. Woolam
SOUTHWEST RESEARCH INSTITUTE
1235 Jefferson Davis Highway
Suite 1406
Arlington, VA 22202

Robert Woolfolk
SRI INTERNATIONAL
1611 N. 10th Street
Arlington, VA 22209

James Worth
OXFORD SUPERCONDUCTING
TECHNOLOGY
600 Milik Street
Carteret, NJ 07008

Otis C. Wright Jr.
ELECTRONIC ASSOCIATES, INC.
185 Monmouth Parkway
W. Long Branch, NJ 07764

Lily Wu
GTE SERVICE CORPORATION
1850 M Street, N.W.
Washington, D.C. 20036

M.K. Wu
UNIVERSITY OF ALABAMA
Department of Physics
Huntsville, AL 35899

Professor M.K. Wu
UNIVERSITY OF ALABAMA
Huntsville, AL 35899

L.N. Yannopoulos
CARNEGIE-MELLON UNIVERSITY
4400 Fifth Avenue
Pittsburgh, PA 15213

Professor Baki Yarer
COLORADO SCHOOL OF MINES
Golden, CO 80401

Ronald L. Yates
DOW CHEMICAL COMPANY
2030 Willard H. Dow Center
Midland, MI 48674

B.G. Yee
GENERAL DYNAMICS
P.O. Box 748
Ft. Worth, TX 76101

Mawlin Yeh
NYNEX
70 West Red Oak Lane
White Plains, NY 10604

Clayton Yeutter
U.S. TRADE REPRESENTATIVE
600 17th Street, N.W.
Washington, D.C. 20508

Edward M Yokley
CELANESE RESEARCH COMPANY
86 Morris Avenue
Summit, NJ 07901

Peter L. Young
BOEING ELECTRONICS COMPANY
P.O. Box 24969
M/S 7J-20
Seattle, WA 98124-6269

Rosa Young
ENERGY CONVERSION DEVICES
1675 West Maple Road
Troy, MI 48084

Stuart A. Young III
YOUNG GROUP
45 Milk Street
Boston, MA 02109

Daniel Yu
UNIVERSAL ENERGY SYSTEMS
4401 Dayton Xenia Road
Dayton, OH 45432

Dr. Gregory J. Yurek
MASSACHUSETTS INSTI.
TECHNOLOGY
H.H. Uhlig Corrosion Laboratory
Materials Science & Engineering
Cambridge, MA 02138

Gregory J. Yurek
MASSACHUSETTS INSTITUTE
OF TECHNOLOGY
77 Massachusetts Avenue
Cambridge, MA 02139

Dr. Peter Zalm
PHILIPS RESEARCH
5600 JA Eindhoven
P.O. Box 80000
NETHERLANDS

David Zavisza
HERCULES INC. RESEARCH CENTER
Wilmington, DE 19894

Bruce Zeitlin
INTERMAGENTICS GENERAL
CORPORATION
P.O. Box 566
New Karner Road
Guilderland, NY 12084

Dr. Hansruedi Zeller
BROWN BOVERI RESEARCH
CENTER
5405 Badem
Dattwil, SWITZERLAND

Bruce B. Zimmerman
U.S. ARMY
HQDA
SARD-TR
Washington, D.C. 20310-3558

John C. Zink
CENTRAL & SOUTH WEST SERVICES
P.O. Box 660164
Dallas, TX 75266-0164

Ed Zschau
BRENTWOOD ASSOCIATES
3000 Sand Hill Road
Building 2, Suite 210
Menlo Park, CA 94025

Alexander Zucker
OAK RIDGE NATIONAL LABORA-
TORY
P.O. Box X
Oak Ridge, TN 37831-6255

Michael A. Zuniga
SCIENCE APPLICATIONS INTERNA-
TIONAL
CORPORATION
803 W. Broad Street
Suite 300
Falls Church, VA 22046

Klaus M. Zwilsky
NATIONAL RESEARCH COUNCIL
2101 Constitution Avenue, N.W.
Washington, D.C. 20418

German de la Fuente
FIBERTEK
510-A Herndon Parkway
Herndon, VA 22070

CHAPTER 14

THE FEDERAL SUPERCONDUCTIVITY INITIATIVE

President Reagan announced, on July 28, 1987, an eleven-point initiative to promote further work in the field of superconductivity and ensure U.S. readiness in commercializing technologies resulting from recent and anticipated scientific advances.

The U.S. has been a leader for years in the field of superconductivity—the phenomenon of conducting electricity without resistance. U.S. private and Government researchers have also been at the forefront of recent laboratory discoveries allowing superconductivity to occur at higher temperatures and with greater current-carrying capacity than was previously possible.

The Federal Government has played a key role in these developments through the funding of basic research. The National Science Foundation (NSF) and the National Aeronautics and Space Administration (NASA) both provided funding for Dr. Paul Chu at the University of Houston in his landmark efforts in raising the temperature at which superconductivity occurs. In addition, the Department of Energy (DOE), which is the principal Federal supporter in the field of superconductivity, has been a leader in the search for the mechanism that produces high-temperature superconductivity and in research into the practical uses of these new materials. The Federal government is currently spending approximately $55 million in superconductivity research, with more than one-half of that reallocated within the last six months.

The President's initiative reflects his belief that it is critical that the U.S. translate our leadership in science into leadership in commerce. While the U.S. private sector must take the lead, the Administration is taking important actions to facilitate and speed the process, including increasing funding for basic research and removing impediments to procompetitive collaboration on géneric research and production and to the swift transfer of technology and technical information from the Government to the private sector.

The President's Superconductivity Initiative has three objectives:

1. Promoting greater cooperation among the Federal government, academia, and U.S. industry in the basic and enabling research that is necessary to continue scientific breakthroughs in superconductivity;

2. Enabling the U.S. private sector to convert more rapidly scientific advances into new and improved products and processes; and

3. Better protecting the intellectual property rights of scientists, engineers, and businessmen working in superconductivity.

The Superconductivity Initiative includes both legislative and administrative proposals. The major components of the Initiative are:

LEGISLATIVE

1. Amending the National Cooperative Research Act (NCRA) to expand the concept of a permissible joint venture to include some types of joint production ventures. This is a particularly important step that would ease the risk of antitrust litigation perceived by U.S. businesses that could otherwise benefit from procompetitive joint ventures. If enacted, it could benefit not just developments using superconductivity, but other high technology products as well.

2. Amending U.S. patent laws to increase protection for manufacturing process patents. This would enable U.S. owners of process patents to obtain damages for infringement where products made with those processes are imported into the U.S.

3. Authorizing Federal agencies to withhold from release under the Freedom of Information Act (FOIA) commercially valuable scientific and technical information generated in Government owned and operated laboratories that, if released, will harm U.S. economic competitiveness.

ADMINISTRATIVE

4. Establishing a *"Wise Man"* Advisory Group on Superconductivity under the auspices of the White House Science Council. This would be a small group of three to five people from industry and academia that would advise the Administration on research and commercialization policies.

5. Establishing a number of *"Superconductivity Research Centers"* (SRCs) and other similar groups that would (1) conduct important basic research in superconductivity and (2) serve as repositories of information to be disseminated throughout the scientific community.

a. The Department of Energy will establish three SRCs, as well as a computer data base:

— Center for Superconductivity Applications at the Argonne National Laboratory;

— Center for Thin Film Applications at the Lawrence Berkeley Laboratory;

— Center for Basic Scientific Information at the Ames Laboratory; and

— Computer Data Base on Superconductivity at the DOE Office of Scientific & Technology Information.

b. The Department of Commerce (DOC) will establish a Superconductivity Center at the National Bureau of Standards (NBS) laboratory in Boulder, Colorado. The center will focus on electronic applications of high temperature superconductivity.

c. The National Aeronautics and Space Administration (NASA) is establishing a coordinating group on superconductivity activities at its office of Aeronautics and Space Technology.

d. The National Science Foundation (NSF) will augment its support for research in high temperature superconductivity programs at three of its materials research laboratories. In addition, NSF is initiating a series of *"quick start"* grants for research into processing superconducting materials into useful forms including wires, rods, tubes, films, and ribbons.

e. The Department of Defense is developing a multi-year plan to ensure use of superconductivity technologies in military systems as soon as possible. DOD will spend nearly $150 million over three years.

DOD will build upon its long experience in superconductivity R&D to systematically: define the engineering parameters for high-temperature superconducting materials; develop the required processing and manufacturing capabilities; and accomplish the necessary development, engineering, and operational prototype testing of superconductors.

Small scale applications with commercial spin-off potential include sensors and electronics. Potential large-scale applications include compact, high-efficiency electric ship drive; electrical energy storage; pulsed power systems; and free electron lasers.

6. Urging all Federal agencies to implement quickly the steps outlined in Executive Order 12591 designed to: (1) transfer technology developed in Federal laboratories into the private sector; and (2) encourage Federal, university, and industry cooperation in research. The White House Science Advisor will report to the President on progress in implementing the Executive Order, particularly with regards to superconductivity.

7. Directing the Patent and Trademark Office to accelerate the processing of patent applications and adjudication of disputes involving superconductivity technologies when requested by the applicants to do so.

8. Directing the NBS to accelerate its efforts to develop and coordinate common standards (e.g. measurement methods, standard reference materials, and supporting technical data) in the U.S. and internationally for superconductors and related devices.

9. Encouraging Federal agencies to continue to reallocate funds into superconductivity basic research, applied research in enabling technologies, and prototype development. Agencies are directed to place a high priority for this area.

10. Requesting that DOD accelerate prototype work in sensor, electronic, and superconducting magnet-based military applications and that the Department of Commerce accelerate development of prototype devices in detection and measurement of weak magnetic fields.

11. Taking advantage of the opportunity presented by the current negotiations for renewing the U.S.-Japan Agreement on Science and Technology to seek reciprocal U.S. opportunities to participate in Japanese government supported research and development, including superconductivity.

In April, the President issued Executive Order 12591 Facilitating Access to Science and Technology directed at encouraging increased commercialization of the U.S. science and technology enterprise.

Within a few months of the initial discoveries, Federal agencies redirected $45 million in fiscal 1987 funds from other R&D to high-temperature superconductors. The scientific breakthroughs prompted a dozen bills during the first session of the 100th Congress, proposals ranging from study commissions to a national program on superconductivity. All reflected, in one way or another, concern over commercialization.

The policy drama reached a peak in the administrations interest in July 1987, when President Reagan brought three ranking cabinet officers to the Federal Conference on Commercial Applications of Superconductivity; in an unprecedented appearance, he announced the 11-point initiative, noted above, for the support of high-temperature superconductors. In a similarly unprecedented move, the Administration closed the meeting to all foreigners except representatives of the press. Although the President's message focused on executive branch actions, he stated that the Administration would also be proposing new legislation.

The following months brought a sense of anticlimax, with no sign of the promised legislative package. Questions of R&D funding then came to the fore, as the end of the fiscal year passed with no resolution of the budget impasse between the President and Congress. Only at the end of the calendar year—several months into fiscal 1988—did Federal agencies know for certain how much money they would have for high-temperature superconductor R&D.

Taken together, Federal agencies spent nearly $160 million for superconductivity R&D in fiscal 1988, over half ($95 million) on the new materials (and the rest for lower-temperature superconductors). The Department of Defense and the energy Department together account for three-quarters of the high-temperature superconductor budget, and received most of the increase. DOE, for instance, will have nearly twice as much high-temperature superconductor money as the National Science Foundation (NSF). Most of the Federal high-temperature superconductor money will go to government laboratories, contractors, and universities that are well removed from the commercial marketplace.

The President's legislative package, which reached Congress in February 1988, did not address R&D funding. Consistent with the Administration's emphasis on indirect incentives for commercialization, the package included provisions that would further liberalize U.S. antitrust policies, and extend the reach of U.S. patent protection.